HEINEMANN MODULAR MATHEMATICS *for* EDEXCEL AS AND A-LEVEL

Decision Mathematics 1

John Hebborn

Heinemann

Edexcel
Success through qualifications

Heinemann Educational Publishers,
a division of Heinemann Publishers (Oxford) Ltd,
Halley Court, Jordan Hill, Oxford, OX2 8EJ

OXFORD MELBOURNE JOHANNESBURG
AUCKLAND BLANTYRE IBADAN GABORONE
PORTSMOUTH NH (USA) CHICAGO

First published 2000

02 01 10 9 8 7 6 5 4 3 2 1

ISBN 0 435 51080 0

Cover design by Gecko Limited

Original design by Geoffrey Wadsley: additional design work by Jim Turner

Typeset and illustrated by Tech-Set Limited, Gateshead, Tyne & Wear

Printed in Great Britain by The Bath Press, Bath

Acknowledgements:

The publisher's and author's thanks are due to the Edexcel for permission to reproduce questions from past examination papers. These are marked with an [E].

The answers have been provided by the authors and are not the responsibility of the examining board.

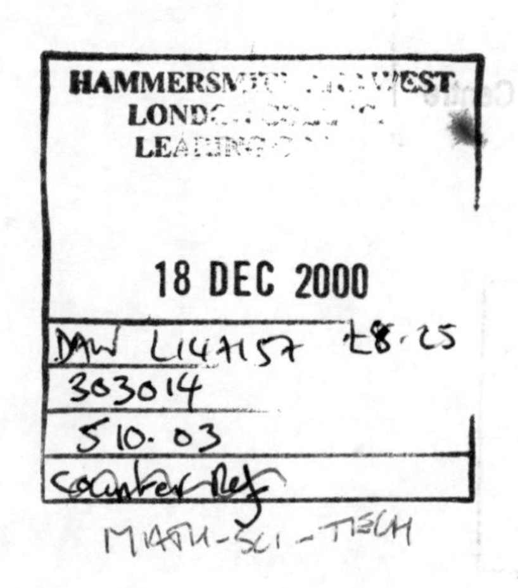

About this book

This book is designed to provide you with the best preparation possible for your Edexcel D1 exam. The series authors are senior examiners and exam moderators themselves and have a good understanding of Edexcel's requirements.

Use this **new edition** to prepare for the new 6-unit specification. Use the first edition (*Heinemann Modular Mathematics for London AS and A-Level*) if you are preparing for the 4-module syllabus.

Finding your way around

To help to find your way around when you are studying and revising use the:

- **edge marks** (shown on the front page) – these help you to get to the right chapter quickly;
- **contents list** – this lists the headings that identify key syllabus ideas covered in the book so you can turn straight to them;
- **index** – if you need to find a topic the **bold** number shows where to find the main entry on a topic.

Remembering key ideas

We have provided clear explanations of the key ideas and techniques you need throughout the book. Key ideas you need to remember are listed in a **summary of key points** at the end of each chapter and marked like this in the chapters:

■ **For all events on a critical path $e_i = l_i$**

Exercises and exam questions

In this book questions are carefully graded so they increase in difficulty and gradually bring you up to exam standard.

- **past exam questions** are marked with an [E];
- **review exercises** on pages 77, 180 and 239 help you practise answering questions from several areas of mathematics at once, as in the real exam;
- **exam style practice paper** – this is designed to help you prepare for the exam itself;
- **answers** are included at the end of the book – use them to check your work.

Contents

Algorithms 1

1.1 What is an algorithm?

When you buy a new piece of equipment, such as a video recorder, the package usually includes a set of step-by-step instructions for installing it or setting it up. Such a set of instructions is called an **algorithm**.

In this chapter, we will be considering algorithms of various kinds, beginning with some well-known algorithms from the everyday world, moving then to some algorithms from elementary mathematics and then to some algorithms that have been developed in the areas of sorting, packing and searching.

Before considering specific algorithms we give a definition of an algorithm and outline some of its essential properties.

- **An algorithm is a set of precise instructions which if followed will solve a problem.**

The word *precise* in the definition is important. It means that there must be *no ambiguity* in any instruction and that after a particular instruction has been obeyed there must be *no ambiguity* as to which instruction is to be carried out next.

Ambiguous – having two or more possible meanings.

Not all sets of instructions constitute an algorithm. A set of instructions would fail to be an algorithm if one of the instructions was ambiguous or if after completing a particular instruction it was not clear which instruction was to be carried out next.

It is also implicit in the definition that the problem will be solved in a finite time so that the set of instructions should include an instruction to stop. This instruction must be reached after carrying out a *finite* number of instructions.

1.2 Some examples of algorithms

There are many things in everyday life that are algorithms but for which we use other more commonplace names. Here are three examples from knitting, cooking and repairing broken china.

Example 1
The instructions for knitting a man's cardigan include the following set of instructions for knitting the rib.

Using No 3 mm needles cast on 115 sts.
1st row — S1, K1 * P1, K1 repeat from * to the last st, K1.
2nd row — S1 * P1, K1, repeat from * to end.
Repeat 1st and 2nd rows 11 times.
(Abbreviations: K knit, P purl, S1 slip one stitch, st stitch)

Example 2
The following recipe was given in a magazine:

Celebration Cornflake Crunch (makes 12–14)

125 g Plain chocolate
250 g Christmas pudding, cooked
60 g Crunchy Nut Cornflakes
Bun tray, lined with paper cases

Melt the chocolate in a bowl over a pan of hot water.
Break the Christmas pudding into pieces and stir into the chocolate.
Fold in the cornflakes and spoon into the paper cases.
Leave to set.

Example 3
On a pack of strong clear adhesive are printed the following instructions for repairing a broken glazed china cup.

Directions for use
1 Make sure surfaces to be joined are clean, dry and free from grease.
2 Apply a thin layer of adhesive to each surface and leave for at least one minute before pressing the surfaces together.

With a little thought you will be able to recall many examples of algorithms from your own experience.

In your earlier courses on mathematics you will have learned how to add, subtract, multiply and divide numbers. In each of these cases you will have used a method – a set of instructions – which, when carried out precisely, will produce the correct answer.

For example, if asked to add 256 and 845 you will probably write:

$$\begin{array}{r} 256+ \\ \underline{845} \\ \underline{1101} \end{array}$$

This does not indicate the order in which you obtained the digits in the answer. A set of instructions would indicate that you work from the right to the left, first adding the units, then the tens and finally the hundreds.

It is not always easy to produce an algorithm. On the other hand it is not necessary to understand the precise logic behind an algorithm to be able to use it.

Consider the usual algorithm for multiplying two numbers, that is:

(multiplicand) × (multiplier)

The **multiplicand** is the number you are multiplying.

1 Multiply successively by each figure of the multiplier taken from right to left.
2 Write these intermediate results one beneath the other, shifting each line one place to the left.
3 Finally add all these rows to obtain the answer.

Example 4

Multiply 178×26.

1 $178 \times 6 = 1068$
  $178 \times 2 = 356$
2 1068
  356
3 4628

You would of course probably write this simply as:

$$\begin{array}{r} 178\times \\ \underline{26} \\ \underline{1068} \\ \underline{3560} \\ \underline{4628} \end{array}$$

There are, however, other methods of multiplying two numbers. This method is sometimes called the Russian peasant's algorithm, although it has also been attributed to the ancient Egyptians. It uses only doubling, halving and addition. Here is a statement of the algorithm:

1 Write multiplier and multiplicand side by side.
2 Make two columns, one under each number, by repeating the following rules until the number under the multiplier is 1:
  (a) Divide the number under the multiplier by 2, ignoring fractions.
  (b) Multiply the number under the multiplicand by 2.

3 Cross out each row in which the number under the multiplier is even.

4 Add up the numbers that remain in the column under the multiplicand.

Example 5

Multiply 178×26.

	Multiplier	Multiplicand
	~~26~~	~~178~~
	13	356
	~~6~~	~~712~~
	3	1424
	1	2848
Total of remaining rows		4628

Using flow charts to represent algorithms

All the algorithms discussed so far have been described in words. Another way of presenting an algorithm is in the form of a **flow chart**. In a flow chart (or **flow diagram**) it is customary to use three kinds of blocks:

- A rectangular block ▭ contains operations to be carried out. These blocks have only one route leading from them.
- A diamond-shaped block ◇ contains questions that can be answered 'yes' or 'no'. These blocks have two alternative routes out, the route taken depending on the answer to the question in the block.
- An oval-shaped block ⬭ indicates a terminus.

Example 1 may be described by flow chart A.

A:

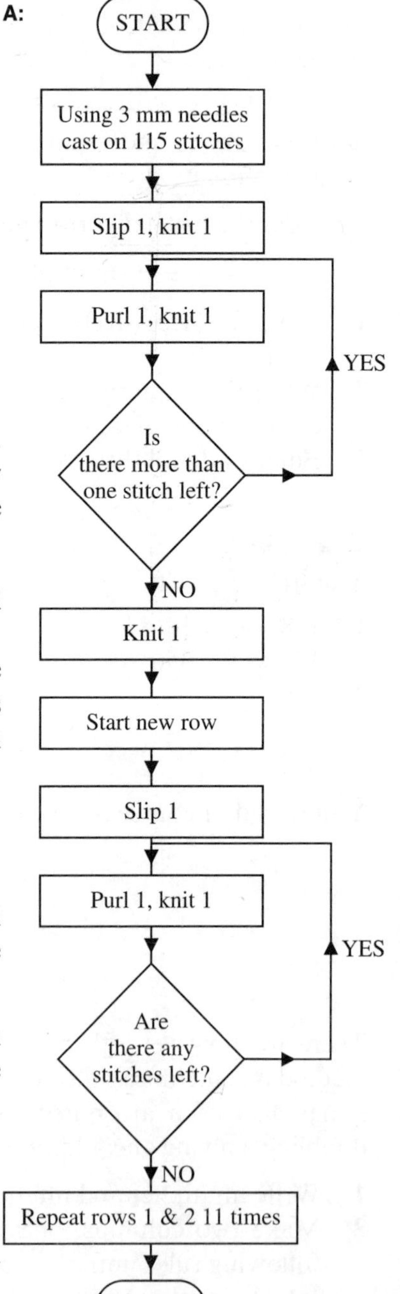

Notice that in this flow diagram there are lines that leave decision blocks and then re-enter the diagram at an earlier point. Lines like these are called **loops**.

You have probably encountered the quadratic equation $ax^2 + bx + c = 0$. The solutions of this equation, if they exist, are obtained from the formula

$$\frac{-b \pm \sqrt{b^2 - 4ac}}{2a}$$

You can solve this equation by using the algorithm given by flow chart B.

B:

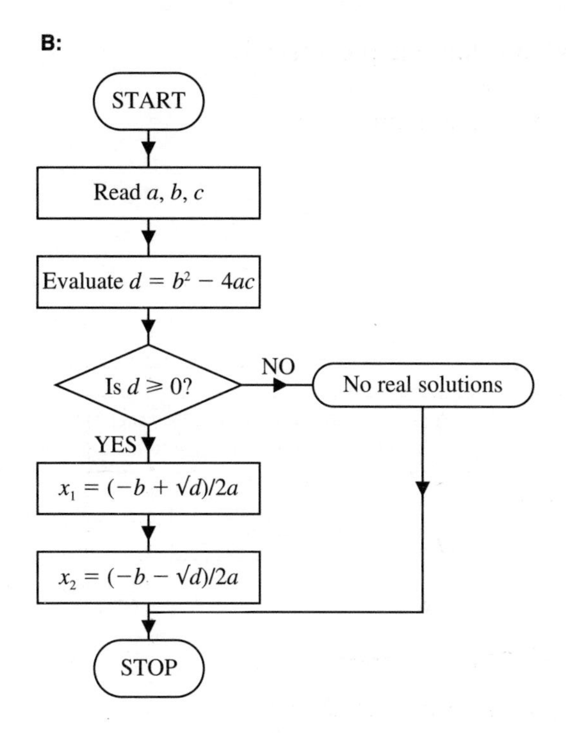

Example 6

Use the algorithm given by flow chart B to find the roots of $x^2 - 21x - 162 = 0$.

Read a, b, c: $a = 1, \quad b = -21, \quad c = -162$

Evaluate d: $d = (-21)^2 - 4(1)(-162) = 441 + 648 = 1089$

d is $\geqslant 0$, $(\sqrt{d} = \sqrt{1089} = 33)$

Evaluate x_1: $x_1 = \dfrac{(-b + \sqrt{d})}{2a} = \dfrac{(21 + 33)}{2} = 27$

Evaluate x_2: $x_2 = \dfrac{(-b - \sqrt{d})}{2a} = \dfrac{(21 - 33)}{2} = -6$

A famous algorithm invented by Euclid for finding the highest common factor of two integers A and B is given by flow chart C. It is called the **Euclidean algorithm**.

Example 7

Use the Euclidean algorithm, given by flow chart C, to find the highest common factor of 35 and 112.

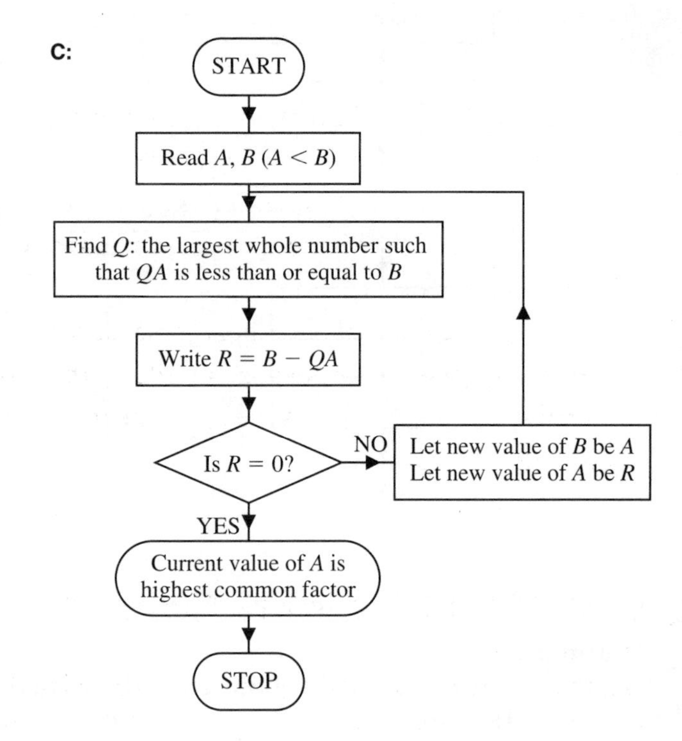

Read A, B: $A = 35$, $B = 112$

Find Q: $3 \times 35 = 105$ and $4 \times 35 = 140$
So, in this case, $Q = 3$

Find R: $R = 112 - (3 \times 35) = 7$

$R \neq 0$, so assign new values to A and B:
$A = 7$, $B = 35$

Find Q: $5 \times 7 = 35$
So, in this case, $Q = 5$

Find R: $R = 35 - (5 \times 7) = 0$

$R = 0$ and the current value of A is 7. The highest common factor of 35 and 112 is 7.

Exercise 1A

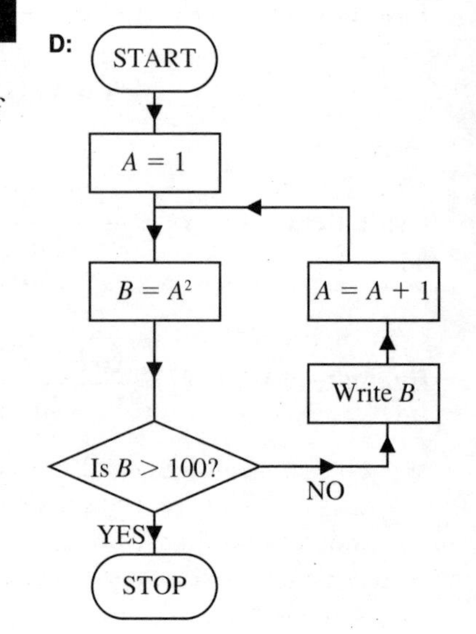

1 Use the algorithm given by flow chart B to find real solutions, if they exist, of:

(a) $x^2 - x - 2 = 0$

(b) $3x^2 - 14x - 5 = 0$

(c) $2x^2 + 3x + 5 = 0$.

2 Use the Euclidean algorithm given in flow chart C to find the highest common factor of:

(a) 39 and 169

(b) 666 and 972.

3 Implement the algorithm given by flow chart D and state what the algorithm actually produces.

4 Implement the algorithm given by flow chart E and state what the algorithm actually produces.

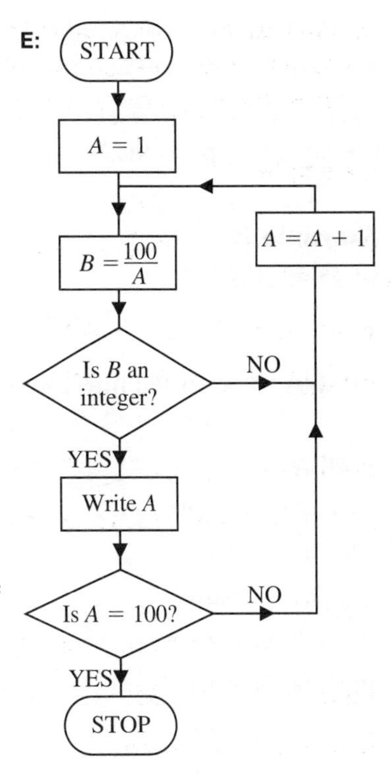

5

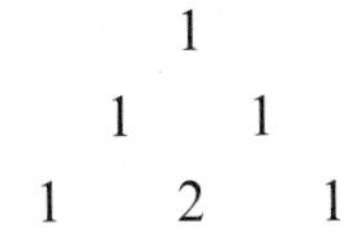

The diagram shows the first three rows of what is known as Pascal's triangle. Construct the next five rows given that:

(i) the border consists entirely of 1s

(ii) each of the other numbers is the sum of the two numbers immediately above it.

6 The first three terms of the Fibonacci sequence are 1, 1, 2. The remaining numbers are obtained by adding the two previous numbers in the sequence. Write down the first 12 numbers of the sequence.

1.3 Sorting algorithms

Suppose a list of students is given in alphabetical order and you want to arrange them in a new order according to the marks they obtained in an examination. This is an example of a **sorting** problem.

In Kruskal's algorithm for finding a minimum spanning tree (discussed in Chapter 3) the weights of the edges give you a list of numbers. You have to work through this list from the smallest to the largest and so the first step is to produce a new list in ascending order. This is again a **sorting** problem.

Given a list of people's surnames in a particular club the most useful directory to produce is one in which these surnames are in alphabetical order. This requires sorting to take place.

Bubble-sort algorithm

One of the simplest sorting algorithms is called **bubble-sort**. Just as bubbles in a fizzy drink rise to the surface, in a bubble sort the largest or smallest number 'rises' to one end in a list.

- **The bubble-sort algorithm makes repeated passes through a list of numbers. On each pass adjacent numbers in the list are compared, and switched if they are in the wrong order. The bubble-sort algorithm terminates when a pass produces no changes to the order of the list.**

In the first pass of the algorithm the first and second numbers are compared, then the second and third, and so on.

Each comparison refers to the list in the order resulting from all previous comparisons.

Example 8

Consider the list

16, 9, 4, 6, 12, 3, 8, 7

which is to be sorted into *ascending* order.

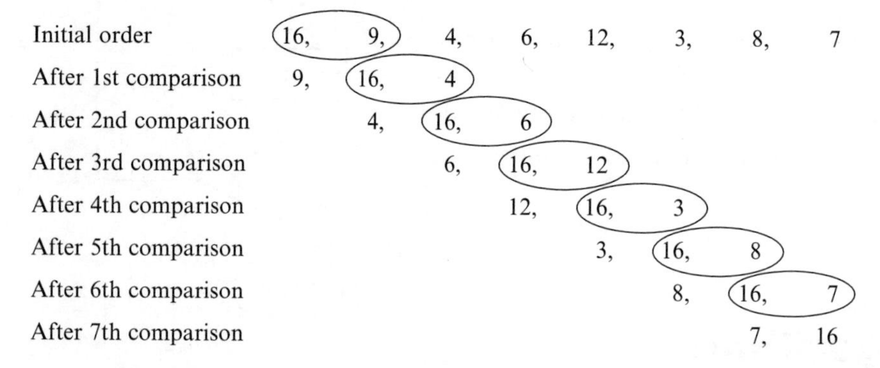

At the end of this **first pass** the largest number, 16, is in its correct position. You can see from the above diagram how the 'bubble' has moved to the right.

Each pass of the bubble-sort algorithm succeeds in placing at least one number in its correct position.

Each pass of the algorithm through the list is called an **iteration**. **Iterate** means repeat.

Having completed the first pass the list is now

9, 4, 6, 12, 3, 8, 7, [16]

The algorithm is now applied to the sublist with the [16] removed. This is called the *second pass*.

The result of the second pass is

4, 6, 9, 3, 8, 7, [12]

So now the two largest numbers of the original list, namely [12] and [16], are in their correct positions.

The result of the *third pass* is

4, 6, 3, 8, 7, [9]

that for the fourth pass is

4, 3, 6, 7, [8]

and the fifth pass gives

3, 4, 6, [7]

A sixth pass produces no change and so the list is sorted, the final list being

3, 4, 6, 7, 8, 9, 12, 16

We can apply the same method to producing a list in *descending* order.

Example 9

Consider the same list

16, 9, 4, 6, 12, 3, 8, 7

and sort it into decreasing order.

The first pass is

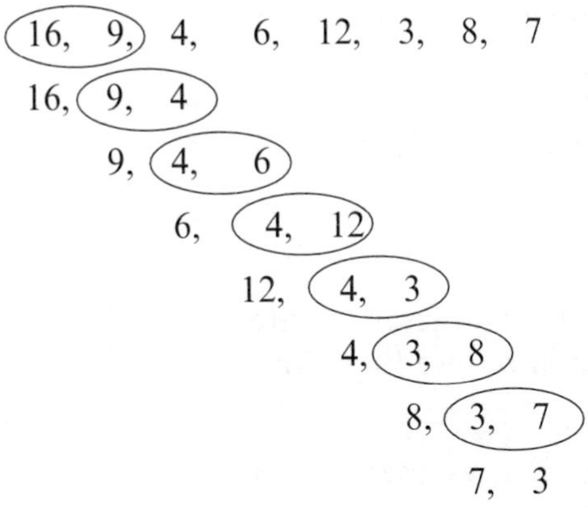

(16, 9,) 4, 6, 12, 3, 8, 7
16, (9, 4)
9, (4, 6)
6, (4, 12)
12, (4, 3)
4, (3, 8)
8, (3, 7)
7, 3

At the end of this pass the smallest number [3] is in its correct place. The other passes give

2nd pass 16, 9, 12, 6, 8, 7, [4]

3rd pass 16, 12, 9, 8, 7, [6]

4th pass 16, 12, 9, 8, 7

The fourth pass produces no change so the list in descending order is 16, 12, 9, 8, 7, 6, 4, 3.

Example 10
The five members of a club have the surnames

Jordan, Smith, Adams, Evans, Kapasi

Use the bubble-sort algorithm to sort this list into alphabetical order.

Denote the names by J, S, A, E, K.

The first pass is

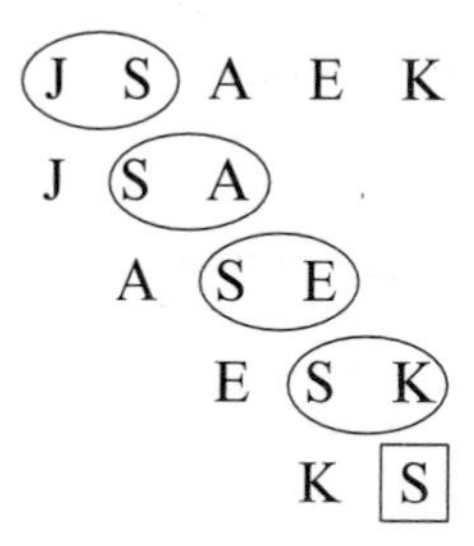

At the end of this pass S (Smith) is in the correct place.

The other passes give

2nd pass   A E J K

3rd pass   A E J

The third pass produces no change so the list in alphabetical order is

A	E	J	K	S
Adams	Evans	Jordan	Kapasi	Smith

Quick-sort algorithm

A more efficient method of sorting, in that it in general requires fewer comparisons, is the **quick-sort algorithm** introduced by Hoare in 1962.

Before describing the algorithm we will define what we mean by the number at the **mid-point** of a list. For a list with an odd number of members the mid-point is obvious. For example, if a list has nine numbers the fifth is the mid-point. So in the list

8, 4, 5, 6, ③, 10, 2, 1, 12

the mid-point or middle number is 3. In a similar way the mid-point of the list

B, A, D, Ⓕ, E, G, H

is the letter F.

For a list with an even number of members we have a choice as to what we call the mid-point. In this book we adopt the convention of choosing the higher numbered member of the two possible choices. For example, for the list

4, 3, 6, 5, 8, 2

we could choose either the 6 or the 5 as the mid-point. We will choose the 5.

In this context *higher numbered* refers to the position of the member within the list. It does not mean *higher valued.*

Let $\lceil x \rceil$ be the smallest integer greater than or equal to x (notice the shape of the brackets). Then whether N, the number of members in a list, is odd or even the mid-point has position $\lceil \frac{1}{2}(N+1) \rceil$.

For $N = 9$ the mid-point is the fifth
$N = 7$ the mid-point is the fourth
$N = 6$ $\lceil \frac{1}{2}(6+1) \rceil = \lceil 3\frac{1}{2} \rceil = 4$. The mid-point is the fourth

The steps in the quick-sort algorithm are as follows.

Step 1 Select a specific number from the list, which will be used as a kind of pivot. Some authors select the first number but in this book the number at the mid-point of the list will be selected.

Let us apply the quick-sort algorithm to sort the list L in ascending order:

L: 16, 9, 4, 6, (12), 3, 8, 7

Using the above definition, the mid-point is the fifth member of the list ($\lceil \frac{1}{2}(8+1) \rceil = \lceil 4\frac{1}{2} \rceil = 5$). The number 12 is therefore our pivot and we indicate this by ringing it.

Step 2 Now write all the numbers *smaller* than 12 to the *left* of 12, reading the original list from left to right. This creates a sublist L_1.

In a similar way write all the numbers *larger* than 12 to the *right* of 12, reading the original list from left to right. This creates a sublist L_2.

Do not reorder these sublists.

For our example the application of step 2 gives

9, 4, 6, 3, 8, 7 , (12), 16

sublist L_1 sublist L_2

Step 3 Apply steps 1 and 2 to each separate sublist until each sublist contains only one number.

(a) Consider L_1. The pivot here is 3 and applying step 2 gives

③, 9, 4, 6, 8, 7

sublist L_3

(b) Sublist L_2 has only one member.

(c) Now consider sublist L_3. The pivot here is 6 and applying step 2 gives

4, ⑥, 9, 8, 7

sublist L_4 sublist L_5

(d) Sublist L_4 has only one member.

(e) Now consider L_5. The pivot here is 8 and applying step 2 gives

7 , ⑧, 9

L_6 L_7

(f) Both sublists L_6 and L_7 have only one number and so the algorithm is complete.

Combining the various steps gives:

16 9 4 6 ⑫ 3 8 7

9 4 6 ③ 8 7 [12] 16

[3] 9 4 ⑥ 8 7

4 [6] 9 ⑧ 7

7 [8] 9

3 4 6 7 8 9 12 16

Starting at the left-hand side the final sorted list is

3, 4, 6, 7, 8, 9, 12, 16

At the end of the sorting process it is a good idea to count the number of items in the final list to ensure that you have not lost any in carrying out the process.

Although all the examples given in this section could easily be sorted by inspection, it is important to have well-defined algorithms to perform the task. Without an explicit method it would not be possible to program computers to sort large data sets.

Exercise 1B

1 The marks obtained in an examination by five students were:

Anne (68), Barry (42), Clare (70), David (45),
Eileen (50) and Greg (55)

Use the bubble-sort algorithm to sort these marks

(a) in ascending order

(b) in descending order.

2 Use the quick-sort algorithm to sort the list

$$6, \ 8, \ 4, \ 5, \ 10, \ 2, \ 9$$

in ascending order.

3 The members of a club have surnames

Monro, Jones, Malik, Wilson and Shah

Use the bubble-sort algorithm to sort these into alphabetical order.

4 Use the bubble-sort algorithm to sort the following list in ascending order:

$$4, \ -3, \ -5, \ 0, \ 2, \ 5, \ -1$$

5 The times recorded by seven competitors for a given distance were:

$$9.8, \ 9.2, \ 9.6, \ 9.7, \ 9.1, \ 8.9, \ 9.0$$

(a) Use the bubble-sort algorithm to sort these times into ascending order.

(b) Use the quick-sort algorithm to sort the times into ascending order.

(c) Which of the methods is the most efficient in this case?

6

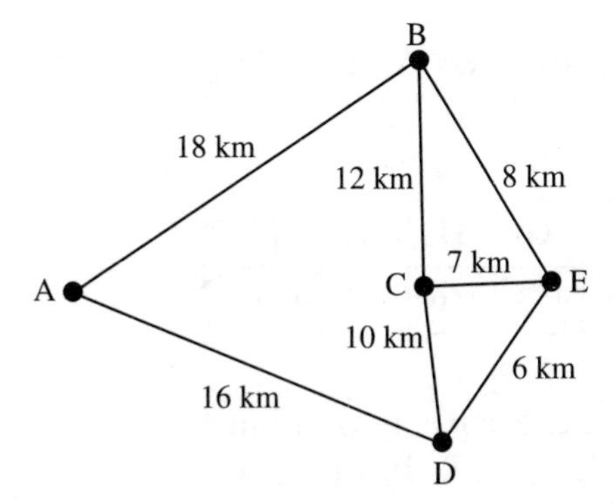

The diagram above shows the roads between five towns and their lengths in kilometres.

(a) Complete the distance table below.

Road	AB	AD	BC	BE	CE	CD	DE
Length (km)	18			8			6

(b) Use the quick-sort algorithm to order the roads in increasing order of lengths.

1.4 Bin-packing algorithms

There is a whole class of problems which may be modelled by the bin-packing problem. Some of these will be included in the exercises.

Consider a set of bins all of the same cross-section and the same height as shown in the figure.

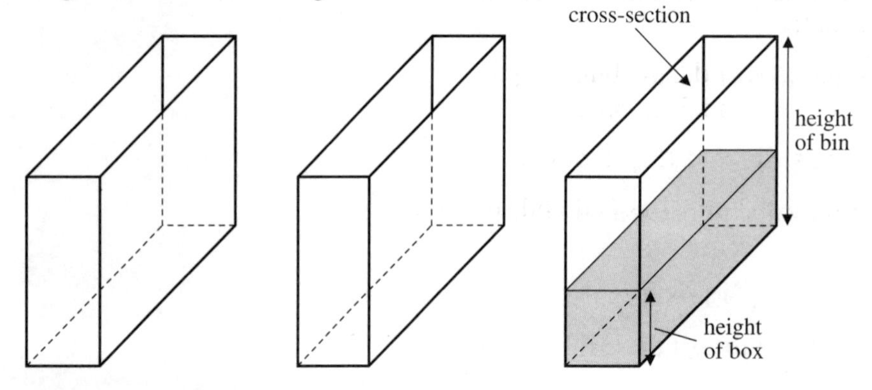

The bin-packing problem considers how to pack into the bins a number of boxes of the same cross-section as the bins but of varying heights, using as few bins as possible.

Example 11

Suppose the bins are 1.5 m tall and you have 10 boxes A, B, C, D, . . . , J with the heights shown in the table.

Box	A	B	C	D	E	F	G	H	I	J
Height (m)	0.8	0.6	0.7	0.5	0.9	0.4	0.3	0.6	0.5	0.6

To save working with decimals, we will define the capacity of a bin as (height $\times$ 10), and the size of a box as (height $\times$ 10). In this case each bin has capacity 15 and the boxes are of sizes:

A(8), B(6), C(7), D(5), E(9), F(4), G(3), H(6), I(5), J(6)

The total size of all the boxes is

$$8 + 6 + 7 + 5 + 9 + 4 + 3 + 6 + 5 + 6 = 59$$

If you divide this by 15, the capacity of a bin, you get $\frac{59}{15} = 3\frac{14}{15}$. This indicates that *at least four bins* are needed. However, because of the sizes of the boxes, it may not be possible to find a solution (packing) using just four bins.

Full-bin combinations

For a problem involving only a few bins and boxes it is possible to look for combinations of boxes that fill a bin. This is not a practical method for larger problems and so other algorithms are considered below.

Example 12

For the data given in Example 11 find full-bin combinations.

In this case you can easily see that:

box A + box C = 8 + 7, so boxes A and C fill bin 1
box B + box E = 6 + 9, so boxes B and E fill bin 2

The next largest boxes remaining are H and J, each of size 6.

box H + box J + box G = 6 + 6 + 3, so boxes H, J and G fill bin 3

The remaining boxes D, I and F give

box D + box I + box F = 5 + 5 + 4 = 14

and so these may be fitted into bin 4. The situation can be summarised by a diagram:

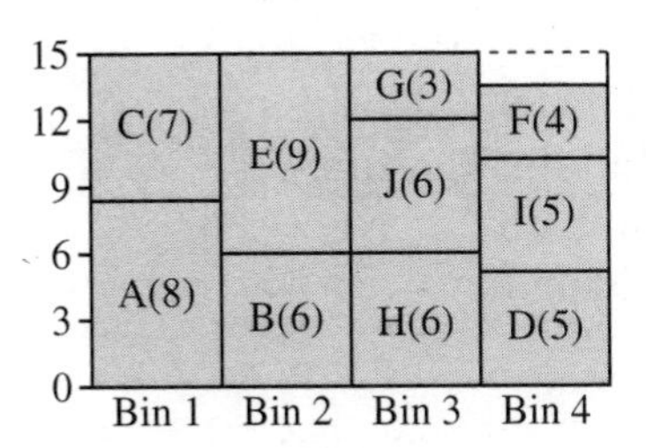

All the boxes can in fact be fitted into four bins.

First-fit algorithm

This is an algorithm that provides a method of dealing with more complicated problems. It may be stated as follows:

- ***Taking the boxes in the order listed,* place the next box to be packed in the *first* available bin that can take that box.**

Example 13

Apply the first-fit algorithm to the data given in Example 11.

Applying the algorithm to this situation gives:

box A(8) into bin 1, leaving space of 7
box B(6) into bin 1, leaving space of 1
box C(7) into bin 2, leaving space of 8
box D(5) into bin 2, leaving space of 3
box E(9) into bin 3, leaving space of 6
box F(4) into bin 3, leaving space of 2
box G(3) into bin 2, leaving space of 0
box H(6) into bin 4, leaving space of 9
box I(5) into bin 4, leaving space of 4
box J(6) into bin 5, leaving space of 9

The situation can be summarised as:

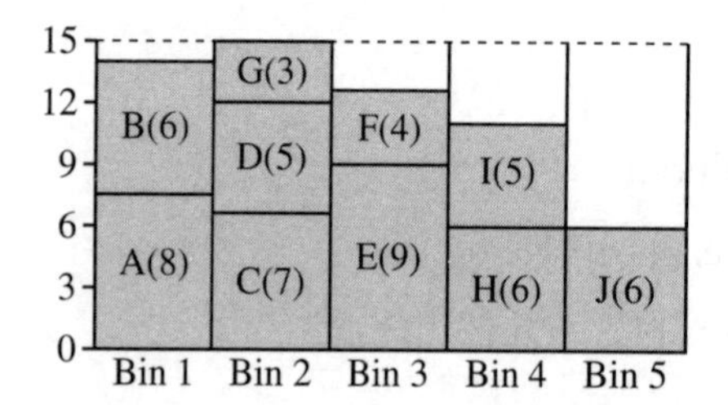

To get a picture of what is happening, try to build up this diagram as you assign boxes to bins.

Notice that the application of this algorithm has produced a solution requiring five bins, although the best (optimum) solution requires only four bins as we have seen above.

For the bin-packing problem there is no known efficient algorithm that will always produce the optimal or best solution. The first-fit algorithm is an example of an algorithm that attempts to find a good solution. Such algorithms are called **heuristic algorithms**.

First-fit decreasing algorithm

From Example 13 you can see that if there is a box of a large size towards the end of the list it will probably have to go into a bin on its own, since any spare capacity of the other bins is likely to be distributed in small amounts. This suggests that the first-fit algorithm is more likely to give the optimal solution if the boxes are reordered in *descending order of size* before allocation to bins is started.

The first-fit decreasing algorithm has two steps.

Step 1 Reorder the boxes in *decreasing* order of size using one of the sorting algorithms.
Step 2 Apply the first-fit algorithm to the reordered list.

Example 14
Apply the first-fit decreasing algorithm to the data given in Example 11.

Reordering the boxes in decreasing order of size gives:

Box	E	A	C	B	H	J	D	I	F	G
Size	9	8	7	6	6	6	5	5	4	3

Applying the first-fit algorithm to this list gives:

box E(9) into bin 1, leaving space of 6
box A(8) into bin 2, leaving space of 7
box C(7) into bin 2, leaving space of 0
box B(6) into bin 1, leaving space of 0
box H(6) into bin 3, leaving space of 9
box J(6) into bin 3, leaving space of 3
box D(5) into bin 4, leaving space of 10
box I(5) into bin 4, leaving space of 5
box F(4) into bin 4, leaving space of 1
box G(3) into bin 3, leaving space of 0

This solution is summarised by:

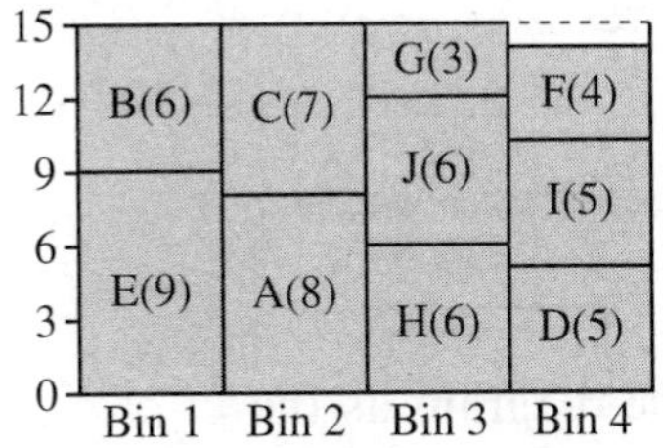

In this case the algorithm does give an optimal solution. Although this algorithm is generally more efficient than the first-fit algorithm, it is not guaranteed to always give an optimal solution.

Example 15
The times, in hours, taken to produce articles A, B, C, ..., K are given in the table.

Article	A	B	C	D	E	F	G	H	I	J	K
Time (hours)	2	2	3	3	4	4	4	5	7	7	7

Determine the numbers of workers required to produce these articles in a 12-hour shift using
(a) the first-fit algorithm
(b) the first-fit decreasing algorithm.
(c) Is it possible to obtain a better solution than either of those given by (a) or (b)?

In this example, each 'bin' is a 12-hour shift for one worker.

(a) The first-fit algorithm applied to this situation gives:

in bin 1 you can pack A(2), B(2), C(3) and D(3), leaving space of 2
in bin 2 you can pack E(4), F(4) and G(4), no space left
in bin 3 you can pack H(5) and I(7), no space left
in bin 4 you can pack J(7), leaving space of 5
in bin 5 you can pack K(7), leaving space of 5

This solution is summarised by the diagram below:

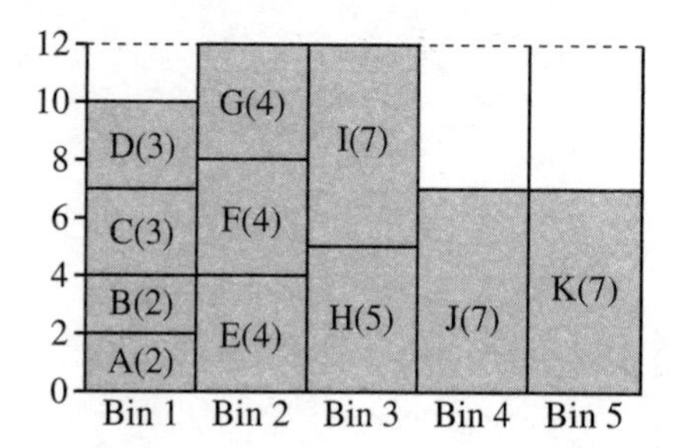

This solution requires five bins, that is five workers to produce the articles.

(b) Reordering the list in decreasing order of size gives:

K	J	I	H	G	F	E	D	C	B	A
7	7	7	5	4	4	4	3	3	2	2

It is particularly useful in this case to construct the diagram as the assignments to bins are made:

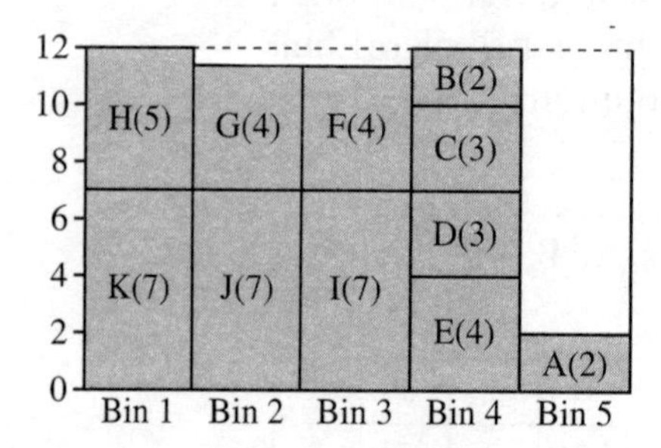

place K(7) in bin 1, leaving space of 5
place J(7) in bin 2, leaving space of 5
place I(7) in bin 3, leaving space of 5
place H(5) in bin 1, leaving space of 0
place G(4) in bin 2, leaving space of 1
place F(4) in bin 3, leaving space of 1
place E(4) in bin 4, leaving space of 8
place D(3) in bin 4, leaving space of 5
place C(3) in bin 4, leaving space of 2
place B(2) in bin 4, leaving space of 0
place A(2) in bin 5, leaving space of 10

Again this solution requires five bins, that is five workers.

(c) You can see from the solution in (b) that there is only A(2) in bin 5. In other words, worker 5 would only be working for 2 hours! The question is, can the assignments to the other bins be reorganised so that A(2) may be fitted in?

Here is an optimal solution:

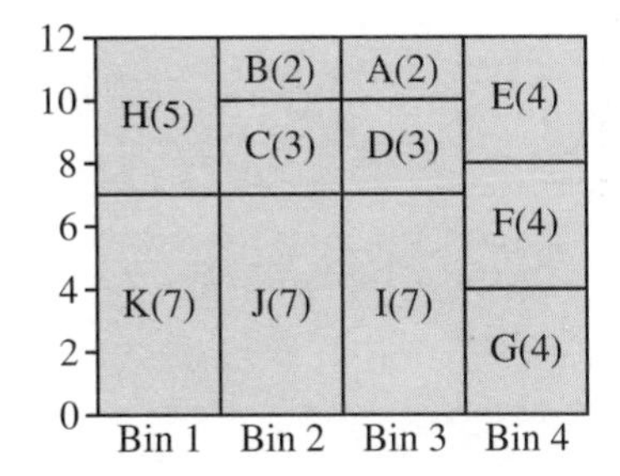

This solution involves only four bins, that is four workers can complete the task.

Exercise 1C

1 A project is to be completed in 13 days. The activities involved in the project and their durations in days are given in the table.

A	B	C	D	E	F	G	H	I	J
3	8	7	5	8	4	5	4	4	4

To determine how many workers are required

(a) apply the first-fit algorithm

(b) apply the first-fit decreasing algorithm.

(c) Is it possible to obtain a better solution than either (a) or (b)?

2 A small ferry that sails between two of the islands in the Hebrides has three lanes each 20 m long on its car deck. The vehicles waiting to be loaded are:

petrol tanker	13 m	small van	3 m	truck	7 m
small truck	6 m	coach	12 m	car	3 m
car	4 m	lorry	11 m		

(a) Can you use the first-fit decreasing algorithm to load all the vehicles on to the ferry?
(b) Can all the vehicles be taken in one trip?

3 A project consists of eight activities whose durations are as follows:

A	B	C	D	E	F	G	H
2	4	3	1	5	4	2	3

Use full-bin combinations to determine the minimum number of workers needed to finish the project in 12 hours.

4 A certain kind of pipe is sold in 10 m lengths. For a particular job the following lengths are required:

2 m, 8 m, 4 m, 5 m, 2 m, 5 m, 4 m

By looking for full-bin combinations, or otherwise, find the number of 10 m lengths required for the job.

5 Joan decided that she wanted to record a number of programmes on the video recorder. The lengths of the programmes were:

45 min, 1 h, 35 min, 15 min, 40 min, 30 min, 50 min, 55 min and 25 min

Help Joan to decide how many 2 h tapes she requires, using
(a) the first-fit algorithm
(b) the first-fit decreasing algorithm
(c) full-bin combinations.

6 120, 78, 100, 90, 60, 38, 80, 26, 150

(a) The list of numbers above is to be sorted into descending order. Perform a quick-sort to obtain the required list. Give the state of the list after each rearrangement and indicate the pivot elements used.

(b) (i) Use the first-fit decreasing algorithm to fit the data into bins of size 200.

(ii) Explain how you decided into which bin to place the number 78.

1.5 Binary-search algorithm

In section 1.3 we considered the problem of sorting a list. Quite often it is necessary to **search** a list to see if it contains a given specific item and if it does to locate that item in the list.

The **binary-search algorithm** is a method for carrying out such a search. To apply this algorithm to a list of names we require the list to be in **alphabetical order**. To apply this algorithm to a list of numbers we require the list to be in **ascending order**.

This algorithm concentrates on the middle item of a reducing list. If the items in the reduced list are numbered $n_1, n_1 + 1, n_1 + 2, \dots, n_2$, then the middle item is numbered $[\frac{1}{2}(n_1 + n_2)]$, using the notation defined in section 1.3. For example, suppose the reduced list is:

(7) EVANS
(8) FRIEND
(9) GOOD
(10) HEAP
(11) JONES
(12) LOVE

then $n_1 = 7$ and $n_2 = 12$.

So $[\frac{1}{2}(n_1 + n_2)] = [\frac{1}{2}(7 + 12)] = [9\frac{1}{2}] = 10$ and the middle item is (10) HEAP.

Suppose we are looking for NAME in a list of names in alphabetical order.

Step 1 Compare NAME with the middle name of the list. There are three possible outcomes of this comparison.
(i) The middle name is NAME and the search is then complete.
(ii) NAME occurs *before* the middle name alphabetically.
(iii) NAME occurs *after* the middle name alphabetically.

Step 2 If (ii) is true, repeat step 1 on the first half of the list.
If (iii) is true, repeat step 1 on the second half of the list.

Stop Either when NAME has been located or when it has been shown that NAME does not appear in the list.

At each stage the number of names to be searched is halved, hence the name of the algorithm.

The two possible ways of ending the search are shown in the examples which follow.

Example 16

In the list below use the binary-search algorithm to locate the names
(a) DIXON
(b) DAVY.

(1) ABBOTT
(2) BROWN
(3) CARR
(4) CASSON
(5) CATER
(6) DANIEL
(7) DIXON
(8) DOMB
(9) FOX
(10) GOUCH
(11) HAMPTON

(a) As there are 11 names in the list the middle name is at $[\frac{1}{2}(11 + 1)] = 6$, i.e. DANIEL.

DIXON occurs *after* DANIEL alphabetically so we consider the reduced list

(7) DIXON
(8) DOMB
(9) FOX
(10) GOUCH
(11) HAMPTON

The middle name in this list occurs at the position given by $[\frac{1}{2}(7 + 11)] = 9$, i.e. FOX.

As DIXON occurs *before* FOX alphabetically we now consider the reduced list

(7) DIXON
(8) DOMB

The middle name in this list occurs at the position given by $[\frac{1}{2}(7 + 8)] = 8$, i.e. DOMB.

As DIXON occurs *before* DOMB alphabetically, we consider the reduced list

(7) DIXON

Hence DIXON occurs in the list and is in the 7th position.

(b) As before the middle name of the original list is DANIEL.
DAVY occurs *after* DANIEL alphabetically so as before we consider the reduced list

(7) DIXON
(8) DOMB
(9) FOX
(10) GOUCH
(11) HAMPTON

The middle name is again (9) FOX.
DAVY occurs *before* FOX alphabetically so we consider the reduced list

(7) DIXON
(8) DOMB

The middle name is again (8) DOMB and DAVY occurs before DOMB alphabetically so we consider the reduced list

(7) DIXON

Applying the algorithm for a final time results in no list being formed. Hence we conclude that DAVY is not in the list.

Example 17

Use the binary-search algorithm to locate the number 10 in the list

$$1,\ 4,\ 8,\ 9,\ 13,\ 15,\ 16,\ 20$$

We begin by forming the list

(1) 1
(2) 4
(3) 8
(4) 9
(5) 13*
(6) 15
(7) 16
(8) 20

There are eight numbers in the list so the middle number is at the location $[\frac{1}{2}(8+1)] = [4\frac{1}{2}] = 5$. The fifth number in the list is 13, shown starred.

We know that 10 is less than 13 and so occurs *before* the fifth number. So we consider the reduced list

(1) 1
(2) 4
(3) 8*
(4) 9

There are now four numbers in the list and so the middle number is at the location $[\frac{1}{2}(1+4)] = [2\frac{1}{2}] = 3$. The third number in the list is 8, shown starred.

As 10 is larger than 8 it occurs after the third number. So we consider the reduced list

(4) 9

The only number in this list is 9. As this is not 10, the number 10 does not occur in the original list.

Exercise 1D

1 Find the middle item in each of the following lists.

(a) (1) Brown
(2) Green
(3) Orange
(4) Red
(5) Pink

(b) (1) 18
(2) 25
(3) 36

(c) (1) AHMED
(2) BENAUD
(3) BUN
(4) CLARKE
(5) McKENZIE
(6) PETRA

(d) (6) BLACK
(7) COOPER
(8) DENNIS
(9) EDNA
(10) FULLER
(11) MOOSE
(12) RICHARDS

(e) (5) DAVIES
(6) FARMER
(7) JONES
(8) KNOWLES
(9) MANNION
(10) SHEER

2 Use the binary-sort algorithm to locate in the list below

(a) BHAVANA

(b) BEN

(c) CHARLES.

(1) ANGELA

(2) BEN

(3) BHAVANA

(4) CINDY

(5) PETROS

3 The winning ticket numbers in a raffle are given in this table:

	Ticket number
1	26
2	50
3	72
4	81
5	102
6	131
7	265
8	280

Charles bought ticket number 135. Use the binary-search method to show that he did not win a raffle prize.

4 The list of Dr Jadali's appointments for Wednesday morning is given below in alphabetical order:

1 Arnold
2 Beattie
3 Cowen
4 Douglas
5 Evans
6 Gould
7 Harper
8 Norman
9 Piper
10 Wild

Use the binary-search method to confirm that Mrs Cowen is on the appointments list.

The doctor starts his surgery at 10 a.m. and each appointment takes 15 minutes. Work out the time of Mrs Cowen's appointment.

SUMMARY OF KEY POINTS

1 An **algorithm** is a set of precise instructions which if followed will solve a problem.

2 **The bubble-sort algorithm**
To sort a list compare adjacent members of the list, moving from left to right, and switch them if they are in the wrong order. Continue this process until a pass produces no change in the list.

3 **The quick-sort algorithm**
Step 1 Select a specific number (the pivot) from the list, say the middle one.
Step 2 Write all numbers smaller than the pivot to the left of the pivot, reading the original list from left to right, and so create a sublist L_1.
Write all numbers larger than the pivot to the right of the pivot, reading the original list from left to right, and so create a sublist L_2.
Step 3 Apply step 1 and step 2 to each sublist until all the sublists contain only one number.

4 **First-fit algorithm**
Taking the boxes in the order listed place the next box to be packed in the *first* available bin that can take that box.

5 **First-fit decreasing algorithm**
Step 1 Reorder the boxes in *decreasing* order of size.
Step 2 Apply the first-fit algorithm to the reordered list.

6 **Binary-search algorithm**
May be applied to search a list of names in alphabetical order or a list of numbers in ascending order.
Step 1 Compare the required item with the middle item in the list. If this is the required item then the search is complete.
(i) If the required item is before the middle item then consider the top half of the list.
(ii) If the required item is after the middle item then consider the bottom half of the list.
Step 2 Apply step 1 to the top half of the list in case (i) or to the bottom half of the list in case (ii).
Stop Either when the required item is located or when it has been shown that the required item is not in the list.

Graphs and networks

2

2.1 Basic definitions

In your GCSE course you will have considered graphs of functions such as $y = 2x + 3$ and $y = 2x^2$. However, the word 'graph' is used in a much wider sense in advanced mathematics and there is in fact an area of mathematics called **graph theory**.

In this chapter we will use the following definition:

- **A graph G consists of a finite number of points (usually called vertices or nodes) connected by lines (usually called edges or arcs).**

Here is an example of a graph, G_1:

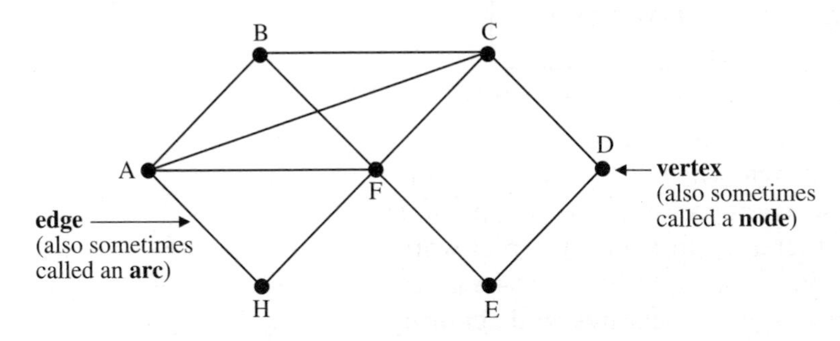

G_1 has seven vertices, A, B, C, D, E, F and H, and 11 edges, AB, AC, AF, AH, BC, BF, CD, CF, DE, EF and FH. Notice that the intersection of AC and BF is not a vertex.

Vertices is the plural of vertex.

2.2 Mathematical modelling

For more than 2000 years physicists and engineers have been using mathematics to solve problems. Today mathematics is also used to solve problems in many other fields, such as biology, economics, geography, management and medicine. Examples of problems that may be solved using mathematics include:

- estimating the height of the leaning tower of Pisa, without climbing it
- predicting the effect of a 30% reduction in income tax, without actually reducing the rate
- predicting the weather.

These problems and many others may be solved using a process called **mathematical modelling**. Mathematical modelling involves:

- **translating** a real-world problem into a mathematical problem or **model**
- **solving** the mathematical problem
- **interpreting** the solution in terms of the real world.

This process of mathematical modelling can be shown as a diagram:

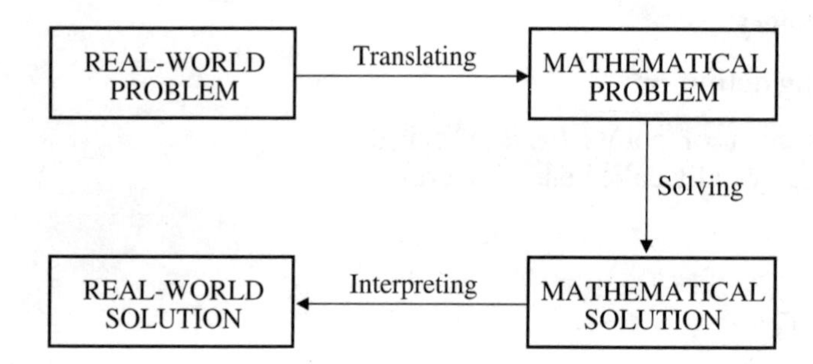

Real-world problems often cannot be translated perfectly into mathematical problems. Even when they can, the resulting mathematical problem may be so complex that it is not possible to solve it. It is often necessary therefore to simplify a real-world problem into one which produces a solvable mathematical problem by concentrating on its essential features and ignoring the rest.

If the model does not produce reliable predictions it may need to be changed in some way, usually by considering the assumptions made in simplification. This is called **refining the model**.

2.3 Modelling using graphs

The following examples involve situations that can be modelled by graphs.

Example 1

Mrs Headley lives in Ashford and wishes to know which towns she can reach directly from Ashford by train. She found the following graph in *Rail routes in Surrey*, and it models the rail network that includes Ashford.

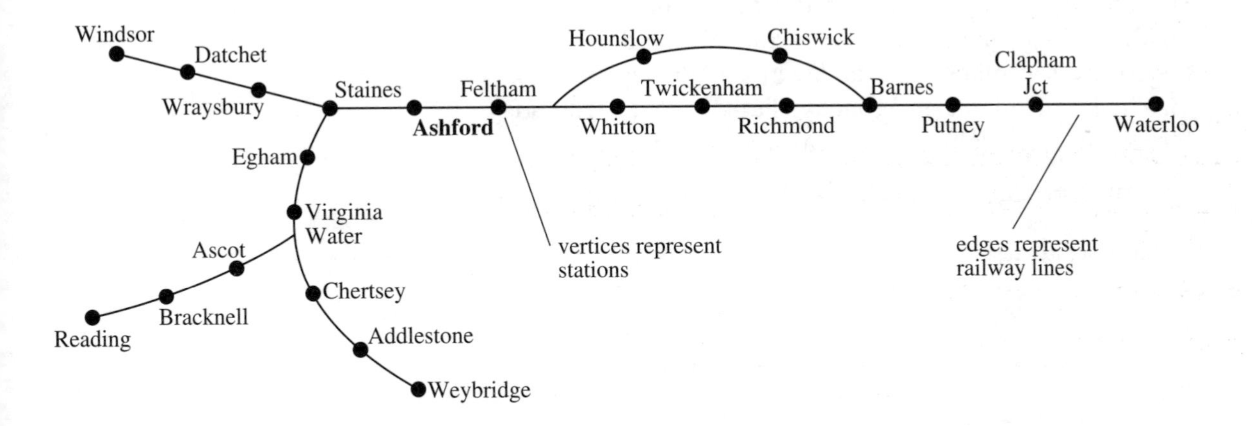

Example 2

In a college there are three faculties, Arts, Science and Management, each of which has a Dean. The heads of the departments in the faculties are responsible to the Dean of their faculty. The Deans are responsible to the Vice Principal (Academic). There is also a Vice Principal (Administration) who is assisted by the Accountant and the Registrar. The two Vice Principals are responsible to the Principal.

This administration can be modelled by the following graph:

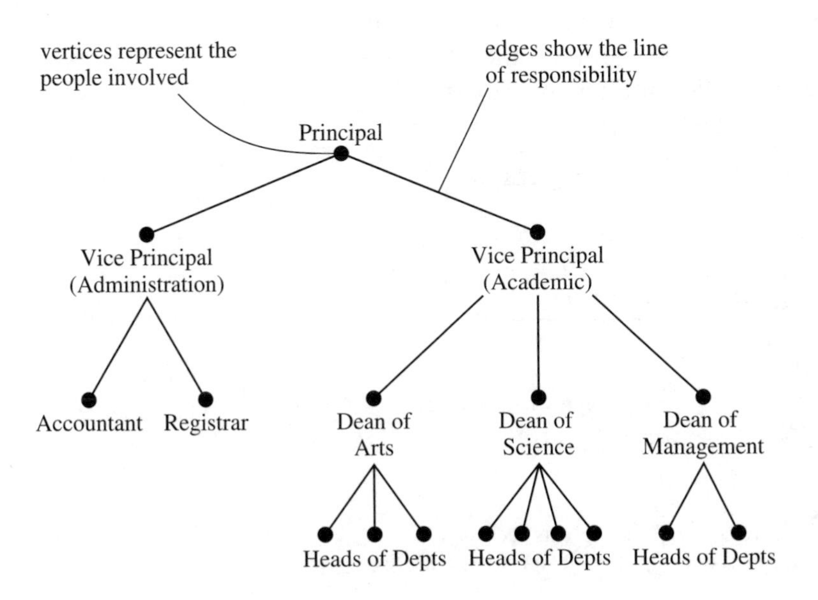

Example 3

A charity shop wishes to find one person per day to manage the shop from Monday to Friday. Five people, Mr Ahmed, Mr Brown, Ms Candy, Ms Davis and Mrs Evans, come forward. They each fill in a form and the following information is obtained:

Mr Ahmed is available on Thursday and Friday
Mr Brown is available on Tuesday and Wednesday
Ms Candy is available on Tuesday and Thursday
Ms Davis is available on Monday, Tuesday and Wednesday
Mrs Evans is available on Tuesday and Friday

This information can be modelled by the following graph:

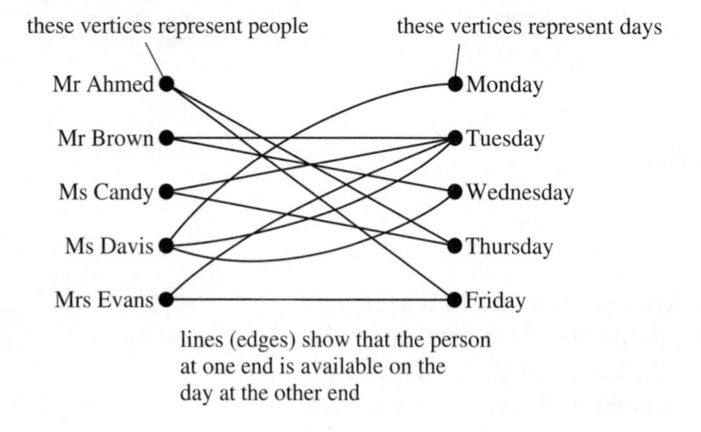

Example 4

A project consists of activities A, B, C, ..., H, which are not all independent. Activities A and B can start any time. Once A is complete C and D can start. When B is complete G can start. E can start only when C is complete. F requires both D and E to be complete before it can start. H can start when F and G are complete. The project is complete when H is complete.

All this information can be modelled by the following graph:

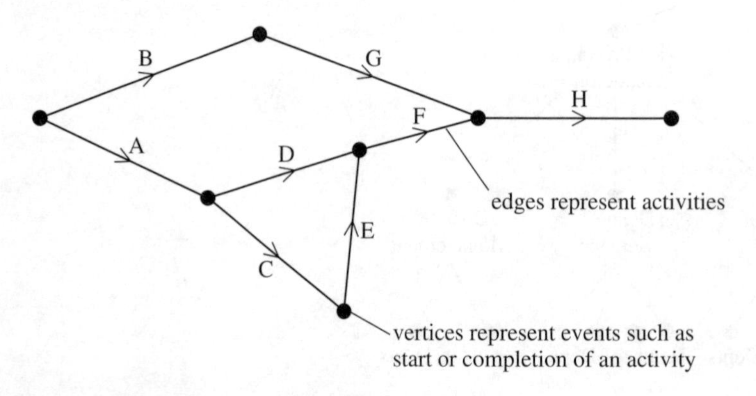

Example 5

The ground-floor plan of a house is shown in the following diagram:

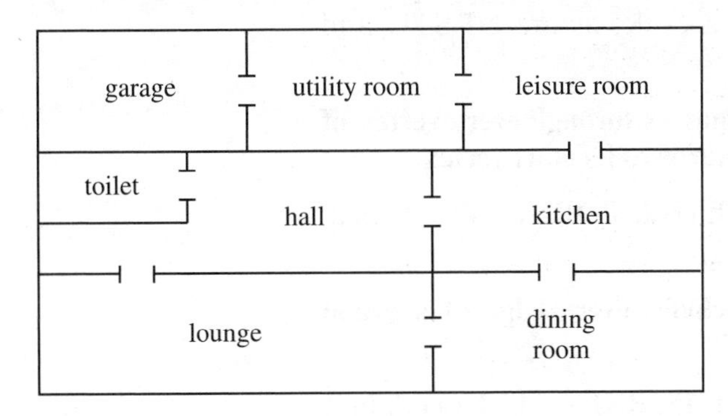

Using vertices to represent rooms and an edge to show two rooms having a connecting door we may represent the plan by the following graph:

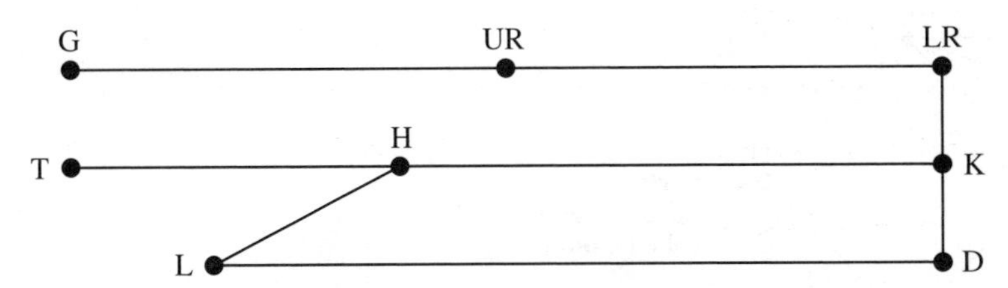

Such a graph is called a **circulation graph** and is often used by architects.

2.4 Definitions of terms used in graph theory

- **A *path* is a finite sequence of edges such that the end vertex of one edge in the sequence is the start vertex of the next.**

For example, in the graph G_2:

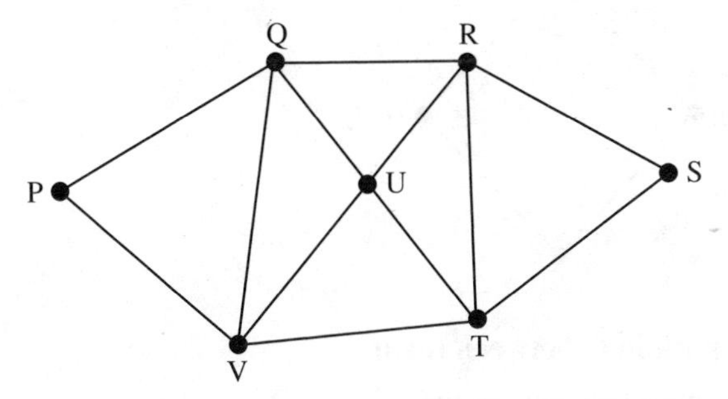

PQRS is a path, as is PVUTR.

- **A *cycle* (or *circuit*) is a closed path, i.e. the end vertex of the last edge is the start vertex of the first edge.**

In graph G_2 on page 31, PQRTVP is a cycle, as are URSTU and PQUTVP.

- **A *Hamiltonian cycle* is a cycle that passes through every vertex of the graph once and only once, and returns to its start vertex.**

A Hamiltonian cycle in the above graph G_2 is PQRSTUVP. (It may not be possible to find such a cycle in a graph.)

- **A *Eulerian cycle* is a cycle that includes every edge of a graph exactly once.**

A Eulerian cycle in the above graph G_2 is PQRSTRUTVUQVP. Not all graphs contain a Eulerian cycle.

- **The *vertex set* is the set of all vertices of a graph.**

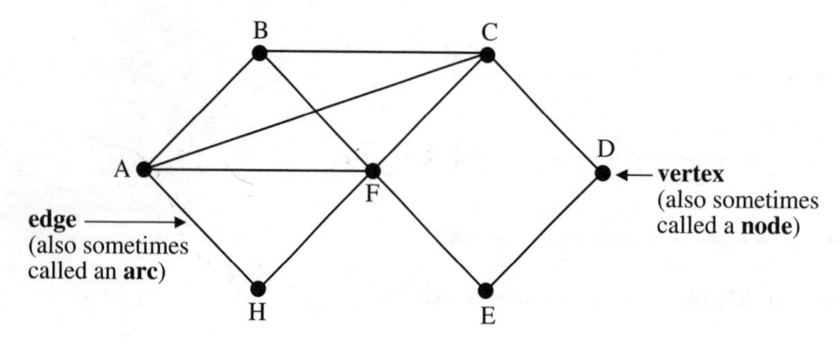

The vertex set for the graph G_1 above is {A, B, C, D, E, F, H}.

- **The *edge set* is the set of all edges of a graph.**

The edge set for the graph G_1 is {AB, AC, AF, AH, BC, BF, CD, CF, DE, EF, FH}.

- **A *subgraph* of a graph is a subset of the vertices together with a subset of edges.**

The following are subgraphs of G_1:

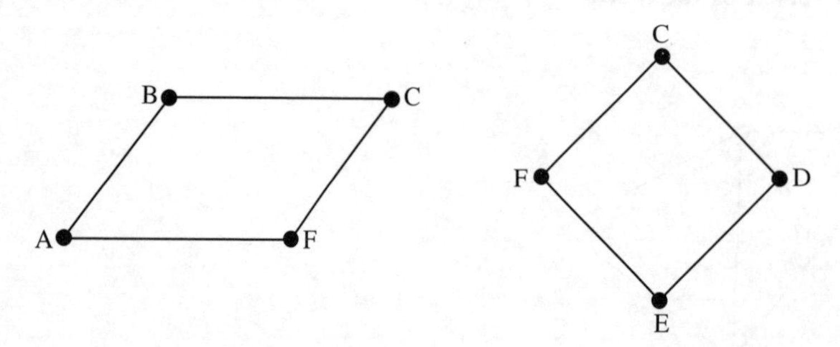

- **Two vertices are *connected* if there is a path in *G* between them.**
- **A graph is *connected* if all pairs of its vertices are connected.**

The graph G_1 is connected but the graph G_3 (below) is not connected as there is no path in G_3 from P to M

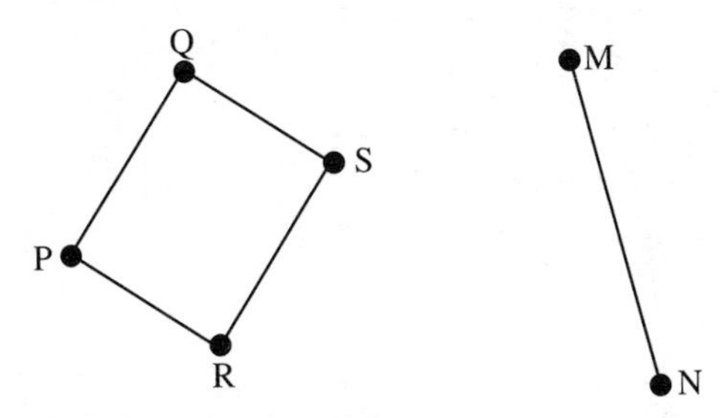

- **A *simple graph* is one in which there are no loops, i.e. no edges with the same vertex at each end, and not more than one edge connecting any pair of vertices.**

The two graphs below are **not simple**.

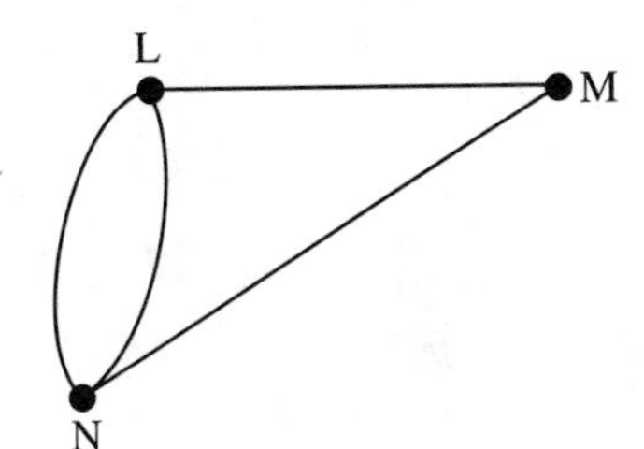

two edges connect L and N

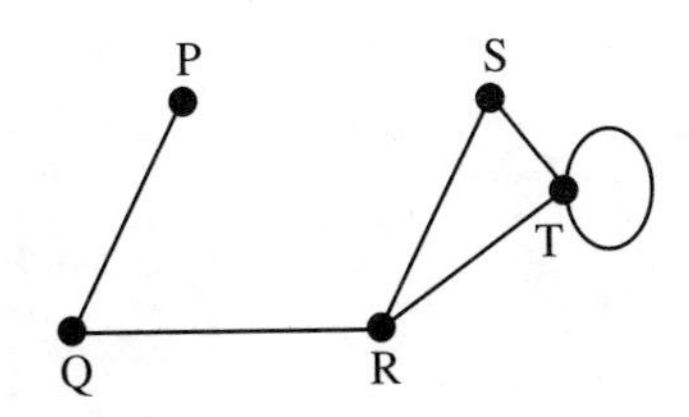

this graph contains a loop

- **The *degree* (or *valency* or *order*) of a vertex is the number of edges connected to it.**

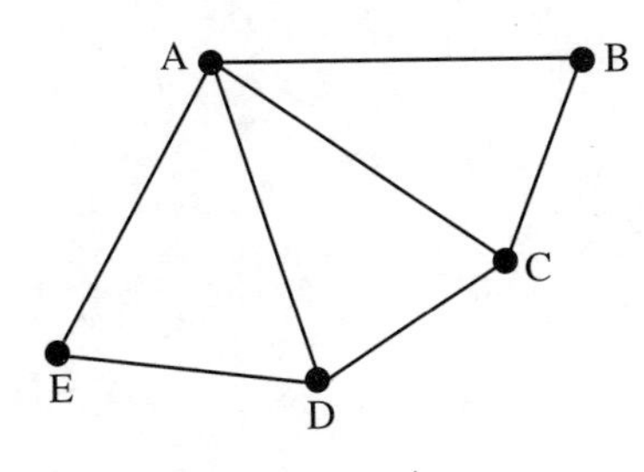

Vertex	Degree
A	4
B	2
C	3
D	3
E	2

Vertices with an odd degree, i.e. C and D, are sometimes called **odd vertices** and vertices with an even degree are sometimes called **even vertices**. A, B and E are all even vertices.

- **If the edges of a graph have a direction associated with them they are known as *directed edges*, and the graph is known as a *digraph*.**

An example of a digraph is shown below.

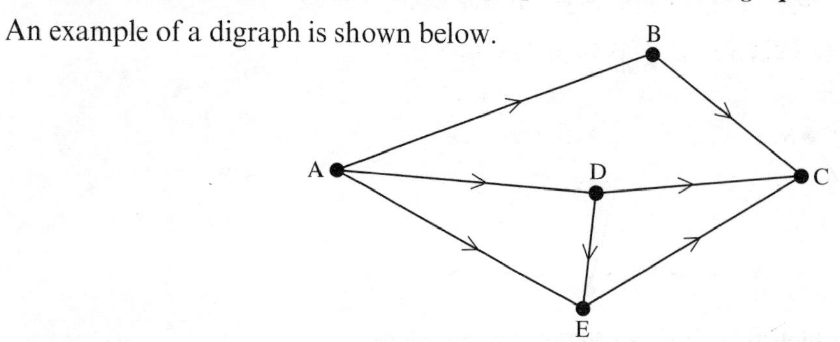

Exercise 2A

1 The plan of the upper floor of a house is shown below. Draw a circulation graph to model this.

stairs
landing
bedroom 2
store room
bedroom 1
shower
bathroom
study

2 Five people, A, B, C, D and E, in an office were asked which of five jobs, 1, 2, 3, 4 and 5, they would be willing to do. Their replies are summarised in the table.

Person	Jobs
A	1, 3, 5
B	1, 4
C	2, 3, 4
D	1, 2, 3
E	1, 5

Represent this information as a graph with people and jobs modelled by vertices and their preferences by edges.

3 There are seven activities, A, B, C, D, E, F and G, involved in a project. The project is complete when all activities are complete. Activity A does not depend on any others. Once it is complete B and C can start. Activities D and E require B to be completed before they begin. Only when both C and D are finished can F start. Activity G requires both E and F to be finished before it can start.
Model this information by a graph using edges to represent activities.

4

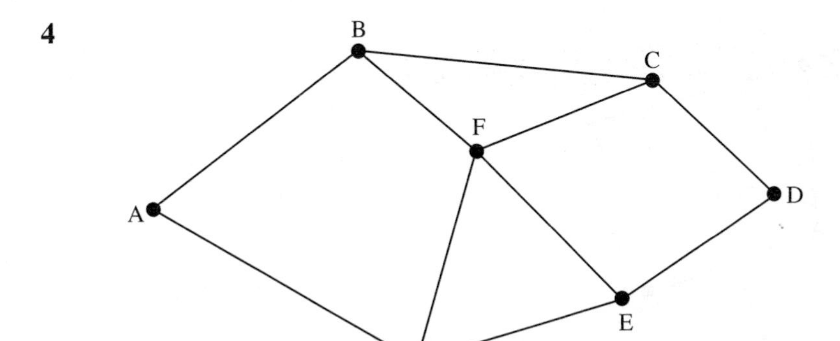

For the above graph:
(a) write down four different paths from A to D via F
(b) write down the degree of each of the vertices
(c) write down two cycles that pass through A and D
(d) write down two Hamiltonian cycles.

5

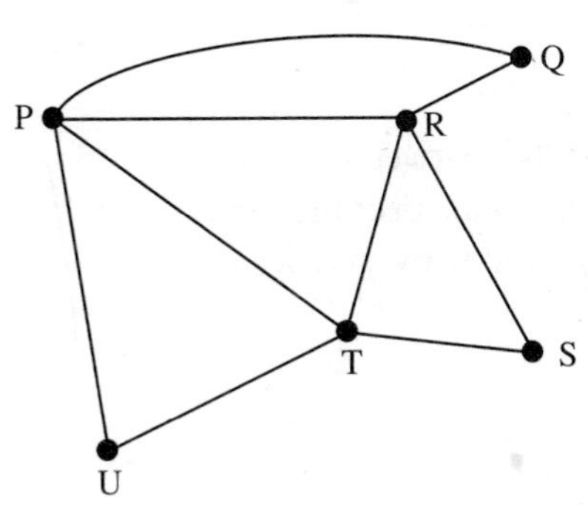

(a) Write down the degree of each vertex in the above graph.
(b) Obtain an Eulerian cycle for the graph.
(c) If the vertex Q together with the edges QP and QR are removed, is it still possible to obtain a Eulerian cycle? Give a reason for your answer.

6

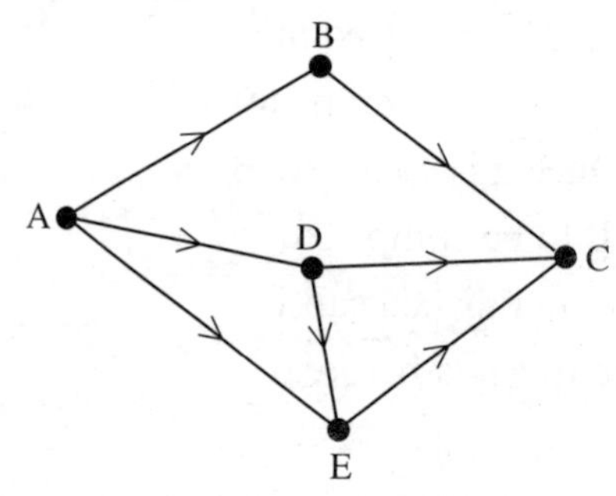

The figure shows a digraph that models the one-way streets between A and C. Write down all the possible routes from A to C for a car.

7

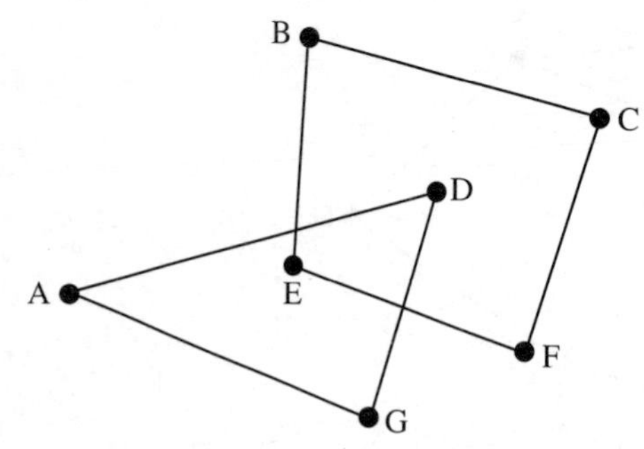

In the above graph the intersections of AD and BE, and DG and EF are not vertices. Is the graph connected or not connected? Give a reason for your answer.

2.5 Other ways of representing graphs

Graphs are very useful for picturing relationships between objects but if there are many objects involved the graphs may become very complicated. Graphs can also be represented by lists or by matrices.

Consider the graph:

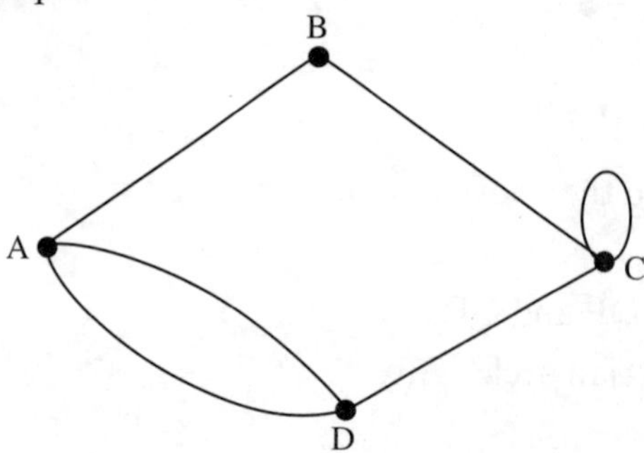

This may be represented by listing the vertex set and the edge set:

vertex set {A, B, C, D}
edge set {AB, AD (twice), BC, CC, CD}
↑
loop

The graph may also be represented by a matrix called the **adjacency matrix**. Each row and each column of the matrix represents a vertex of the graph. The numbers in the matrix give the number of edges joining the pair of vertices. By convention a loop is regarded as two edges.

The adjacency matrix for the above graph is:

	A	**B**	**C**	**D**
A	0	1	0	2
B	1	0	1	0
C	0	1	2	1
D	2	0	1	0

The first line arises from the fact that A is joined to B by one edge and A is joined to D by two edges.

When we have a digraph we have to take into account the directions of the edges and only include in the matrix the number of edges in the given direction. For example, the digraph

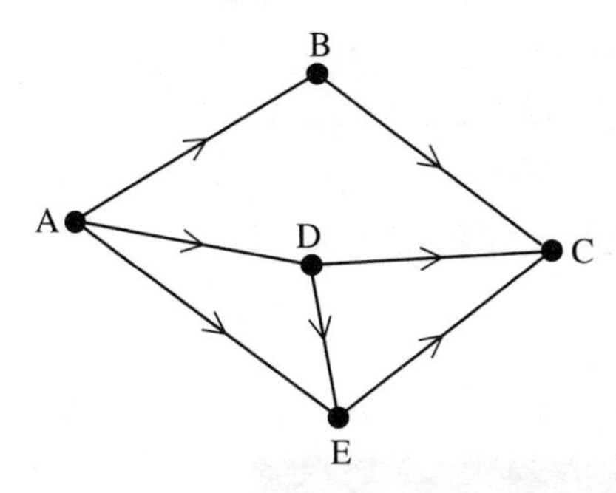

may be represented by the matrix

To

From		**A**	**B**	**C**	**D**	**E**
	A	0	1	0	1	1
	B	0	0	1	0	0
From	**C**	0	0	0	0	0
	D	0	0	1	0	1
	E	0	0	1	0	0

The vertex set is {A, B, C, D, E} and the edge set is {AB, AD, AE, BC, DC, DE, EC}. In this case the order of letters is important.

It is possible to have a graph in which only some edges are directed, for example:

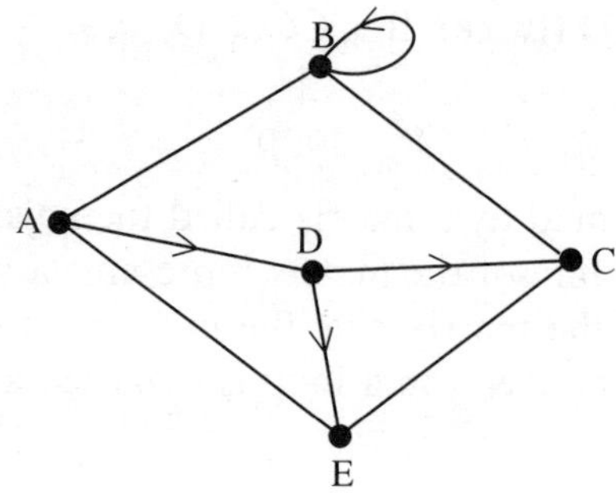

The undirected edges, that is edges without arrows, are assumed to be bidirectional. The adjacency matrix for this graph is:

To

From		A	B	C	D	E
	A	0	1	0	1	1
	B	1	1	1	0	0
From	**C**	0	1	0	0	1
	D	0	0	1	0	1
	E	1	0	1	0	0

Notice that the loop at B is directed so the entry at BB is 1, not 2. The vertex set for this graph is {A, B, C, D, E} and the edge set is {AB, BA, AD, AE, EA, BB, BC, CB, CE, EC, DC, DE}.

Representing a graph by a matrix allows the information stored in the graph to be manipulated by a computer.

Exercise 2B

1 For each of the following graphs write down the vertex set and the edge set.

(a) (b)

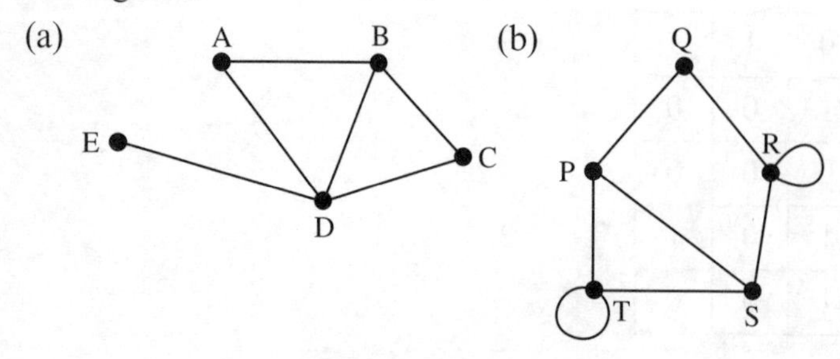

2 Write down the adjacency matrix for each of the graphs in question 1.

3 Produce an adjacency matrix for the circulation graph in Example 5 in section 2.3.

4 Write down the adjacency matrix for this digraph.

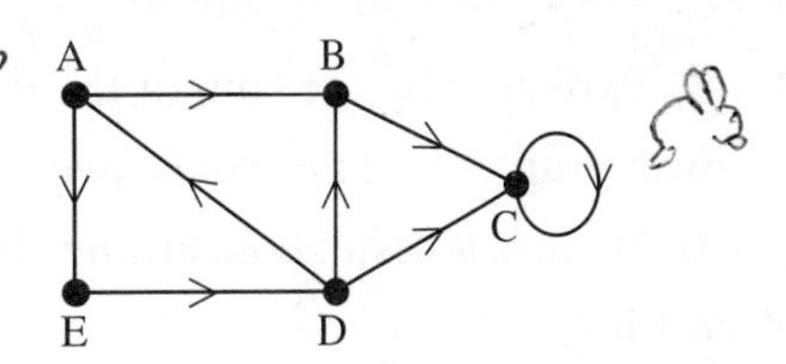

5 Write down the vertex set and edge set for the graph in question 4.

6 Write down the adjacency matrix for this digraph.

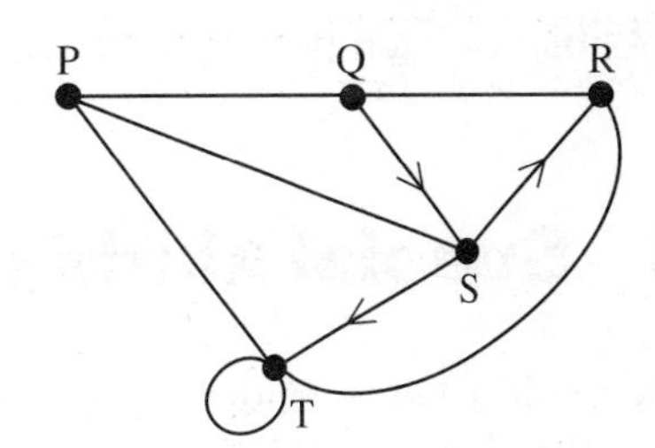

7 Draw a pictorial representation of the graph given by the adjacency matrix

		To				
		A	**B**	**C**	**D**	**E**
From	**A**	2	1	1	0	1
	B	1	0	1	1	1
	C	1	1	0	0	1
	D	0	1	0	2	0
	E	1	1	1	0	2

8 Draw a pictorial representation of the digraph given by the adjacency matrix

		To				
		P	**Q**	**R**	**S**	**T**
From	**P**	0	1	0	0	1
	Q	0	0	1	0	0
	R	0	0	0	0	0
	S	0	1	1	0	0
	T	0	0	1	1	0

9 Produce an adjacency matrix to represent the digraph given by

vertex set {X, Y, Z, V, W}

edge set {VW, WV, VX, XV, VY, VZ, WW, XW, XZ, YX, ZX}

Draw a picture of the graph.

10 (a) Explain why the sum of the degrees of all the vertices in a given graph must always be even.

(b) Deduce a result concerning the number of odd vertices in a graph.

(c) There are seven people in a group. Show that it is not possible for each person to be friends with exactly five others.

2.6 Special kinds of graphs

In this section we consider some special kinds of graphs that will occur in the sections and chapters that follow.

Trees

■ **A *tree* is a *connected* graph with *no* cycles.**

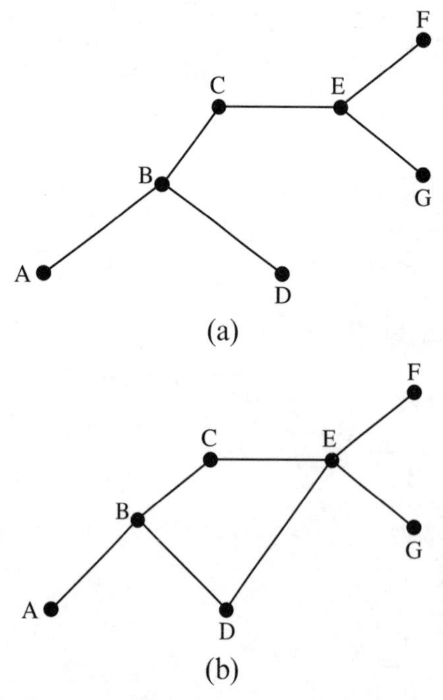

(a)

(b)

Graph (a) is a tree. The graph (b) is not a tree as it contains a cycle, BCEDB.

Here are some examples of trees.

A family tree

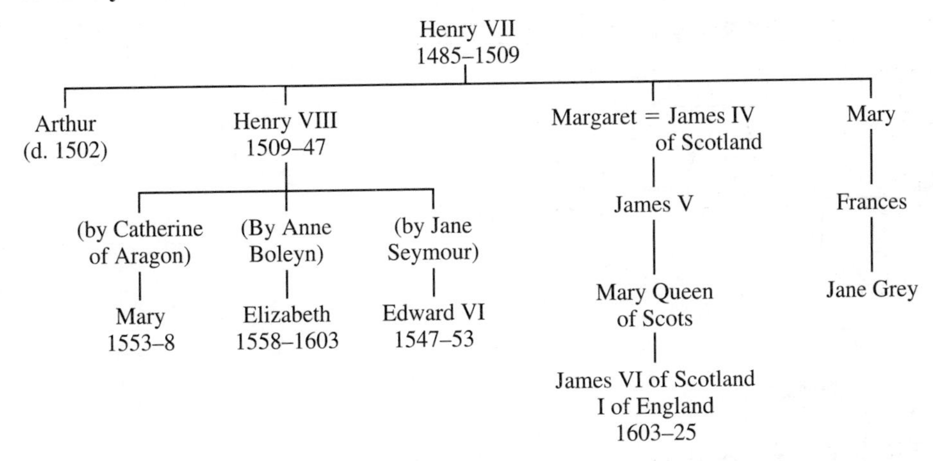

A probability tree

You have probably come across probability trees in your GCSE course. Here is the tree for the sampling of two balls without replacement from a bag containing three red balls and nine blue balls.

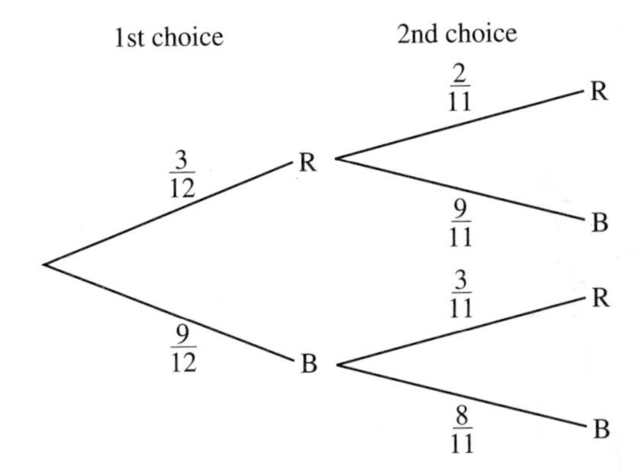

A river system

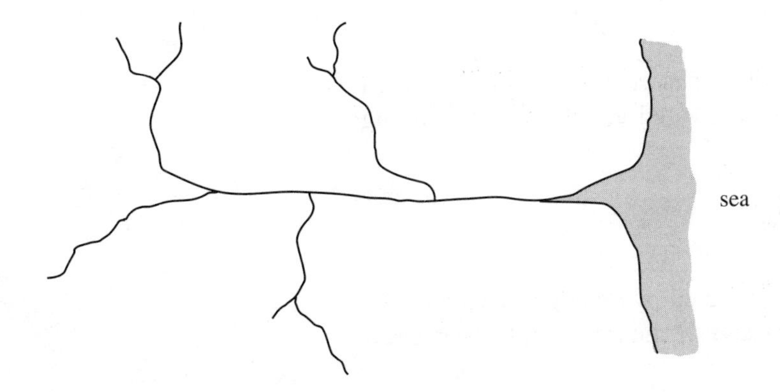

A storage system in a computer

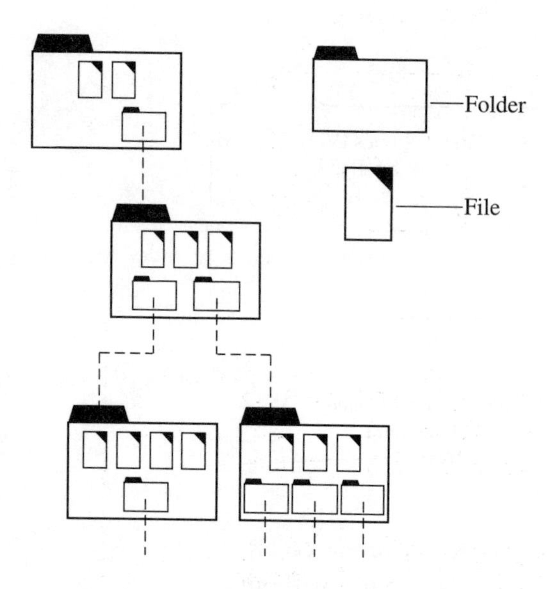

■ **A *spanning tree* of a graph *G* is a subgraph that includes all the vertices of *G* and is also a tree.**

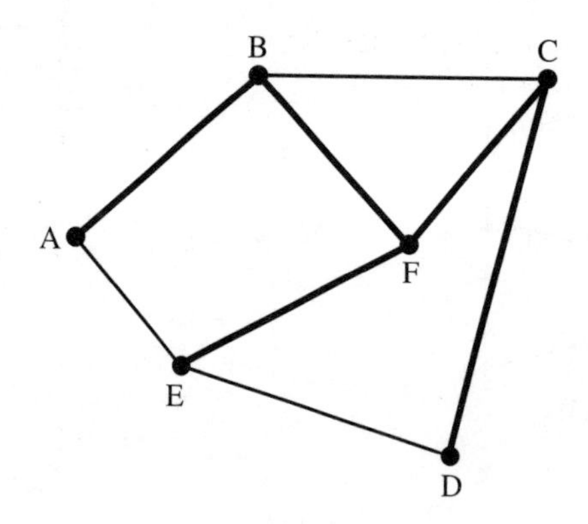

A spanning tree in the above graph is shown by the bold lines. In Chapter 3 you will see that every connected graph has a spanning tree.

Complete graph

A graph in which every vertex is connected by an edge to each of the other vertices is called a **complete graph**. If the graph has n vertices it is denoted by K_n.

Some examples of complete graphs are:

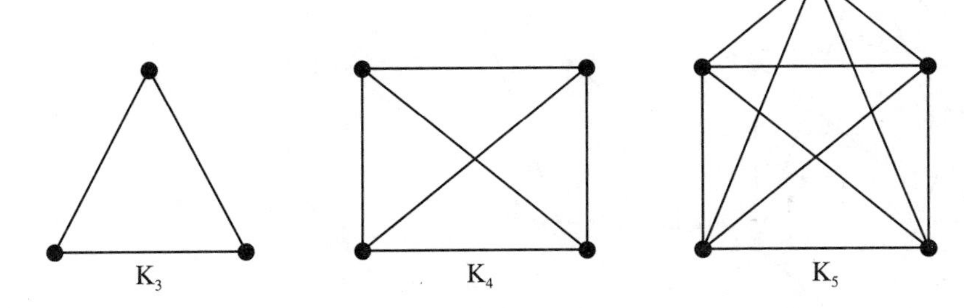

Bipartite graph

The graph that modelled the information in Example 3 in section 2.3 had a very special property: its edges only joined people with days; there were no edges joining two people or two days. Such a graph is called a **bipartite graph.**

- **A bipartite graph consists of two sets of vertices, X and Y. The edges only join vertices in X to vertices in Y, not vertices within a set.**

Sometimes it is useful to talk about complete bipartite graphs.

- **If there are r vertices in X and s vertices in Y and every vertex in X is joined to every vertex in Y then the graph is called $K_{r,s}$.**

For example, suppose there are three houses A, B and C which have to be connected to supplies of water W and electricity E.

The two sets are $X = \{A, B, C\}$
$Y = \{W, E\}$

The graph modelling this situation is

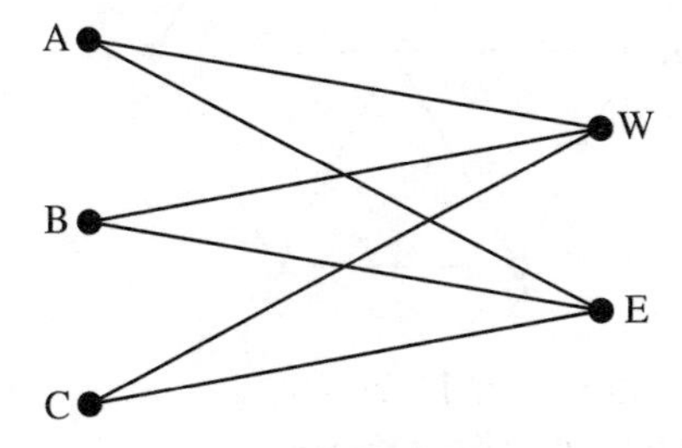

Here $r = 3$ and $s = 2$ and the graph is $K_{3,2}$.

Planar graph

A graph that can be drawn in a plane in such a way that no two edges meet each other, except at a vertex to which they are both incident, is called a **planar graph**.

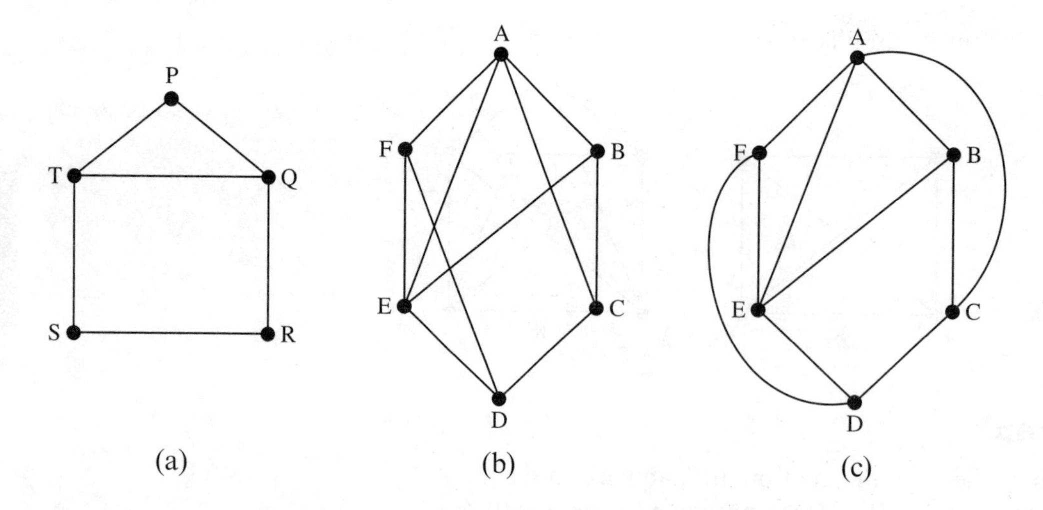

The graph in (a) is obviously planar. The graph in (b) is not obviously planar but by redrawing the graph as shown in (c) it can be seen to be planar.

Isomorphic graphs

Two graphs G_1 and G_2 are **isomorphic** if they have the same number of vertices and the degrees of corresponding vertices are the same. The following two graphs are isomorphic:

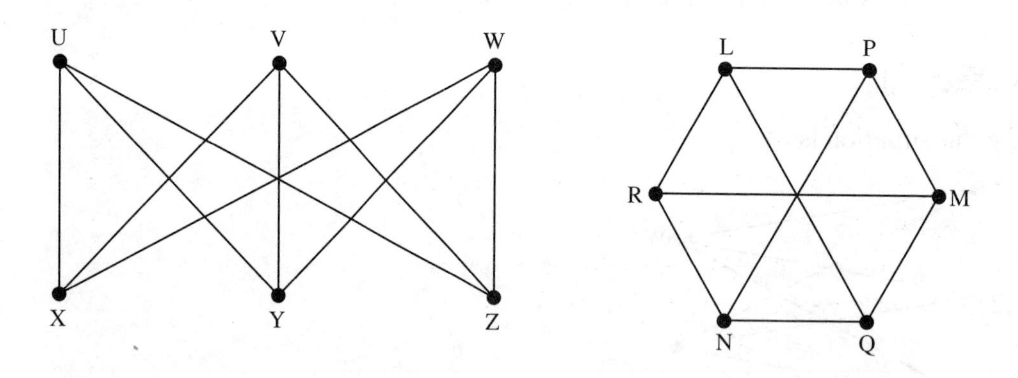

The corresponding vertices are (U, L), (V, M), (W, N), (X, P), (Y, Q) and (Z, R).

Networks

If a graph has a number associated with each edge, usually called its **weight**, then the graph is called a **weighted graph** or **network**.

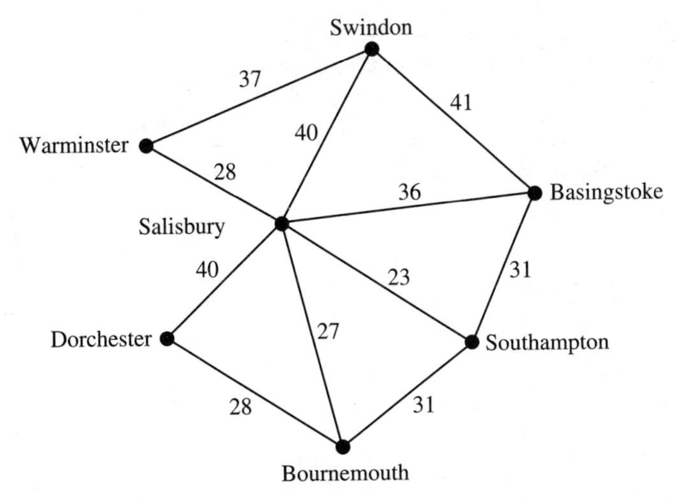

The above diagram shows a typical network. It shows several towns and the distances, in miles, between them.

In a triangle ABC the following inequality always applies:

$$\text{length AB} + \text{length AC} \geqslant \text{length BC}$$

This is called the **triangle inequality**.

If in a graph the inequality

$$\text{weight BC} \leqslant \text{weight AB} + \text{weight AC}$$

holds for all sets of three vertices {A, B, C}, then the network is said to satisfy the **triangle inequality**.

If the weights are times or costs the triangle inequality may not be satisfied. It is even possible for it not to be satisfied when the weights are distances. For example,

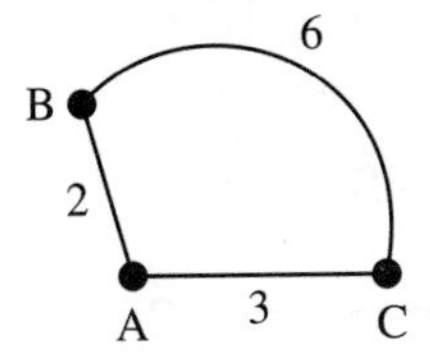

represents the situation where the distance from B to C via the bypass is 6 miles, whereas using the direct routes BA and AC the distance is only 5 miles.

The network of towns shown above may also be represented by a **distance matrix**, in which the entries are the distances between the two towns. We then have:

	W	Sw	Sa	Ba	So	Bo	D
W	0	37	28	—	—	—	—
Sw	37	0	40	41	—	—	—
Sa	28	40	0	36	23	27	40
Ba	—	41	36	0	31	—	—
So	—	—	23	31	0	31	—
Bo	—	—	27	—	31	0	28
D	—	—	40	—	—	28	0

Exercise 2C

1 State which of the following graphs are trees.

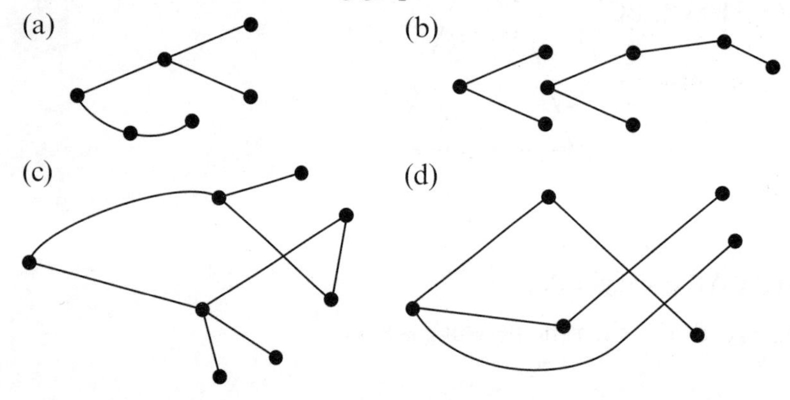

2 A company is divided into two sections: manufacturing and marketing. Each section has a head who reports to the managing director. In each section there are three departments, each with a head who reports to the section head. Each department head is responsible for four workers. Model this company using a tree.

3 For the graph shown draw eight spanning trees.

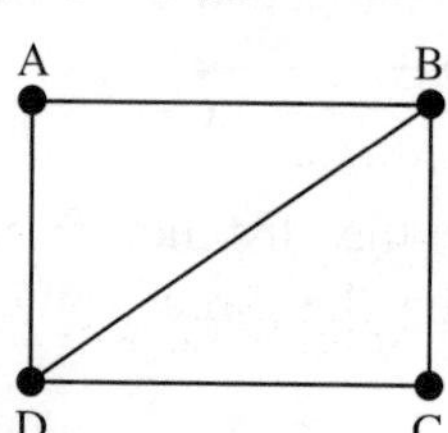

4 Draw the bipartite graph $K_{2,4}$.

5 Which, if either, of the following graphs are bipartite? Give reasons for your answers.

(a)

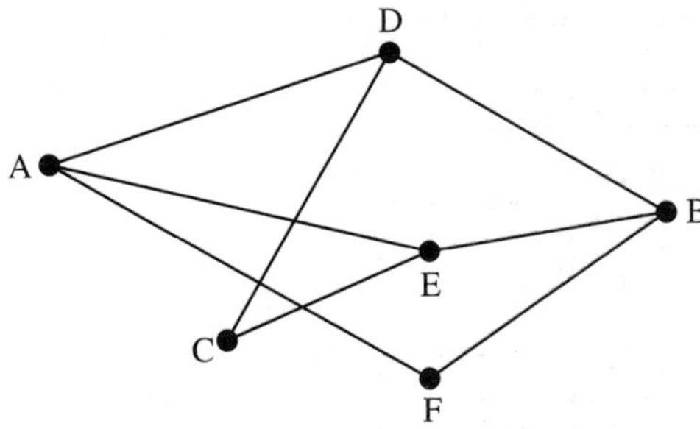

(b)

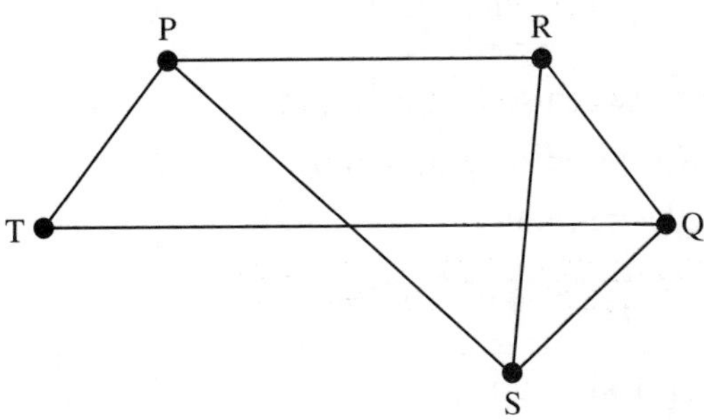

6 Which, if either, of the following networks satisfy the triangle inequality? Give reasons for your answers.

(a)

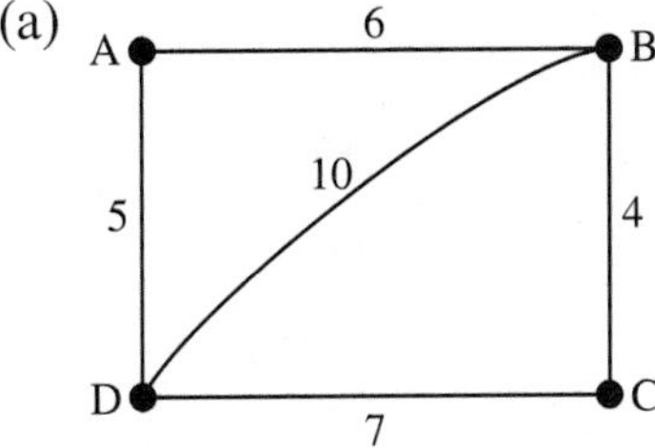

(b)

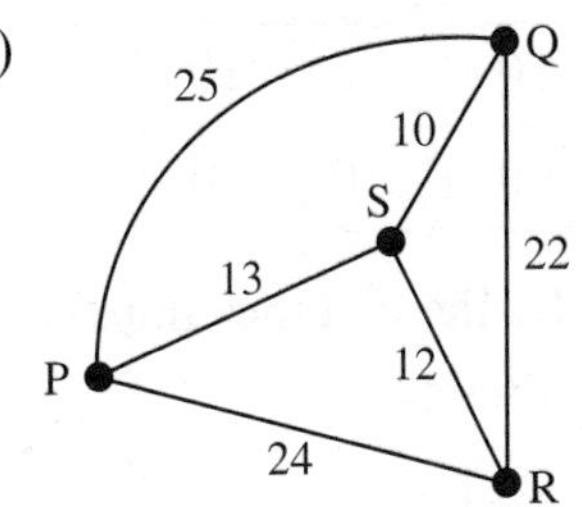

7

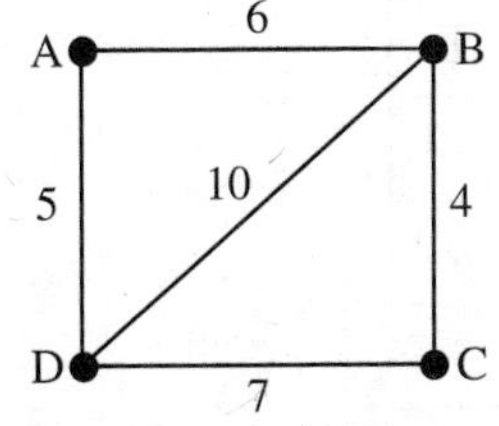

The network shown is the same graph as in question 3 but now with weights on the edges.

(a) For each of the spanning trees obtained in question 3 work out the total weight of the edges.

(b) Which tree has the least weight and which the greatest weight?

8 The table below gives the distances between some towns. Use the information to draw a network.

	A	B	C	D	E	F
A	0	6	—	—	—	7
B	6	0	9	8	—	10
C	—	9	0	5	—	—
D	—	8	5	0	6	—
E	—	—	—	6	0	9
F	7	10	—	—	9	0

9 The times, in minutes, taken to travel between some towns are given in the table below. Draw the corresponding graph and show that it does not satisfy the triangle inequality.

	A	B	C	D	E
A	0	40	—	100	90
B	40	0	30	—	20
C	—	30	0	50	60
D	100	—	50	0	10
E	90	20	60	10	0

10 Draw the directed network given by the distance matrix below, where the distances are given in kilometres.

From \ To	A	B	C	D	E	F
A	—	20	—	—	25	30
B	—	—	10	—	—	—
C	—	—	—	14	—	—
D	—	—	—	—	—	—
E	—	—	—	30	—	—
F	—	—	16	18	12	—

SUMMARY OF KEY POINTS

1 A **graph** G consists of a finite number of points (usually called vertices or nodes) connected by lines (usually called edges or arcs).

2 A **path** is a finite sequence of edges such that the end vertex of one edge in the sequence is the start vertex of the next, and in which no vertex appears more than once.

3 A **cycle** (or circuit) is a closed path, i.e. the end vertex of the last edge is the start vertex of the first edge.

4 A **Hamiltonian cycle** is a cycle that passes through every vertex of the graph once and only once, and returns to its start vertex.

5 A **Eulerian cycle** is a cycle that includes every edge of a graph exactly once.

6 The **vertex set** is the set of all vertices of a graph.

7 The **edge set** is the set of all edges of a graph.

8 A **subgraph** of a graph is a subset of the vertices together with a subset of edges.

9 Two vertices are **connected** if there is a path between them.

10 A graph is **connected** if all pairs of its vertices are connected.

11 A **simple graph** is one in which there is no edge with the same vertex at each end, i.e. no loops, and not more than one edge connecting any pair of vertices.

12 The **degree** (or valency or order) of a vertex is the number of edges connected to it.

13 If the edges of a graph have a direction associated with them they are known as **directed edges**, and the graph is known as a **digraph**.

14 A **tree** is a connected graph with no cycles.

15 A **spanning tree** of a graph G is a subgraph that includes all the vertices of G and is also a tree.

16 A **bipartite graph** consists of two sets of vertices, X and Y. The edges only join vertices in X to vertices in Y, not vertices within a set.

17 If there are r vertices in X and s vertices in Y and every vertex in X is joined to every vertex in Y then the graph is called $K_{r,s}$.

Algorithms on graphs 3

In Chapter 1 algorithms of various kinds were considered and in Chapter 2 we looked at graphs and modelling with graphs. These two areas are brought together in this chapter in which we consider some of the algorithms that have been developed for solving real-life problems that can be modelled by graphs.

3.1 The minimum spanning tree (or minimum connector)

- **A minimum spanning tree of a connected and undirected graph is a spanning tree such that the total length of its edges is as small as possible. (This is sometimes called a minimum connector.)**

To help us understand this mathematical definition let us look at an example that occurs in real life.

A sixth-form college has five sites, each of which has a computer. The network below shows the distances, in kilometres, between each pair of sites. The computers must be connected by underground cables. Note that if no arc is shown connecting a pair of nodes this means that because of underground rock formations no cable can be laid between these two computers.

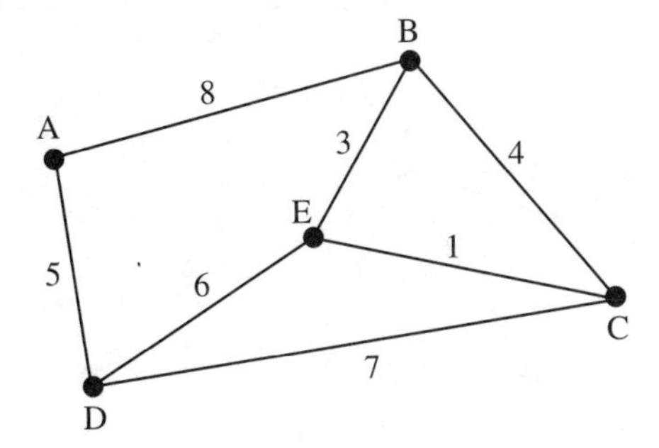

We wish to find the minimum length of cable required to connect the computers, bearing in mind that computers can communicate via other computers.

What we require to solve this problem is a minimum spanning tree for this network. It will then be possible to communicate between any two computers and the sum of the weights of the edges will give the minimum length of cable required.

3.2 Kruskal's algorithm for finding a minimum spanning tree

In 1956 Kruskal published an algorithm for obtaining a minimum spanning tree for a network. Kruskal's algorithm builds a minimum spanning tree by adding one edge at a time (and associated vertices) to a subgraph. At each stage the edge of smallest available weight is chosen **provided that it does not create a cycle with edges already chosen**.

The algorithm may be formally stated as follows.

Step 1 Sort the edges in ascending order of weight.

Step 2 Select the edge of least weight.

Step 3 Select from edges not previously selected the edge of least weight that does not form a cycle together with the edges already included.

Step 4 Repeat step 3 until selected edges form a spanning tree.

Note that the subgraph need not be connected at intermediate stages, although the final graph is connected as it is a tree. Where two edges have the same weight an arbitrary choice may be made. There may be more than one minimum spanning tree but all the minimum spanning trees will have the same minimum total weight.

Kruskal's algorithm chooses the edge of smallest weight at each stage without reference to the edges already chosen as long as no cycles are formed. An algorithm that makes an optimal choice at each stage is called a **greedy algorithm**.

Example 1

Use Kruskal's algorithm to obtain a minimum spanning tree for this network.

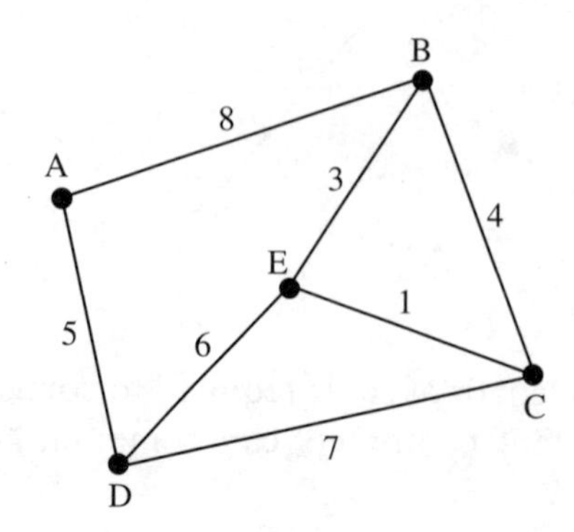

Step 1 By inspection the weights of the edges in ascending order are 1 (EC), 3 (EB), 4 (BC), 5 (AD), 6 (ED), 7 (CD), 8 (AB).

Step 2 The edge of least weight is edge EC, of weight 1. EC is now included.

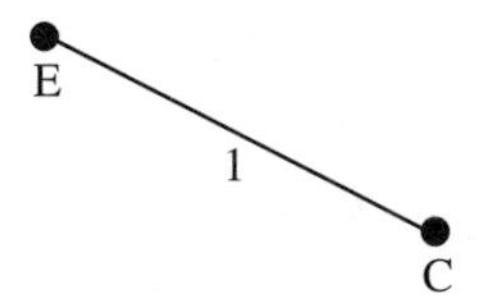

Step 3 (i) The edge of smallest weight available is edge EB, of weight 3. EB is now included.

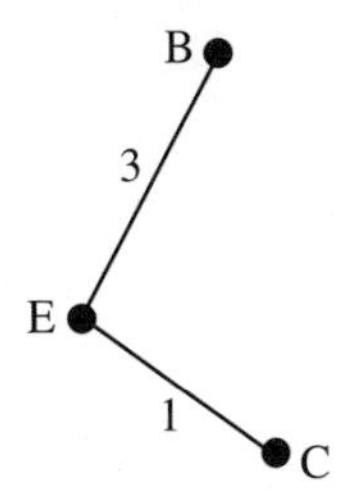

(ii) The edge of smallest weight available is edge BC, of weight 4. However, this would form a cycle with included edges so it is not chosen, i.e. it is excluded.

(iii) The edge of smallest weight available is edge AD, of weight 5. This does not form a cycle and so is included.

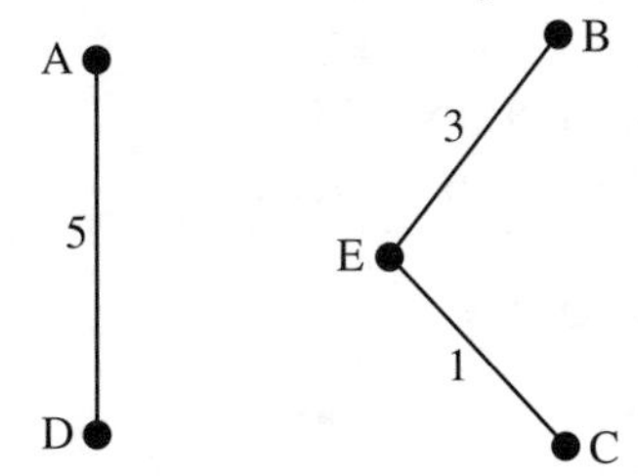

(Notice that at this stage we have a disconnected graph – this does not matter.)

(iv) The edge of smallest weight available is edge DE, of weight 6. This does not form a cycle and so is included.

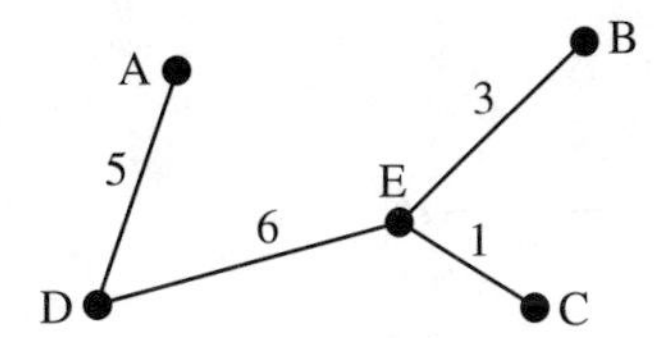

Step 4 All vertices are now in the solution so we stop.

We have now found a minimum spanning tree with a total weight of $6 + 5 + 3 + 1 = 15$. This minimum spanning tree is shown in the diagram in step 3(iv).

Example 2

Using Kruskal's algorithm find as many minimum spanning trees as possible for the following network:

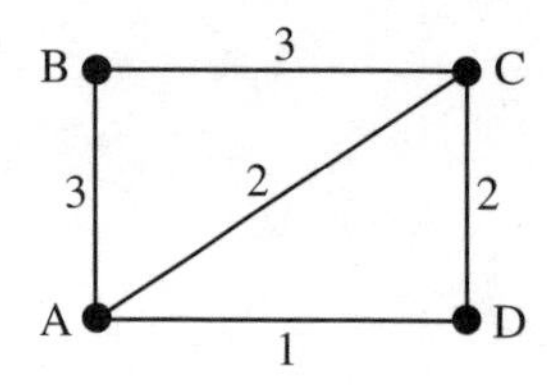

Step 1 By inspection the weights of the edges in ascending order are 1 (AD), 2 (AC), 2 (CD), 3 (AB), 3 (BC).

Step 2 Choose AD.

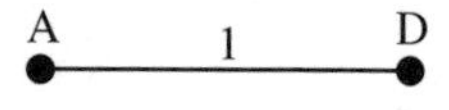

Step 3 (i) Choose either AC or CD

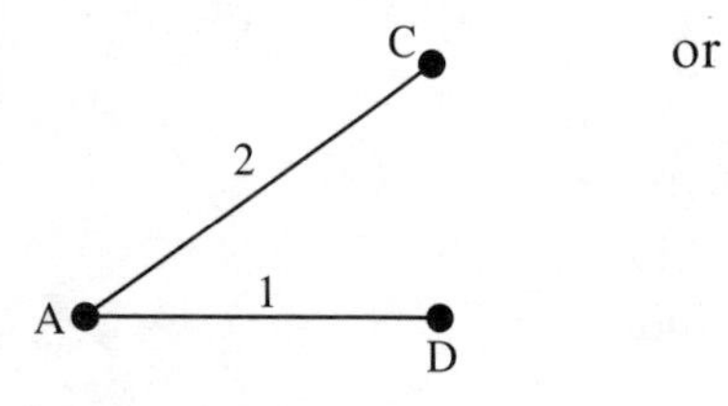

or

(ii) Cannot choose CD as this forms a cycle.

Cannot choose AC as this forms a cycle.

Choose AB

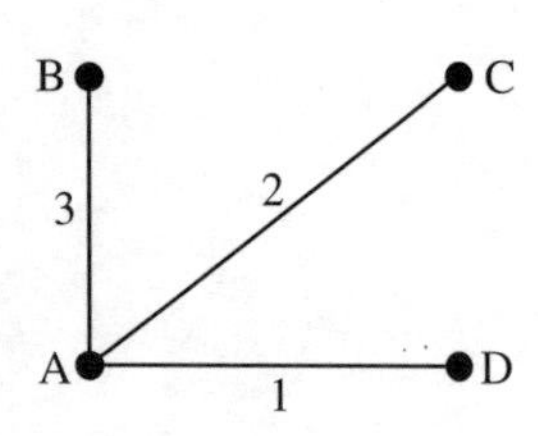

Choose AB

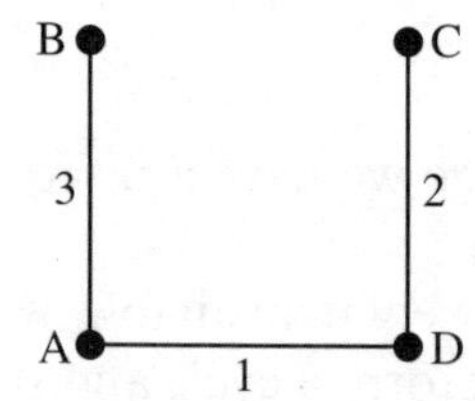

or choose BC

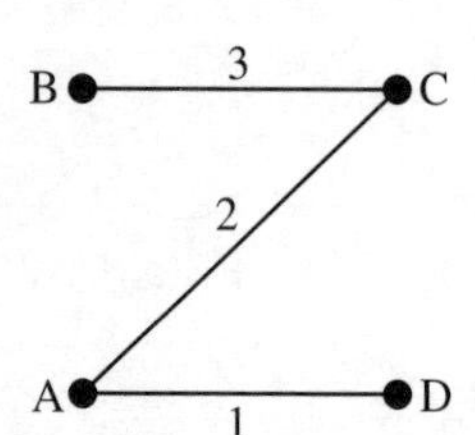

or choose BC

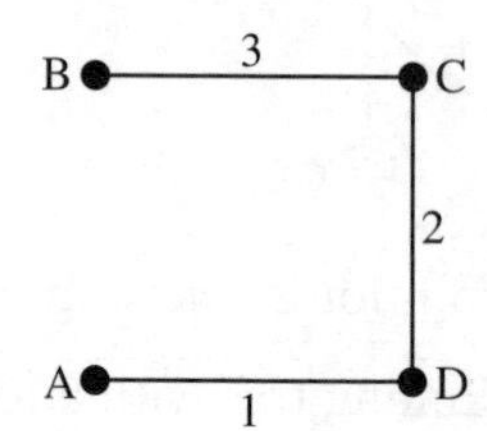

Notice that all four minimum spanning trees have weight 6.

Exercise 3A

1 Use Kruskal's algorithm to find a minimum spanning tree for this network:

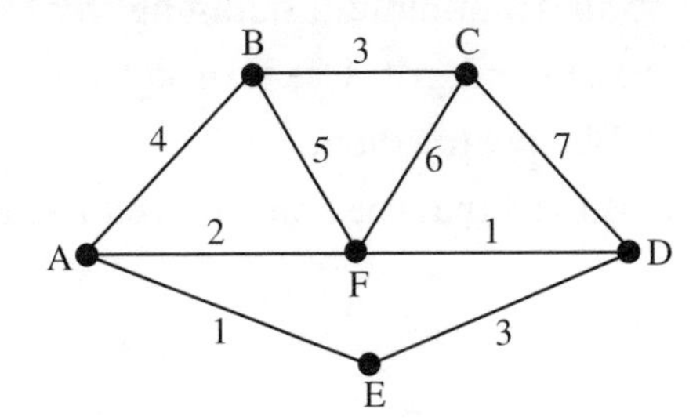

2 Using Kruskal's algorithm find as many minimum spanning trees as possible for this network:

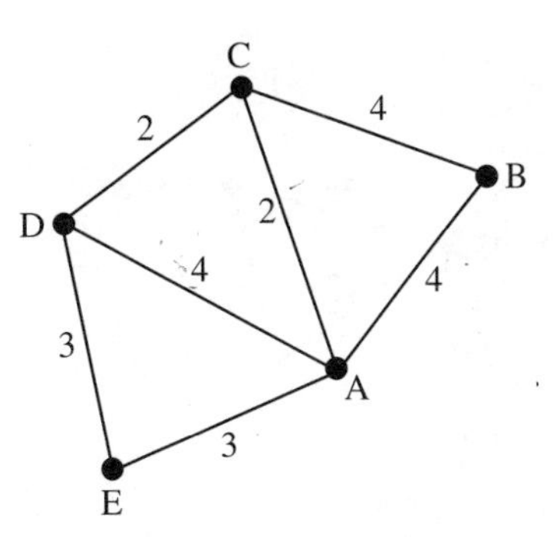

3

	A	B	C	D	E
A	—	10	—	15	17
B	10	—	16	—	20
C	—	16	—	12	18
D	15	—	12	—	24
E	17	20	18	24	—

(a) Draw a pictorial representation of the above distance matrix.

(b) Use Kruskal's algorithm to find a minimum spanning tree and give its total weight.

4

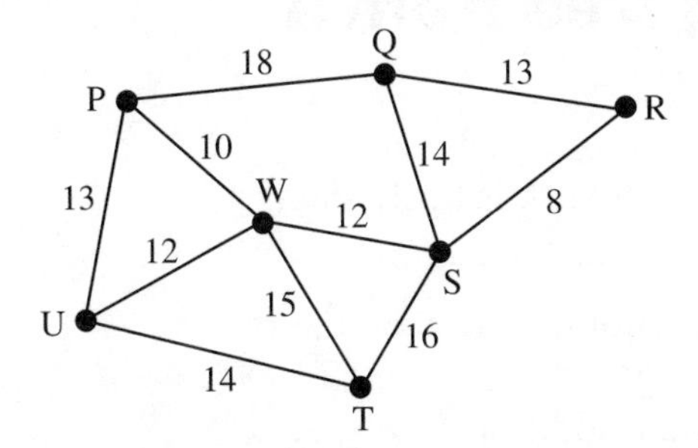

The network above represents the major roads between seven towns. The numbers on the arcs give the distances, in km, between the towns.

The council wishes to keep the towns connected during the winter. They wish therefore to find a minimum spanning tree in order to decide which roads to salt and grit. Use Kruskal's algorithm to determine a suitable tree for them.
What is the minimum length of road that the council need to salt and grit?

5

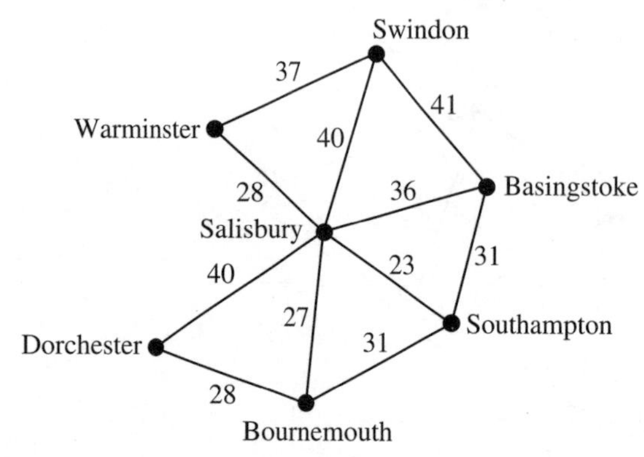

The diagram above shows some places in the south of England. The numbers on the arcs are the distances, in miles, between them. A cable TV company based in Salisbury wishes to link all the places using the minimum length of cable. Use Kruskal's algorithm to find this minimum length of cable. In your solution you should indicate clearly the order in which places are added to your minimum spanning tree. (Specimen Paper Q2 D1 1996)

3.3 Prim's algorithm for finding a minimum spanning tree from a network

Kruskal's algorithm is easy to apply when the network is fairly small but it has two drawbacks:

1 it is necessary to sort the edges in ascending order of their weights before making any selection of edges
2 it is necessary at each stage to check that addition of a given edge does not form a cycle.

Each of these points requires an algorithm within the algorithm. It is quite difficult in a large and complex network to check for cycles.

To overcome these drawbacks, in 1957 Prim proposed another algorithm for finding a minimum spanning tree in a network. Prim's algorithm builds a minimum spanning tree by adding one vertex (and an associated edge) at a time to a connected subgraph.

The algorithm may be formally stated as follows:

Step 1 Choose a starting vertex.

Step 2 Choose the vertex nearest to the starting vertex and add this vertex and the associated edge to the tree.

Step 3 Connect to the tree of connected vertices that vertex that is nearest to **any** vertex in the connected set.

Step 4 Repeat step 3 until all vertices are connected to the tree.

A common error is to misapply step 3. This is often misread or misunderstood to read 'Connect to the tree of connected vertices that vertex that is nearest to the last vertex connected'. The next connection is **not** necessarily to the last vertex connected but to **any** vertex in the connected set, as long as it is the nearest.

Example 3

Use Prim's algorithm to find a minimum spanning tree for this network:

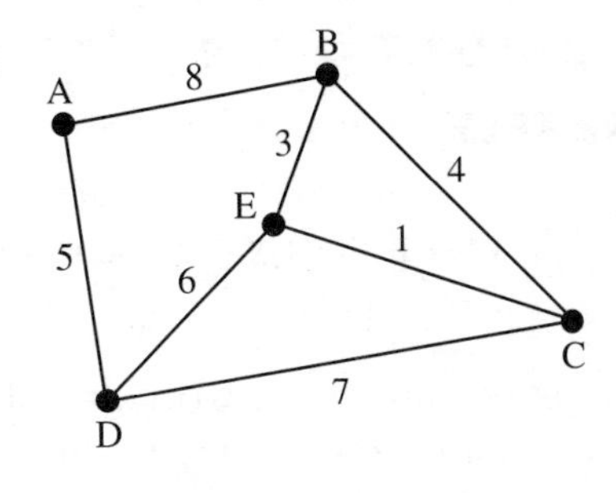

Step 1 Choose vertex A.

Step 2 Nearest vertex to A is vertex D. Add vertex D and edge AD.

Step 3 (i) Vertex nearest to either A or D is E. Add vertex E and edge DE.

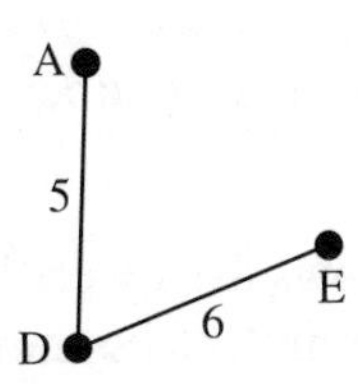

(ii) Vertex nearest to A or D or E is C. Add vertex C and edge EC.

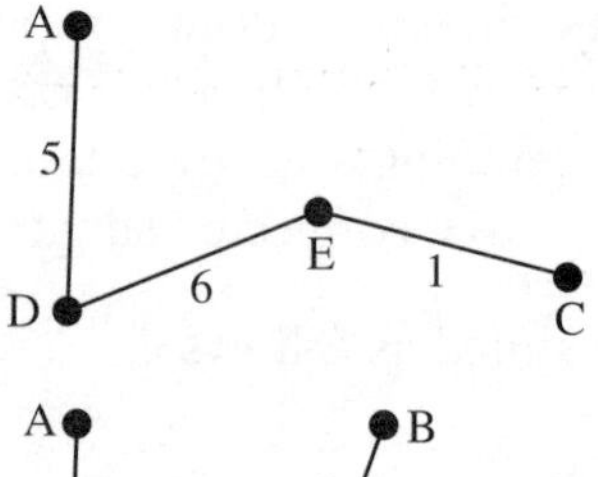

(iii) Vertex nearest to A or D or E or C is B. Add vertex B and edge EB.

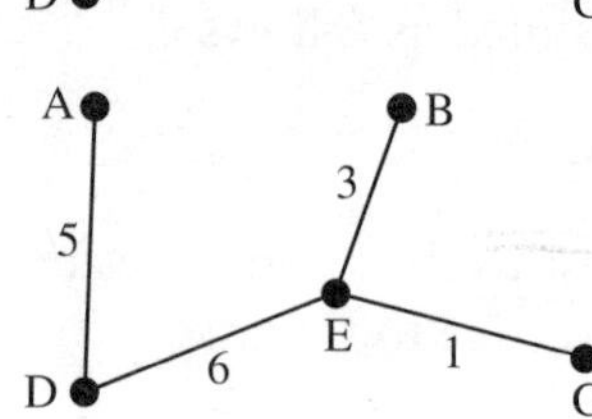

Step 4 All vertices are now connected so we stop.

The network in this example is the same as the one to which we applied Kruskal's algorithm in Example 1. The resulting minimum spanning tree is the same as that obtained in Example 1 as there is only one minimum spanning tree for this network. The weight of the minimum spanning tree is $6 + 5 + 3 + 1 = 15$.

3.4 Prim's algorithm for finding a minimum spanning tree from a distance matrix

Here is how to use Prim's algorithm for a matrix representation of a graph.

Step 1 Start with the matrix representing the network and choose a starting vertex. Delete the *row* corresponding to that vertex.

Step 2 Label with 1 the *column* corresponding to the start vertex, and ring the smallest undeleted entry in that *column*. (There may be a choice of entry to ring.)

Step 3 Delete the *row* corresponding to the entry that you have just ringed.

Step 4 Label (with the next label number) the *column* corresponding to the vertex that you have just ringed.

Step 5 Ring the lowest undeleted entry in *all* labelled columns. (There may be a choice here.)

Step 6 Repeat steps 3, 4 and 5 until all rows are deleted. The ringed entries represent the edges in the minimum connector.

Example 4

Use Prim's algorithm to find a minimum spanning tree for the network represented by the matrix below.

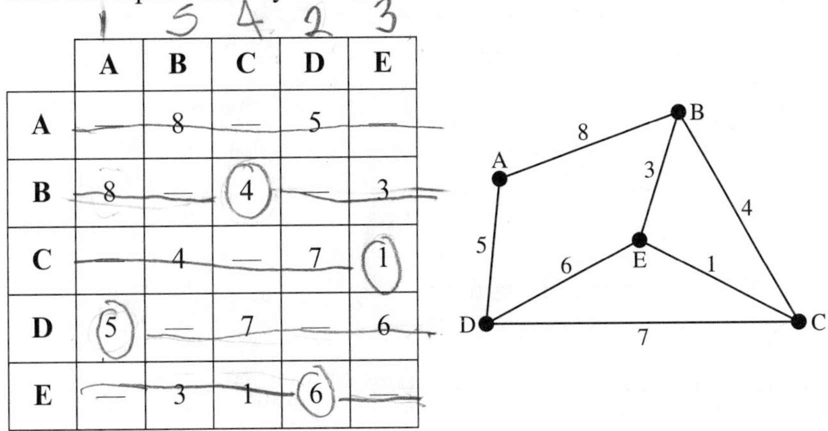

	A	B	C	D	E
A	—	8	—	5	—
B	8	—	4	—	3
C	—	4	—	7	1
D	5	—	7	—	6
E	—	3	1	6	—

Step 1 Choose vertex A and delete the row corresponding to A.

Step 2 Label column A with 1. Ring the smallest entry in column A. This is the 5 in row D.

Step 3 Delete row D.

	1				
	A	**B**	**C**	**D**	**E**
A	—	8	—	5	—
B	8	—	4	—	3
C	—	4	—	7	1
D	(5)	—	7	—	6
E	—	3	1	6	—

Step 4 Label column D with 2.

Step 5 Ring the smallest entry in either column A or column D. This is the 6 in row E.

Step 3 Delete the row corresponding to E.

	1			2	
	A	**B**	**C**	**D**	**E**
A	—	8	—	5	—
B	8	—	4	—	3
C	—	4	—	7	1
D	(5)	—	7	—	6
E	—	3	1	(6)	—

Step 4 Label column E with 3.

Step 5 Ring the smallest entry in columns A, D or E. This is the 1 in row C.

Step 3 Delete the row corresponding to C.

	1			2	3
	A	**B**	**C**	**D**	**E**
A	—	8	—	5	—
B	8	—	4	—	3
C	—	4	—	7	1
D	5	—	7	—	6
E	—	3	1	6	—

Step 4 Label column C with 4.

Step 5 Ring the smallest entry in columns A, C, D or E. This is the 3 in row B.

Step 3 Delete the row corresponding to B.

	1		4	2	3
	A	**B**	**C**	**D**	**E**
A	—	8	—	5	—
B	8	—	4	—	3
C	—	4	—	7	1
D	5	—	7	—	6
E	—	3	1	6	—

All rows have now been deleted. The edges of the minimum spanning tree are DA, ED, CE and BE, as before, with total weight $5 + 6 + 1 + 3 = 15$.

Exercise 3B

1 Use Prim's algorithm to find a minimum spanning tree for the following network:

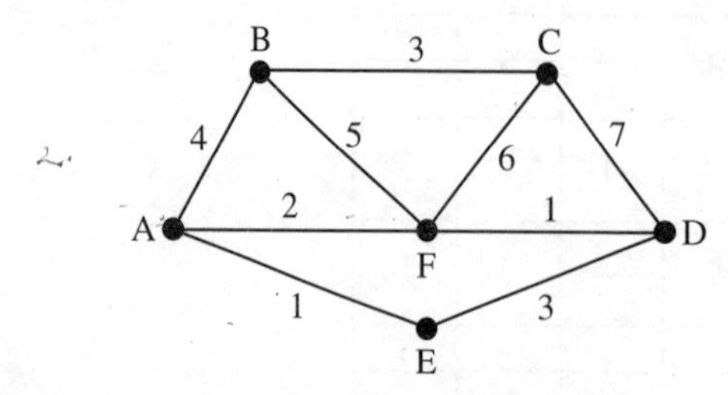

Show clearly the order in which you added vertices to your tree. Start at vertex A.

2 Using Prim's algorithm find as many minimum spanning trees as possible for the following network. Start by choosing vertex C.

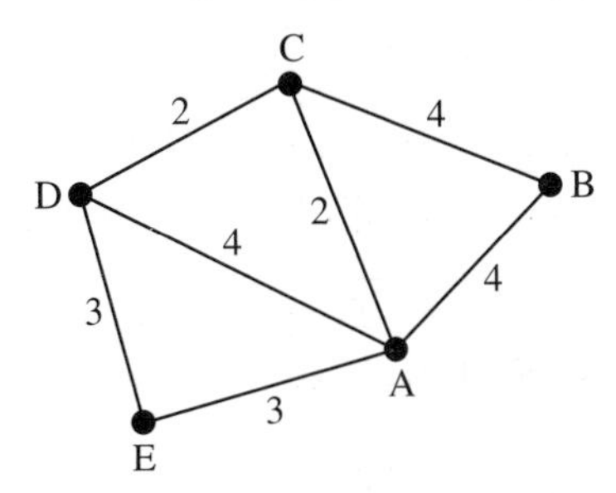

3 (a) Write down the distance matrix that represents the network in question 1.

(b) Use the matrix form of Prim's algorithm to find a minimum spanning tree. State the edges involved in your minimum spanning tree and give its total weight.

4

	A	B	C	D	E
A	—	10	—	15	17
B	10	—	16	—	20
C	—	16	—	12	18
D	15	—	12	—	24
E	17	20	18	24	—

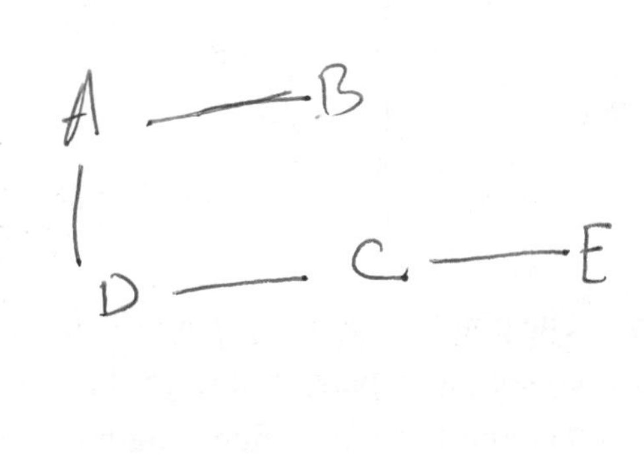

Use the matrix form of Prim's algorithm to find a minimum spanning tree for the network represented by the above distance matrix. Draw your minimum spanning tree and give its total weight.

5 Use both forms of Prim's algorithm to find a minimum spanning tree for the network shown below. Start with vertex P.

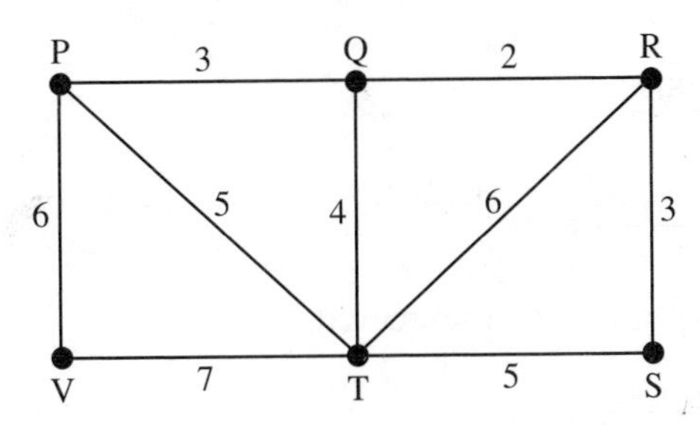

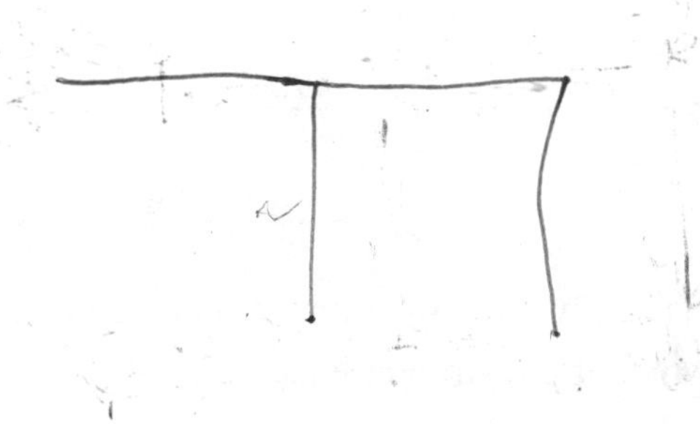

6 The table gives the distances, in miles, between six places in Ireland. Use the matrix form of Prim's algorithm to find a minimum spanning tree connecting these places.

	Athlone	Dublin	Galway	Limerick	Sligo	Wexford
Athlone	—	78	56	73	71	114
Dublin	78	—	132	121	135	96
Galway	56	132	—	64	80	154
Limerick	73	121	64	—	144	116
Sligo	71	135	80	144	—	185
Wexford	114	96	154	116	185	—

7

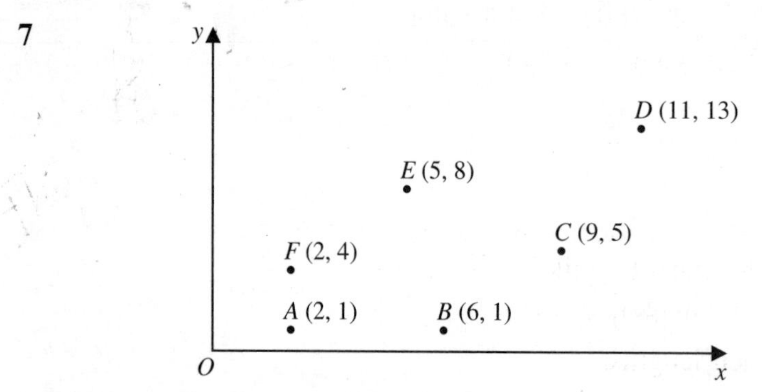

The points A, B, C, D, E and F on the above graph may be connected in pairs by straight lines. Use Prim's algorithm to find a connector of minimum length for the six points. Draw a minimum spanning tree and obtain its length.

8

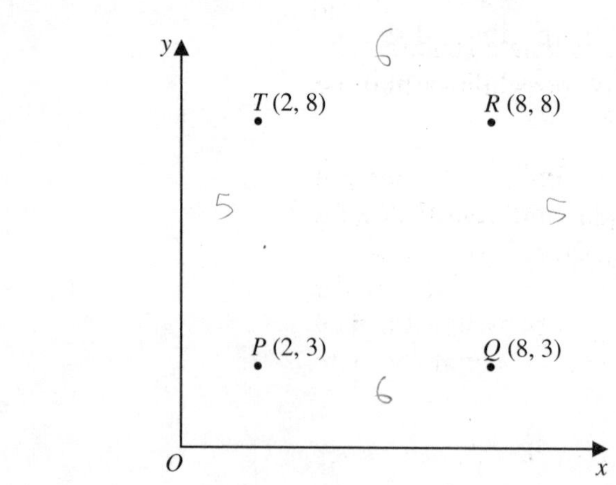

The points P, Q, R and T on the above graph may be connected in pairs by straight lines.

(a) Use Prim's algorithm to find the minimum connector for the four points. State its length.
(b) Locate a fifth point S, such that the minimum connector for the five points P, Q, R, T and S is shorter than that found in (a). State the length of this new connector.

3.5 Dijkstra's algorithm for finding the shortest path

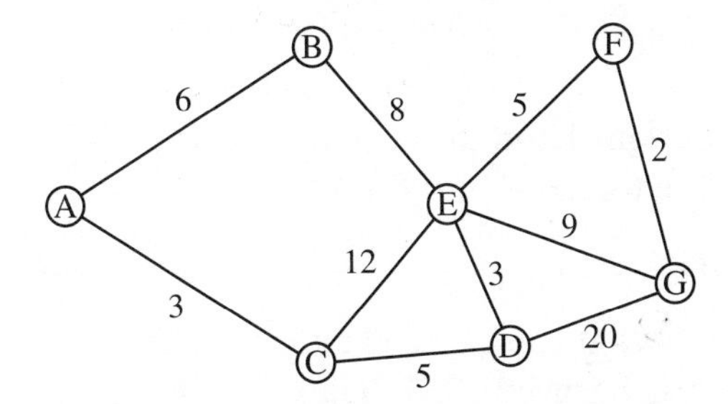

The network above models seven villages and the roads between them. The weights on the edges are the distances (in miles). We want to find the shortest path, or route, between villages A and G.

Here are some paths from A to G with their lengths:

ABEFG	$(6 + 8 + 5 + 2) = 21$
ACDG	$(3 + 5 + 20) = 28$
ACDEFG	$(3 + 5 + 3 + 5 + 2) = 18$
ABEG	$(6 + 8 + 9) = 23$

Listing all possible paths and their lengths will obviously provide the solution in due course. This method is called **complete enumeration**. However, in real-life problems the number of possibilities may be very large, making this an impractical method.

In 1959 Dijkstra developed an algorithm for finding the shortest path through a network between two given vertices. Dijkstra's algorithm is an example of a **labelling algorithm**. It obtains the shortest route from the initial vertex to any other vertex in the network. At each iteration (stage) a fresh vertex is assigned a **final label**. This label gives the shortest distance from the initial vertex to this vertex.

The algorithm works gradually through the network, one new vertex receiving a final label at each stage. The working values of other vertices are also improved at each stage. These working values give the shortest distances found so far to unlabelled vertices.

It is convenient, when applying the algorithm by hand, to record working values and final labels (or values), one for each vertex, thus:

order of labelling	final value
working values	

This is illustrated in the following example.

It should be remembered that this algorithm, like all algorithms, is designed for computer use on large networks.

Dijkstra's algorithm can be formally stated as follows.

Step 1 Label the start vertex 0 (in the top right-hand box) and mark it as the first vertex labelled (in the top left-hand box).

1	0

Step 2 Update working values for all vertices Y that can be reached directly from the vertex X that has just been labelled. The rule for updating the working value of a vertex Y is:

> Take the weight of the arc connecting the vertex Y to the vertex X that has just been labelled and add it to that label. If the vertex Y has no working value then the result becomes the working value. If the vertex Y already has a working value then the result replaces it if it is lower.

It seems odd on the first pass to refer to the vertex just labelled, but the reason for doing so will become clear later.

Step 3 Out of all unlabelled vertices with working values, choose that with the lowest working value. (There may be a choice here, in which case it is a free choice.) Label it with that working value and record the order in which it has been labelled (second, third, fourth, etc.).

Step 4 Repeat steps 2 and 3 until the destination vertex is labelled.

Step 5 The label on the destination vertex is the shortest distance to that vertex.

The shortest route is found by tracing back as follows:

If vertex N lies on the route, then vertex M is the previous vertex *if*:

label at N – label at M = weight of arc MN

Repeat until the route is traced back to the initial vertex.

It is tempting to try to avoid tracing back by keeping a record of the route while labelling, but this is much less efficient.

Example 5

Use Dijkstra's algorithm to find the shortest route from A to G in the network below.

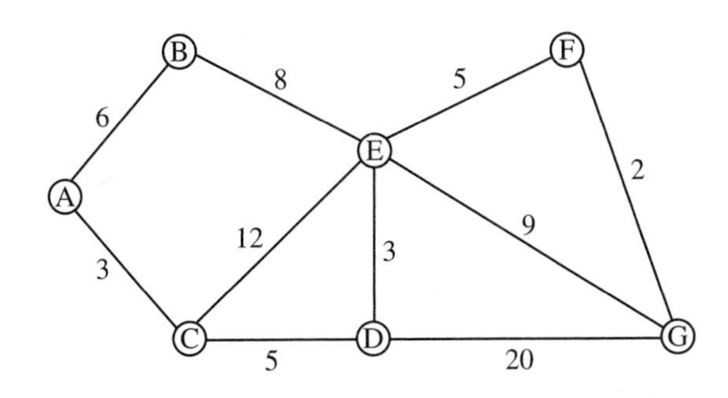

Step 1 Label start vertex A with 0 and number it 1 as the first vertex labelled.

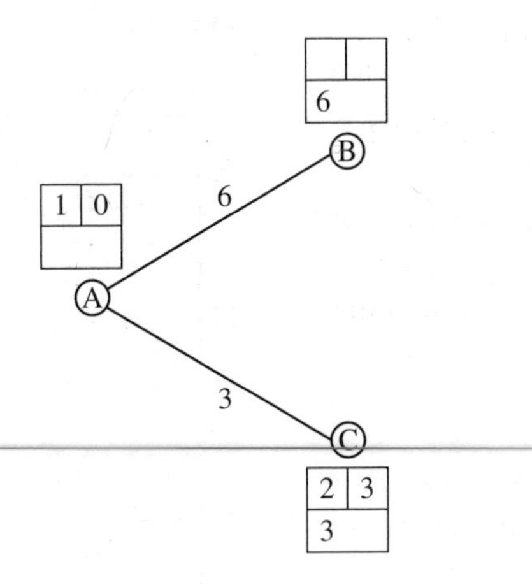

(i) **Step 2** Vertices B and C can be reached directly from A. The working value for the distance along the path to B is $0 + 6 = 6$. The working value for the path to C is $0 + 3 = 3$.

Step 3 The smallest working value is 3 at C, so label C with 3 and number it as 2, the second vertex labelled.

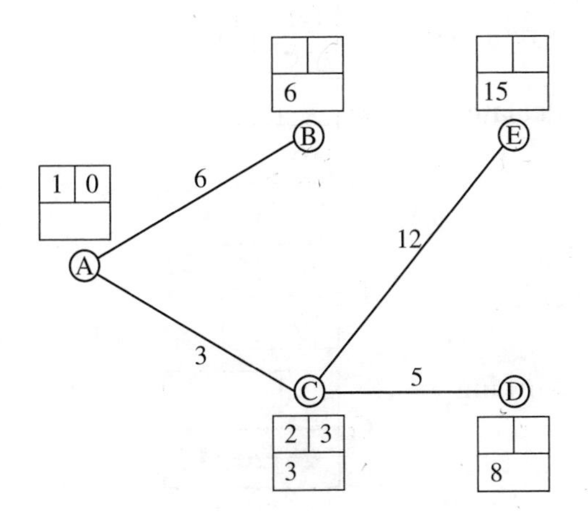

(ii) **Step 2** Vertices D and E can be reached directly from C, just labelled. The working value for D is $3 + 5 = 8$ (label of C + weight of CD). The working value of E is $3 + 12 = 15$.

Step 3 The working values are now B(6), D(8) and E(15). The smallest is 6 at B, so label B with 6 and number it as 3.

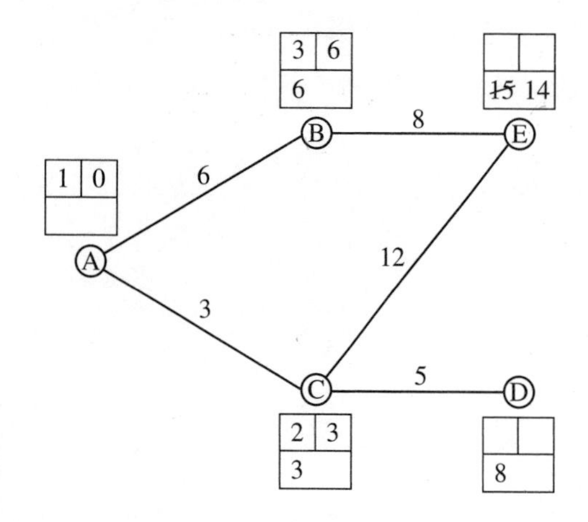

(iii) **Step 2** Vertex E can be reached directly from B, just labelled. Now label B + weight BE = 6 + 8 = 14. As this is smaller than 15, E's existing working value, we replace the 15 by 14.

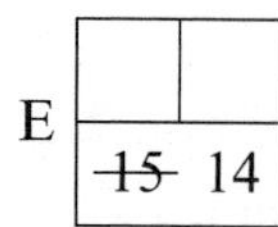

Step 3 The working values are now D(8) and E(14). The smallest is 8 at D, so label D with 8 and number it as 4.

(iv) **Step 2** Vertices E and G can be reached directly from D, just labelled. E already has a working value of 14, but label D + weight DE = 8 + 3 = 11, and so 14 is replaced by 11.
G has no working value, and so we give it one calculated from label D + weight DG = 8 + 20 = 28.

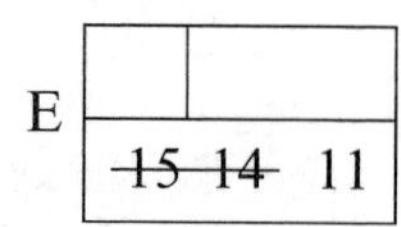

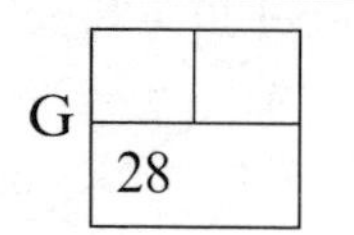

Step 3 The working values are now E(11) and G(28). The smallest is 11 at E, so label E with 11 and number it as 5.

(v) **Step 2** Vertices F and G can be reached directly from E, just labelled. G already has a working value of 28, but label E + weight EG = 11 + 9 = 20, and so 28 is replaced by 20.
F has no working value, and so we give it one calculated from label E + weight EF = 11 + 5 = 16.

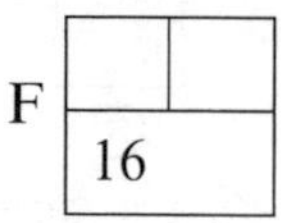

Step 3 The working values are now F(16) and G(20). The smallest is 16 at F, so label F with 16 and number it as 6.

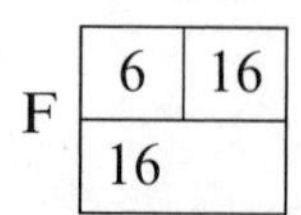

(vi) **Step 2** Only G remains to be reached. G already has a working value of 20, but label F + weight FG = 16 + 2 = 18, and so 20 is replaced by 18.

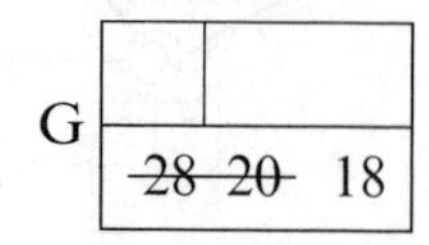

Step 3 The only (and therefore the smallest) unlabelled vertex is G, and this is now labelled and numbered 7. Since we have labelled the destination vertex we now stop repeating steps 2 and 3, and go to step 5.

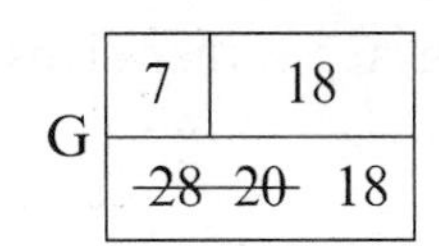

Step 5 The length of the shortest route from A to G is 18, which is the label of G.

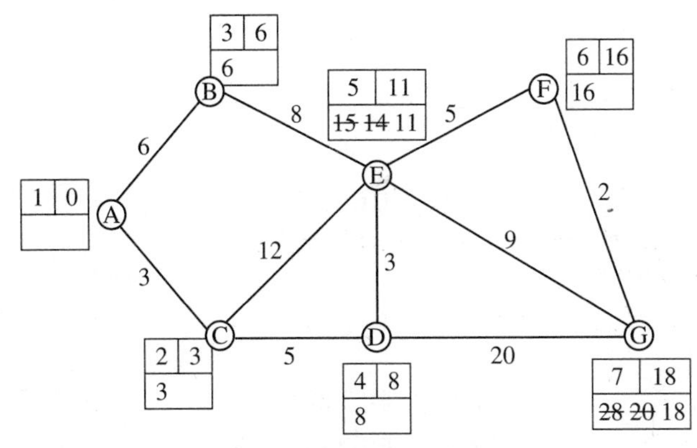

To find the shortest route work **backwards** from G:

label G – label F = 18 – 16 = 2 = weight FG
label F – label E = 16 – 11 = 5 = weight EF
label E – label D = 11 – 8 = 3 = weight DE
label D – label C = 8 – 3 = 5 = weight CD
label C – label A = 3 – 0 = 3 = weight AC

So FG, EF, DE, CD and AC are on the shortest route, which is therefore A → C → D → E → F → G.

For all other edges: label Y – label X ≠ weight XY.

For example, take DG: label G – label D = 18 – 8 = 10, but the weight of DG is 20.

Dijkstra's algorithm can also be applied to directed networks. In doing so we must be careful in step 2 to consider only those vertices Y that can be reached directly from X **by an allowed edge**, i.e. X to Y is allowed (X •——>——• Y).

Example 6

Use Dijkstra's algorithm to find the shortest route from A to D in the following **directed** network:

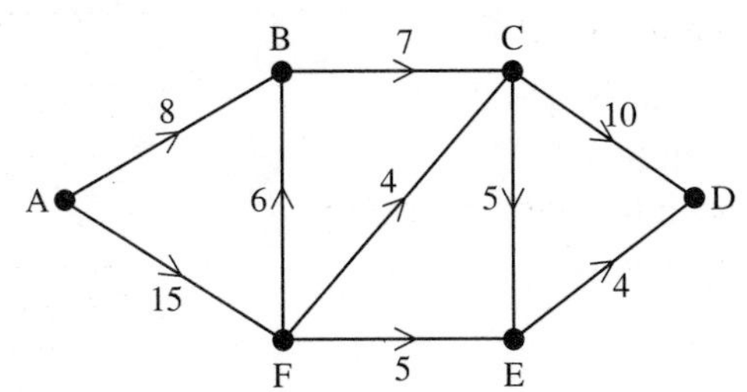

Applying step 1, then steps 2 and 3 once gives

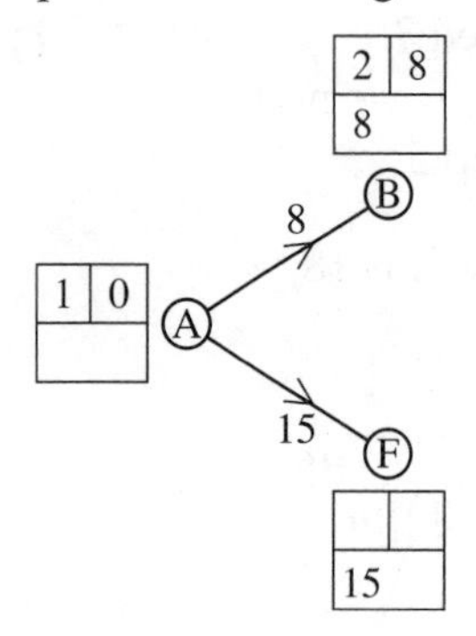

In the next stage only C can be reached directly from B (just labelled) and so we obtain

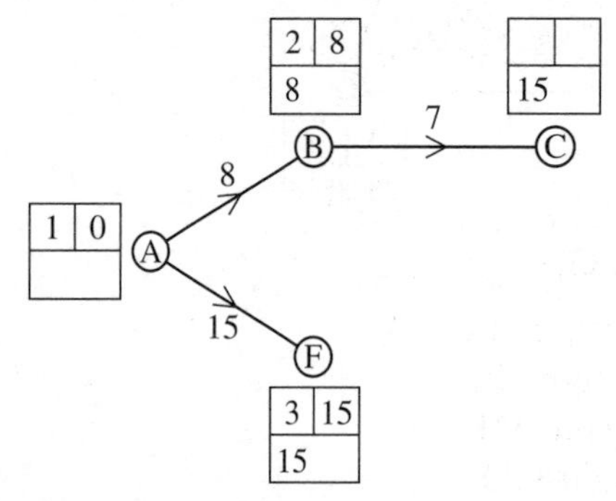

At step 3 we choose F; we could also have chosen C. It does not matter which we choose.

The completed calculation is shown below:

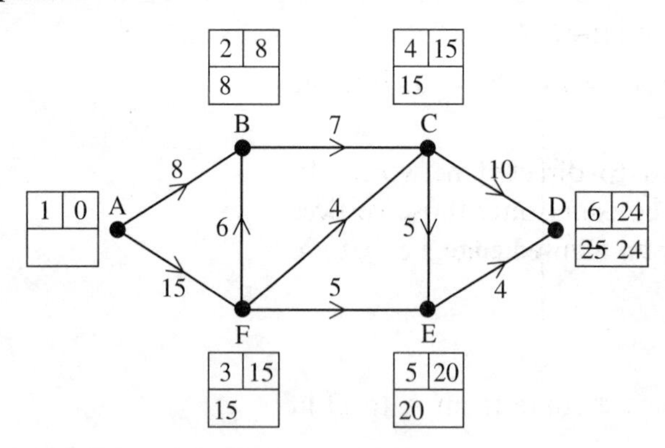

To find the shortest route or routes work backwards from D:

$$\text{label D} - \text{label E} = 24 - 20 = 4 = \text{weight ED}$$
$$\text{label E} - \text{label F} = 20 - 15 = 5 = \text{weight FE}$$
$$\text{label F} - \text{label A} = 15 - 0 = 15 = \text{weight AF}$$

ED, FE and AF are therefore on a shortest route, which is $A \to F \to E \to D$, with length 24.

However, there is another shortest route since

label E – label C = 20 – 15 = 5 = weight CE
label C – label B = 15 – 8 = 7 = weight BC
label B – label A = 8 – 0 = 8 = weight AB

So ED, CE, BC and AB are also on a shortest route, which is A $\rightarrow$ B $\rightarrow$ C $\rightarrow$ E $\rightarrow$ D, with length 24.

You may check for yourself that if the arrows are removed, indicating that one may go either way along each edge, then there is a single shortest route ABFED of length 23.

Exercise 3C

1 By using Dijkstra's algorithm find a shortest path from S to T through each of the networks given in (a), (b) and (c). In each case state the shortest path or paths and give the lengths of these.

(a)

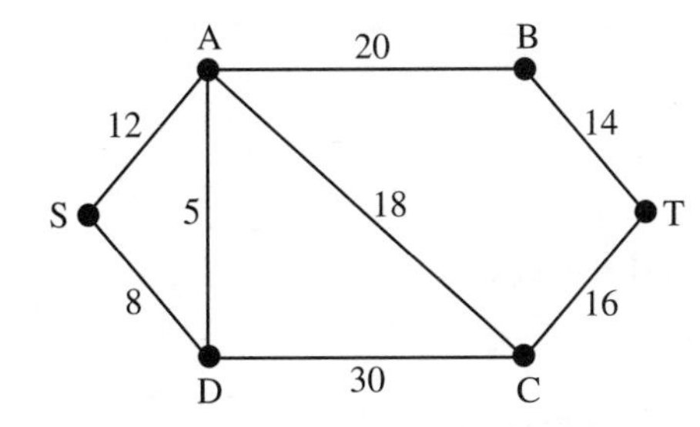

(b)

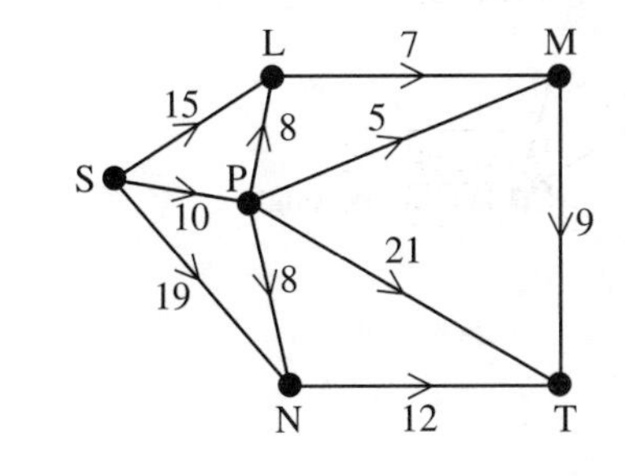

(c)

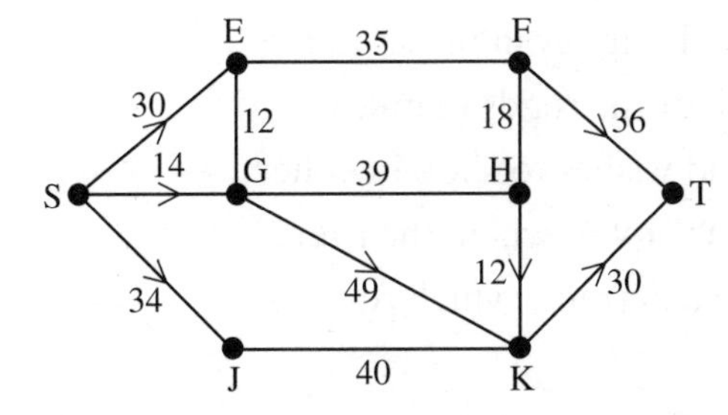

2

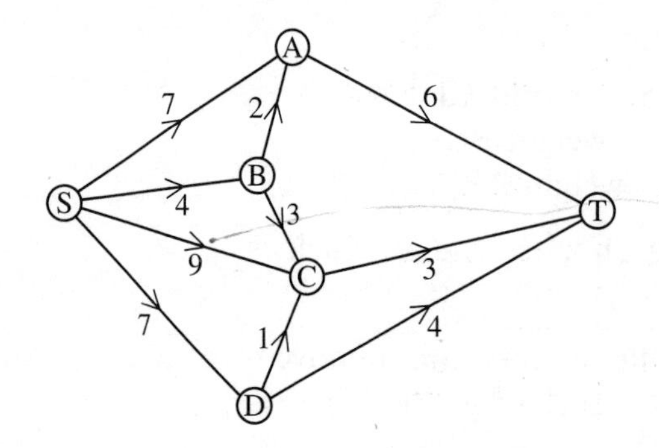

The network above models possible air routes between S and T via certain intermediate cities. The numbers on the arcs indicate the fares in units of £100. The journeys can only be made in the direction indicated. Use the shortest path algorithm to find the route between S and T for which the total fare is a minimum. Your solution should indicate clearly the order in which the nodes (cities) receive their permanent labels.

(Specimen Paper Q5 D1 1996)

3

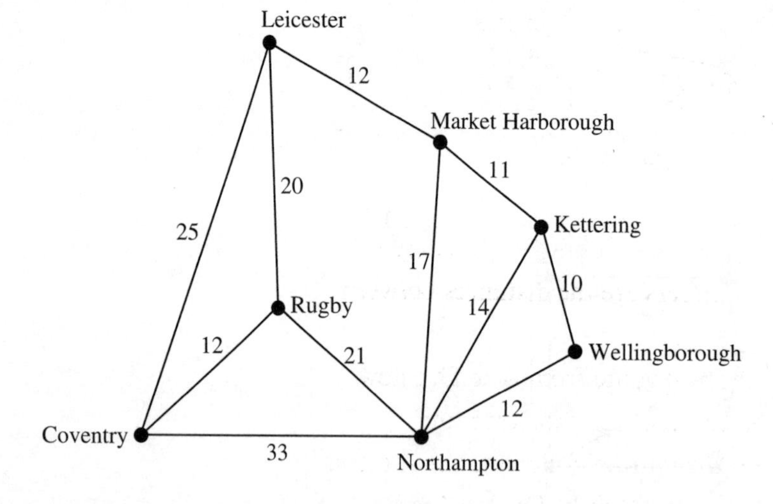

The figure above represents the roads joining seven places in the Midlands. The numbers give the length of the roads in miles. Mrs Brown lives in Wellingborough and wishes to shop in either Coventry or Leicester. She will go to the city to which the route is shortest. Use Dijkstra's algorithm to determine which route is the shortest.

4 The matrix below gives the fares, in pence, for direct bus journeys between towns P, Q, R, S, T and V. Blanks indicate that there is no direct service.

	P	Q	R	S	T	V
P		55				60
Q	55		65	140	50	27
R		65		60		
S		140	60		80	
T		50		80		67
V	60	27			67	

(a) Draw a network to represent this information.

(b) Use Dijkstra's algorithm to find the cheapest route from P to S.

5

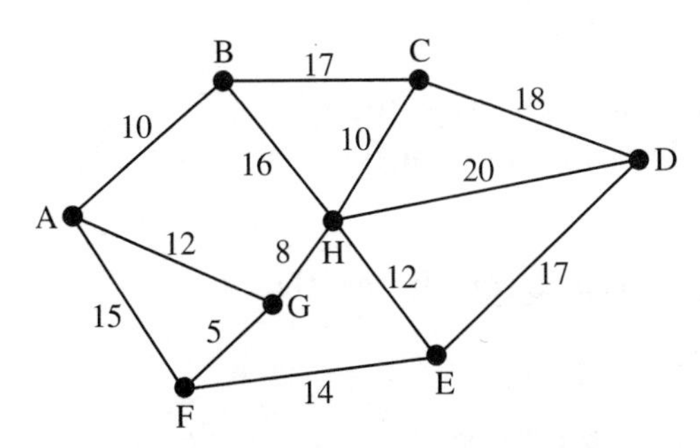

The diagram above represents the roads joining eight villages A, B, C, D, E, F, G and H. The numbers are the distances between the villages in km.

(a) Use Dijkstra to find the shortest route from A to D. Show all your working and find the length of the shortest route.

(b) A driver usually completes the journey following the shortest route driving at an average speed of 60 km/h. The local radio reports an accident at village G and warns drivers of a delay of 10 minutes.

Describe how to modify your approach to (a) to find the new quickest route, explaining how to take account of this information. What is the quickest route and how long will it take?

3.6 Planar and non-planar graphs

Included among the special graphs defined in Chapter 2 were **planar graphs**. These are graphs that can be drawn in a plane in such a way that no two edges meet each other, except at a vertex to which they are both incident.

In printed circuits parts of networks are printed on one side of a non-conducting plate. Since the wires are not insulated they must not cross and the corresponding graphs must be planar.

It is useful therefore to have available some way of deciding if a given graph is planar or not. The algorithm described below does this but it can only be applied to graphs that have a Hamiltonian cycle. Recall that a **Hamiltonian cycle** is a cycle that passes through every vertex of the graph once and only once and returns to its starting vertex.

A planarity algorithm

Step 1 Identify a Hamiltonian cycle in the graph.

Step 2 Redraw the graph so that the Hamiltonian cycle forms a regular polygon and all edges are drawn as straight lines in the polygon.

Step 3 Choose any edge PQ and decide this will stay inside the polygon.

Step 4 Consider any edges that cross PQ.
(i) If it is possible to move all these outside without producing crossings, go to step 5.
(ii) If it is not possible to move all these outside without producing crossings, then the graph is non-planar.

Step 5 Consider each remaining crossing inside and see if any edge may be moved outside to remove it, without creating a crossing outside.

Step 6 Stop when all crossings inside have been considered.
(i) If there are no crossings inside or outside then the graph is planar.
(ii) If there is a crossing inside that cannot be removed then the graph is non-planar.

The implementation of the algorithm is illustrated in the following two examples.

Example 7

Use the planarity algorithm to show that the following graph is planar.

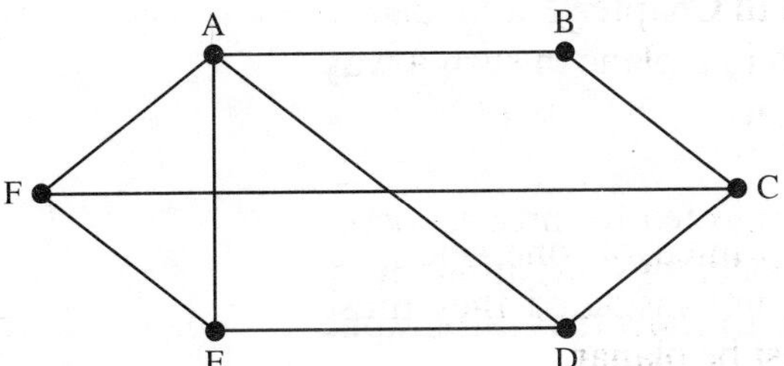

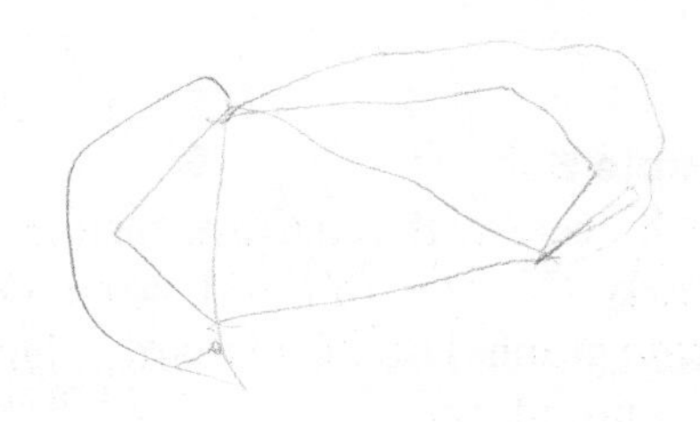

Step 1 There is an obvious Hamiltonian cycle in this graph, ABCDEFA.

Step 2 This cycle is already a polygon and all edges are inside so there is no need to redraw it.
It is useful to identify the circuit in some distinctive way. Here we will do this by making the edges of the circuit dotted lines.

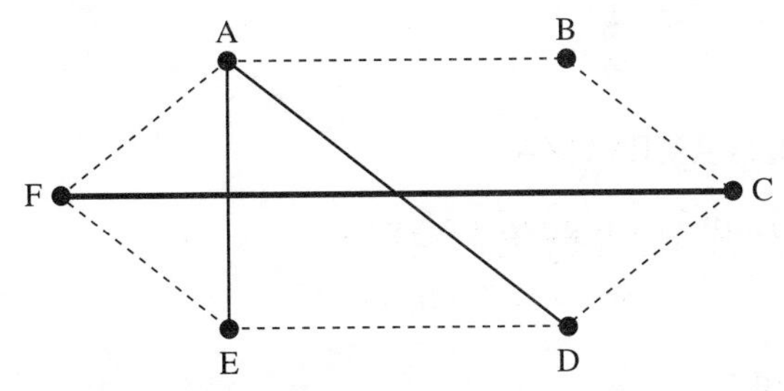

Step 3 We choose edge AD to remain inside and to indicate this we draw this edge as a dotted line.

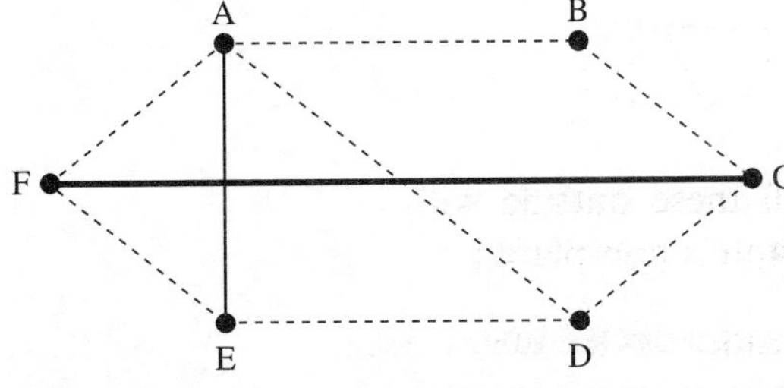

Step 4 Since CF crosses AD we move it outside. We then obtain

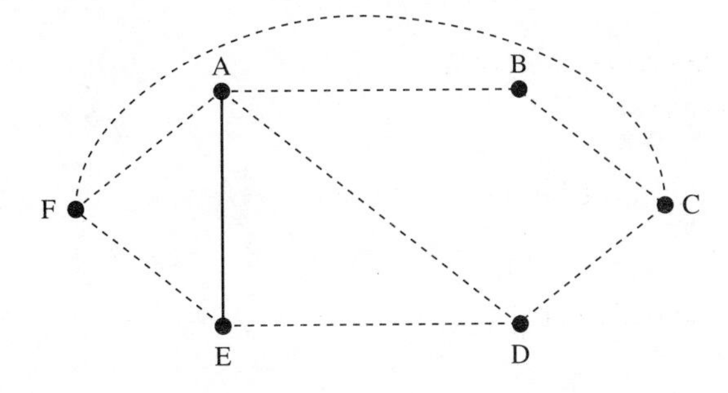

We now have a graph with no crossings inside or outside the Hamiltonian circuit. The graph is therefore planar. The diagram in step 4 is of course a plane drawing of the original graph.

Example 8

Three houses A, B and C are each to be connected to three services: electricity (X), gas (Y) and water (Z). Model this situation by a bipartite graph. Use the planarity algorithm to show that this graph, $K_{3,3}$, is not planar.

The bipartite graph that models this situation is shown below.

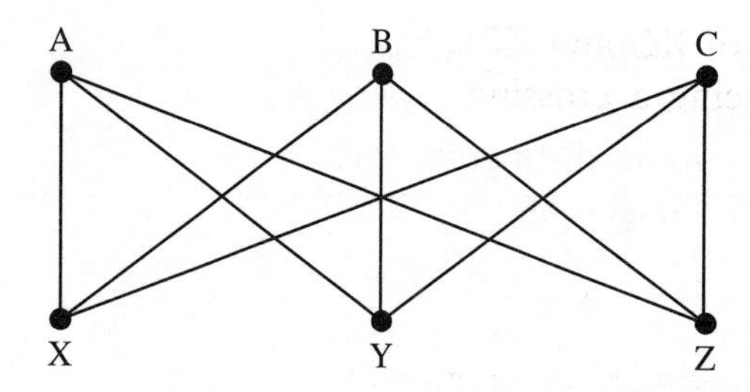

Step 1 A Hamiltonian circuit for the graph is AXBYCZA.

Step 2 Redrawing the graph so that the circuit is a regular polygon we obtain

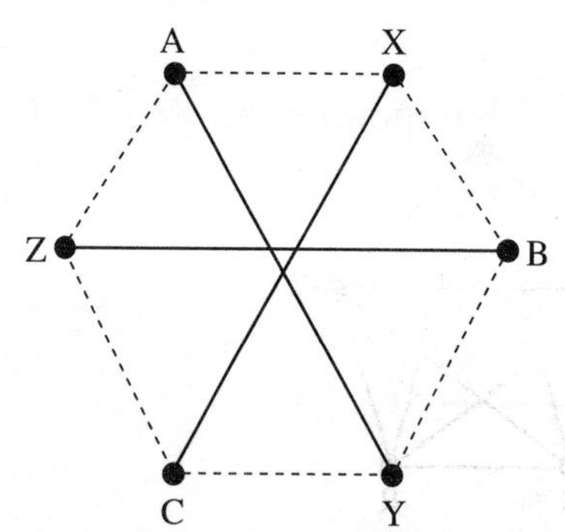

Step 3 We choose edge CX and indicate this as a dotted line.

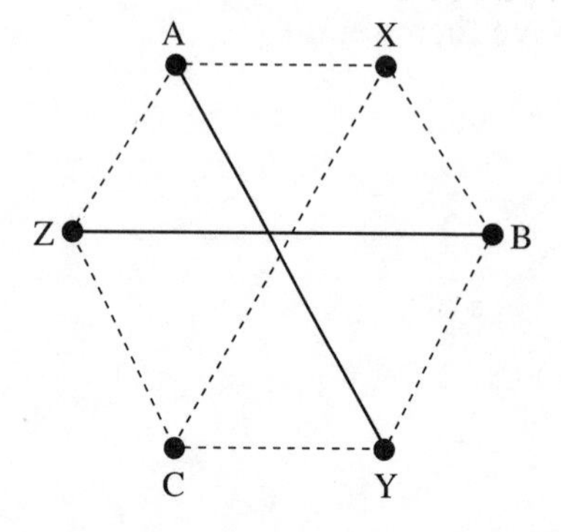

Step 4 Move AY outside as it crosses CX.

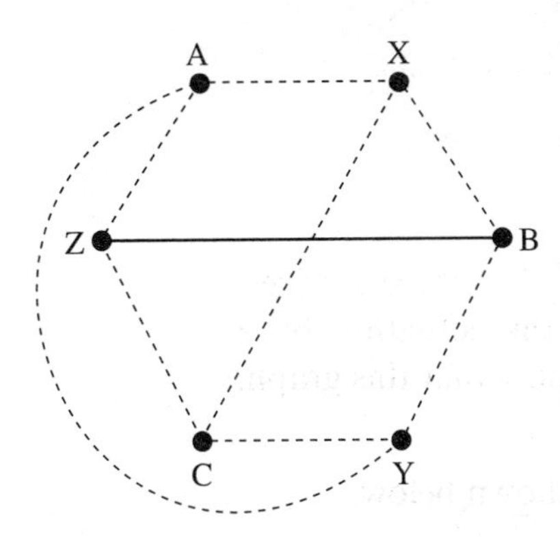

There is one remaining crossing inside, that of XC and ZB. It cannot be moved outside without producing a crossing and so the graph is non-planar.

Exercise 3D

1 Show that (a) is a planar and (b) is non-planar.

((a) is K_4 and (b) is K_5.)

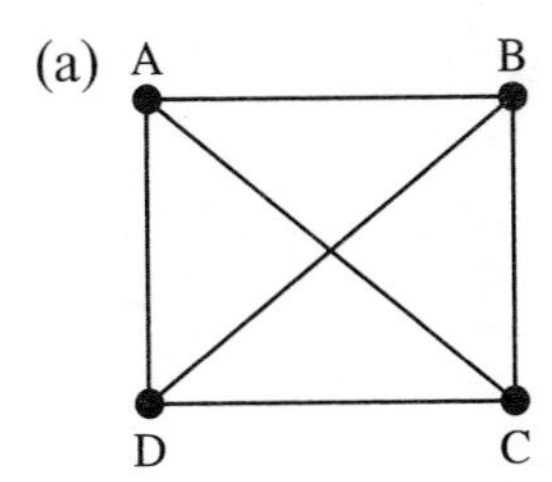

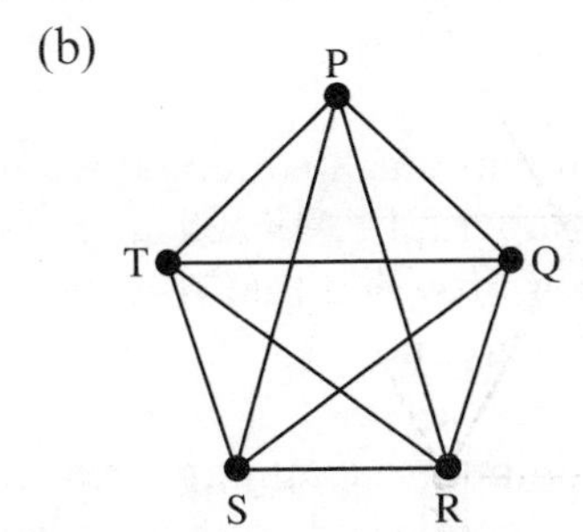

2 Use the planarity algorithm to decide if the graphs below are planar or not.

(a)

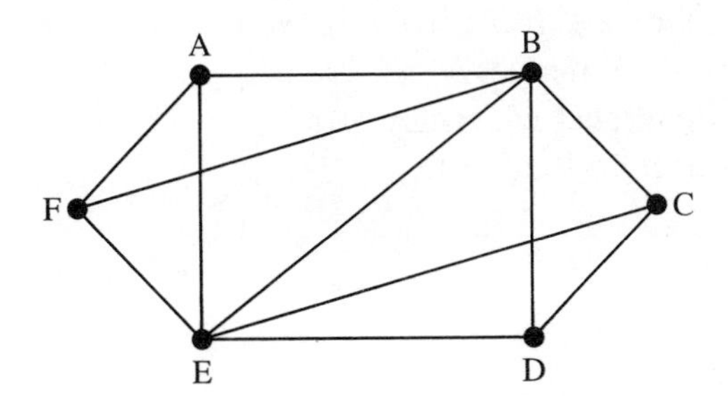

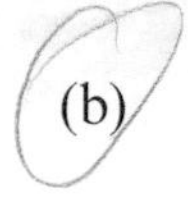
(b)

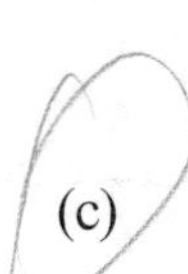
(c)

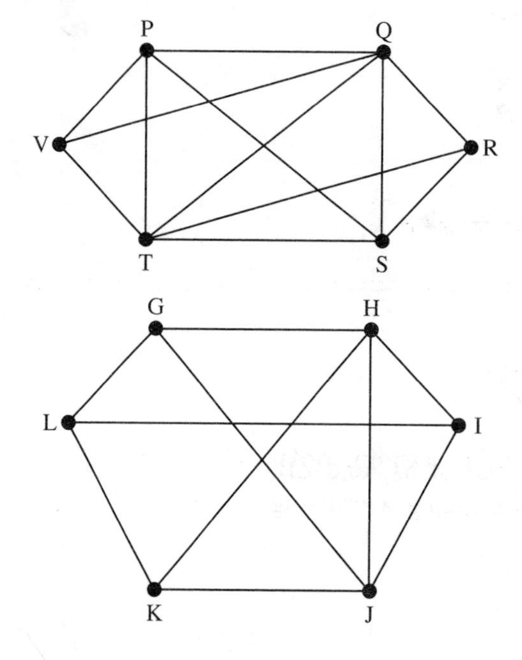

SUMMARY OF KEY POINTS

1 A **minimum spanning tree** of a connected and undirected graph is a spanning tree such that the total length of its edges is as small as possible. (This is sometimes called a minimum connector.)

2 **Kruskal's algorithm** builds a minimum spanning tree by adding one **edge** at a time to a subgraph. At each stage the edge of smallest weight is chosen provided that it does not create a cycle with edges already chosen.

3 **Prim's algorithm** builds a minimum spanning tree by adding one **vertex** at a time to a connected subgraph. The new vertex to be added is the one which is nearest to any vertex already in the subgraph.

4 **Dijkstra's algorithm** obtains the shortest route from the initial vertex to any other vertex in a network. At each stage a fresh vertex is assigned a **final label** which gives the shortest distance from the initial vertex to that vertex.

Review exercise 1

1 Use the bubble-sort algorithm to sort, in ascending order, the list:

27 15 2 38 16 1

giving the state of the list at each stage. [E]

2 (a) Use the bubble-sort algorithm to sort, in descending order, the list:

25 42 31 22 26 41

giving the state of the list on each occasion when two values are interchanged.

(b) Find the *maximum* number of interchanges needed to sort a list of six pieces of data using the bubble-sort algorithm. [E]

3 8, 4, 13, 2, 17, 9, 15

This list of numbers is to be sorted into ascending order. Perform a quick sort to obtain the sorted list, giving the state of the list after each rearrangement. [E]

4 111 103 77 81 98 68 82 115 93

(a) The list of numbers above is to be sorted into descending order. Perform a quick-sort to obtain the sorted list, giving the state of the list after each rearrangement and indicating the pivot elements used.

(b) (i) Use the first-fit decreasing bin-packing algorithm to fit the data into bins of size 200.

(ii) Explain how you decided in which bin to place the number 77. [E]

5 Trishna wishes to video eight television programmes. The lengths of the programmes, in minutes, are:

75 100 52 92 30 84 42 60

Trishna decides to use 2-hour (120 minute) video tapes only to record all of these programmes.

(a) Explain how to use a first-fit decreasing bin-packing algorithm to find the solution that uses the fewest tapes and determine the total amount of unused tape.

(b) Determine whether it is possible for Trishna to record an additional two 25-minute programmes on these 2-hour tapes, without using another video tape. [E]

6 A DIY enthusiast requires the following 14 pieces of wood as shown in the table.

Length in metres	0.4	0.6	1	1.2	1.4	1.6
Number of pieces	3	4	3	2	1	1

The DIY store sells wood in 2 m and 2.4 m lengths. He considers buying six 2 m lengths of wood.

(a) Explain why he will not be able to cut all of the lengths he requires from these six 2 m lengths.

(b) He eventually decides to buy 2.4 m lengths. Use a first-fit decreasing bin-packing algorithm to show how he could use six 2.4 m lengths to obtain the pieces he requires.

(c) Obtain a solution that only requires five 2.4 m lengths. [E]

7 *Note:* This question uses the modulus function. If $x \neq y$, $|x - y|$ is the positive difference between x and y, e.g. $|5 - 6.1| = 1.1$. The algorithm described by the flow chart on page 79 is to be applied to the five pieces of data below.

$U(1) = 6.1$, $U(2) = 6.9$, $U(3) = 5.7$, $U(4) = 4.8$, $U(5) = 5.3$

(a) Obtain the final output of the algorithm using the five values given for $U(1)$ to $U(5)$.

(b) In general, for any set of values $U(1)$ to $U(5)$, explain what the algorithm achieves.

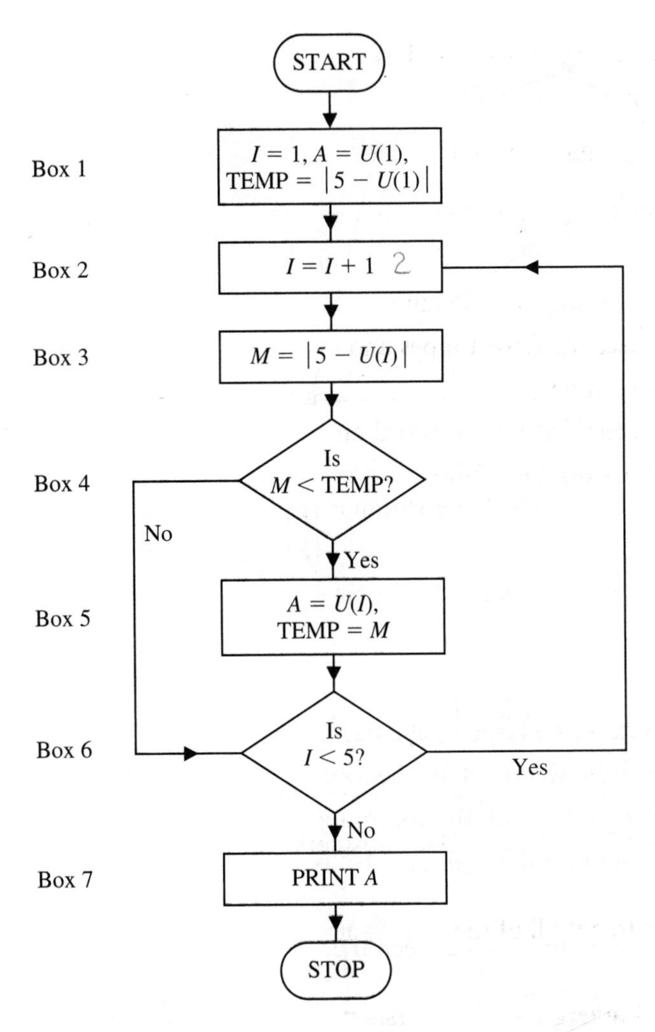

(c) If Box 4 in the flow chart is altered to

'Is $M > \text{TEMP}$?'

state what the algorithm now achieves. [E]

8 (a) State briefly:

(i) Prim's algorithm

(ii) Kruskal's algorithm.

(b) Find a minimum spanning tree for the network shown at the top of page 80 using:

(i) Prim's algorithm, starting with vertex G

(ii) Kruskal's algorithm.

In each case write down the order in which you made your selection of arcs.

(c) State the weight of a minimum spanning tree.

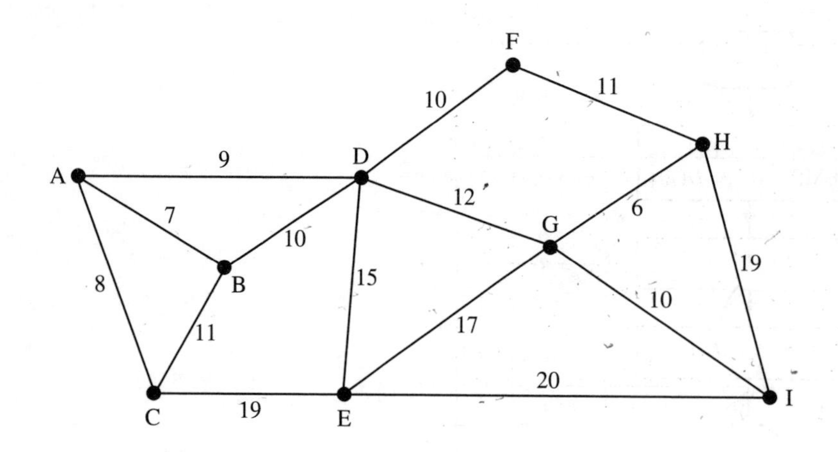

(d) State, giving a reason for your answer, which algorithm is preferable for a large network. [E]

9 It is intended to network five computers at a large theme park. There is one computer at the office and one at each of the four different entrances. Cables need to be laid to link the computers. Cable laying is expensive, so a minimum total length of cable is required.
The table shows the shortest distances, in metres, between the various sites.

	Office	Entrance 1	Entrance 2	Entrance 3	Entrance 4
Office	—	1514	488	980	945
Entrance 1	1514	—	1724	2446	2125
Entrance 2	488	1724	—	884	587
Entrance 3	980	2446	884	—	523
Entrance 4	945	2125	587	523	—

(a) Starting at Entrance 2, demonstrate the use of Prim's algorithm and hence find a minimum spanning tree. You must make your method clear, indicating the order in which you selected the arcs in your final tree.
(b) Calculate the minimum total length of cable required. [E]

10

	A	B	C	D	E	F
A	—	124	52	87	58	97
B	124	—	114	111	115	84
C	52	114	—	67	103	98
D	87	111	67	—	41	117
E	58	115	103	41	—	121
F	97	84	98	117	121	—

The table shows the distances, in mm, between six nodes A to F in a network.

(a) Use Prim's algorithm, starting at A, to solve the minimum connector problem for this table of distances. You must explain your method carefully and indicate clearly the order in which you selected the arcs.

(b) Draw a sketch showing the minimum spanning tree and find its length. [E]

11

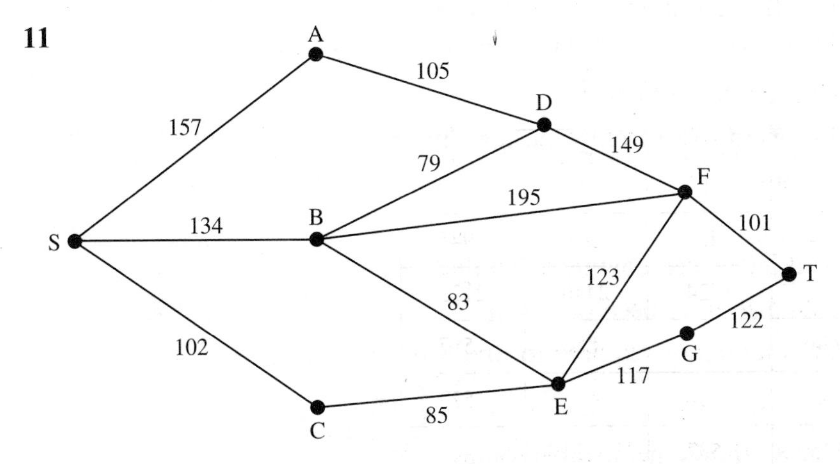

The network above shows the distances, in miles, between nine cities. Use Dijkstra's algorithm to determine the shortest route, and its length, between cities S and T. You must indicate clearly:

(i) the order in which the vertices are labelled

(ii) how you used your labelled diagram to decide which cities to include in the shortest route. [E]

12

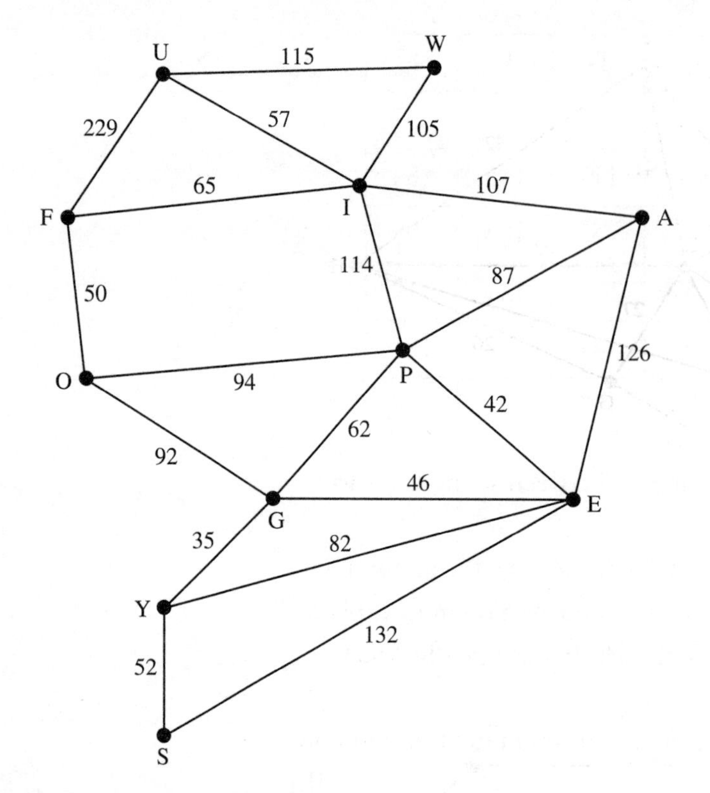

The network above represents the distances, in miles, between eleven places A, E, F, G, I, O, P, S, U, W, and Y.

(a) Use Dijkstra's algorithm to find the shortest route from W to S. State clearly:

(i) the order in which you labelled the vertices

(ii) how you determined the shortest route from your labelling

(iii) the places on the shortest route

(iv) the shortest distance.

(b) Explain how part (a) could have been completed so that the distance from A to S could also have been obtained without further calculation. (You are not required to find this distance.) [E]

13 The network at the top of page 83 shows the possible routes between cities A, B, C, D, E, F, G and H. The number on each arc gives the cost, in pounds, of taking that part of the route. Use Dijkstra's algorithm to determine the cheapest route from A to H and its cost. Your solution must indicate clearly how you have applied the algorithm.

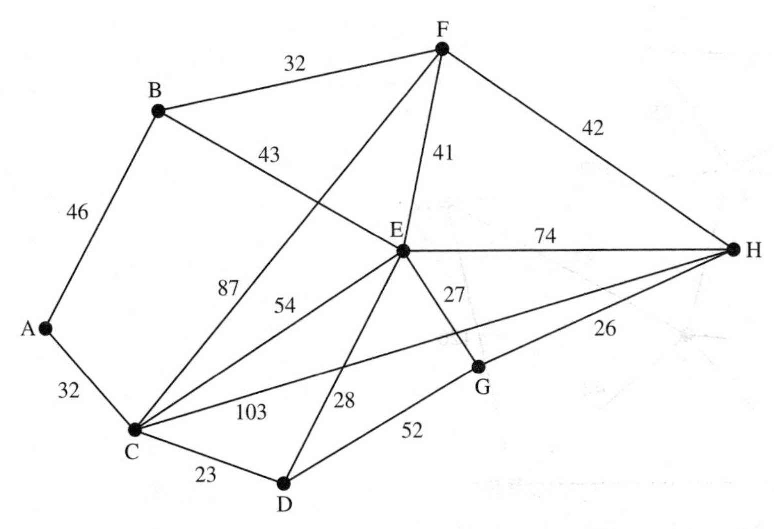

State clearly:

(a) the order in which the vertices are labelled

(b) how you used your labelled diagram to decide on the cheapest route. [E]

14

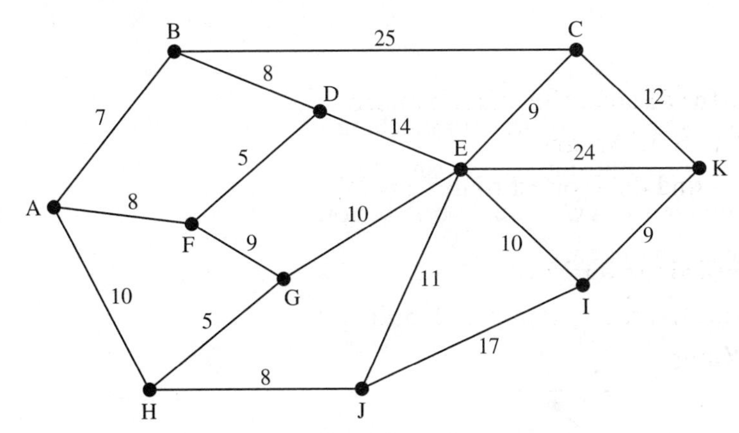

A weighted network is shown above. The number on each arc indicates the weight of that arc.

(a) Use Dijkstra's algorithm to find a path of least weight from A to K.

State clearly:

(i) the order in which the vertices were labelled

(ii) how you determined the path of least weight from your labelling.

(b) List all alternative paths of least weight.

(c) Describe a practical problem that could be modelled by the above network and solved using Dijkstra's algorithm. [E]

15

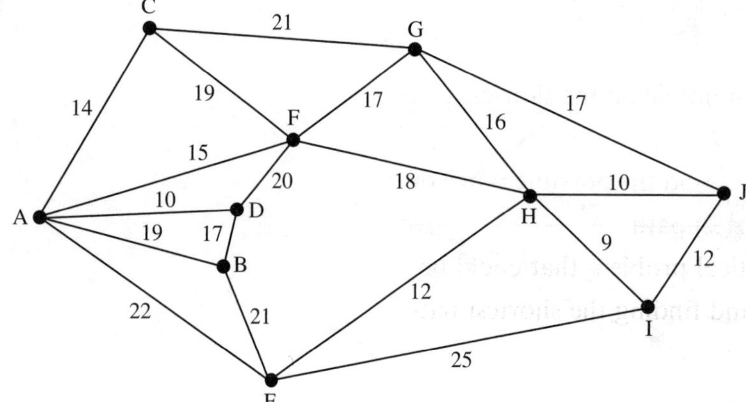

The network above models the roads linking ten towns A, B, C, D, E, F, G, H, I and J. The number on each arc is the journey time, in minutes, along that road.

Alice lives in town A and works in town J.

(a) Use Dijkstra's algorithm to find the quickest route for Alice to travel to work each morning.

State clearly:

(i) the order in which all the vertices were labelled

(ii) how you determined the quickest route from your labelling.

(b) On her return journey from work one day Alice wishes to call in at the supermarket located in town C. Explain briefly how you would find the quickest route in this case. [E]

16

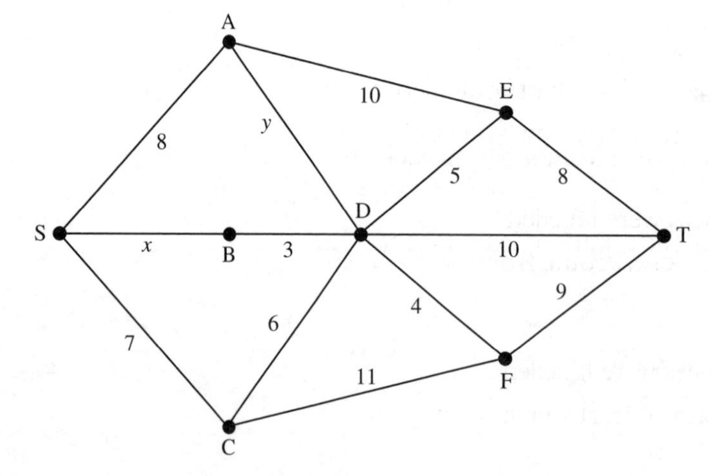

A weighted network is shown above.

Given that the shortest path from S to T is 17 and that $x \geqslant 0$, $y \geqslant 0$:

(a) (i) explain why A and C cannot lie on the shortest path

(ii) find the value of x.

(b) Given that $x = 12$ and $y \geqslant 0$, find the possible range of values for the length of the shortest path.

(c) Give an example of a practical problem that could be solved by drawing a network and finding the shortest path through it. [E]

17

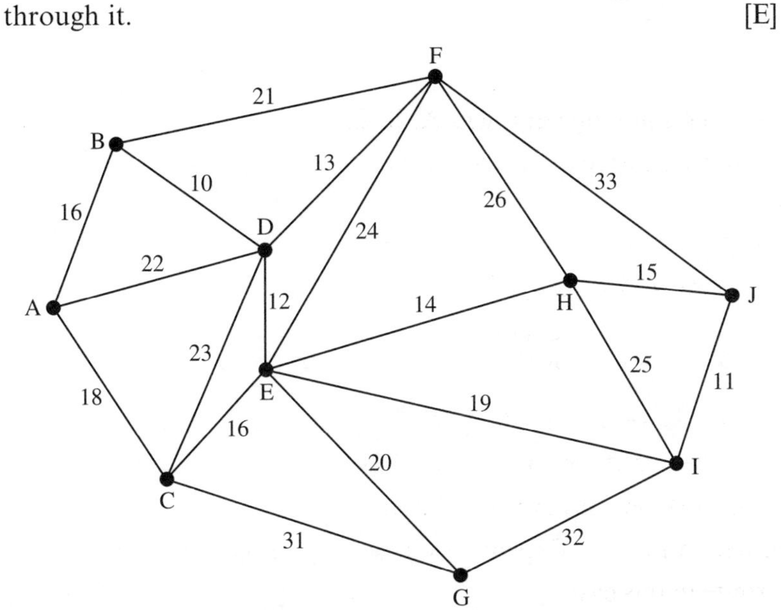

The network above represents the journey time, in minutes, between ten Midland towns.

(a) Use Dijkstra's algorithm to find a quickest route between A and J. Your solution must indicate clearly how you applied the algorithm, including:

(i) the order in which the vertices were labelled

(ii) how you determined your quickest route from your labelling.

(b) Is the route you have found the only quickest route? Give a reason for your answer. [E]

The route inspection problem 4

In Chapter 3 we considered some algorithms for solving problems that can be modelled by networks.

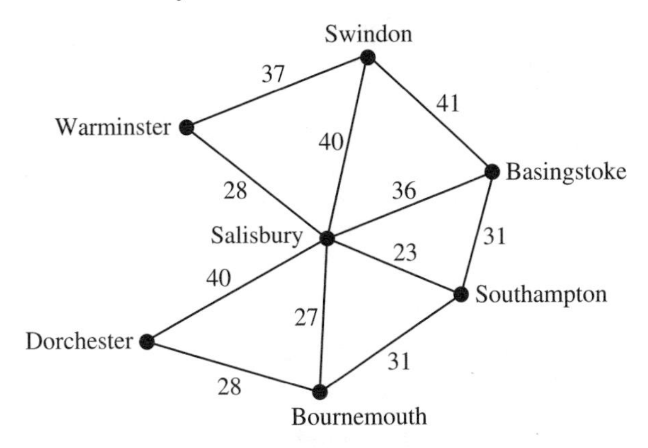

The above network shows some places in the south of England and the roads between them.

There are two classical problems concerning routes in such a network which remain to be considered. The first of these is the problem of visiting **each place once** and returning to the starting point, covering a minimum distance in doing so. This is known as the **Travelling Salesman Problem** and will be considered in Book D2. The other problem concerns the roads rather than the places. The problem is to traverse **every road just once**, in either direction, and return to the starting point. This is known as the **route inspection problem** and is considered in this chapter. Before the algorithm is considered we will look in a little more detail at some relevant properties of graphs.

4.1 Valencies and traversable graphs

- **A *traversable graph* is one that can be drawn without removing your pen from the paper and without going over the same edge twice.**

Suppose we start at vertex X and finish at vertex Y. For the moment let us assume that X and Y are **different** vertices. One edge is used when we leave X and each time we return to X we must arrive and leave by new edges.

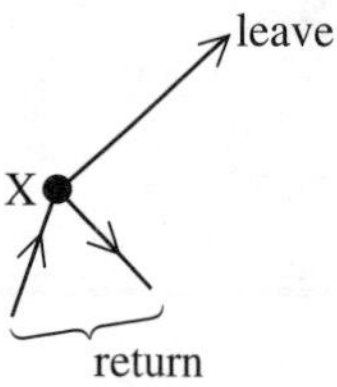

Hence the vertex X must have an **odd valency**. We usually say it is an **odd vertex**.

Similarly, Y must be an **odd vertex**, since we use two new edges each time we pass through Y and one edge when we finish.

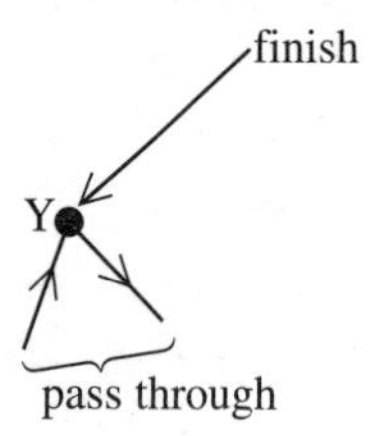

All the remaining vertices in the graph must be **even**, since every time we pass through an intermediate vertex V (not X or Y) we use **two** edges.

- **A route starting and finishing at different vertices X and Y is only possible if *X and Y are odd vertices and all the other vertices are even*. If a graph has this property it is called semi-Eulerian.**

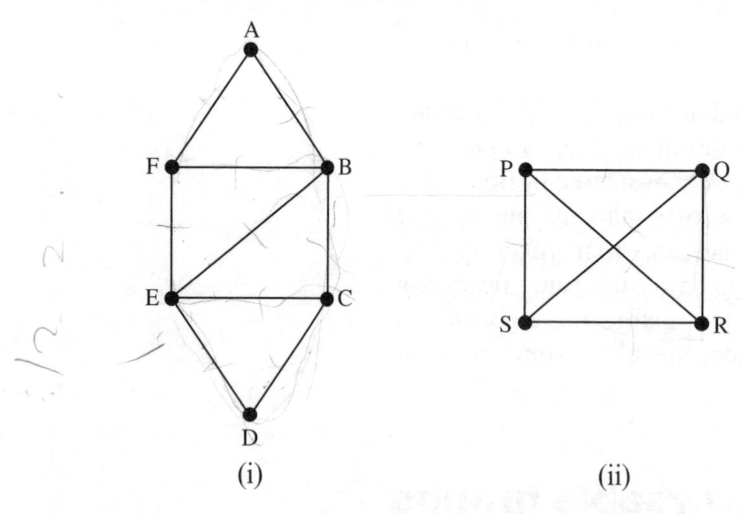

The graph (i) above is semi-Eulerian. It can be drawn by starting at F and finishing at C using the route FABCDEFBEC.

Graph (ii) above is not semi-Eulerian since each of the four vertices has valency 3.

- **If the start and finish vertices are the same, that is X = Y, then in the light of the above it can be seen that for such a route to exist *all the vertices must be even.* If a graph has this property then it is called *Eulerian.***

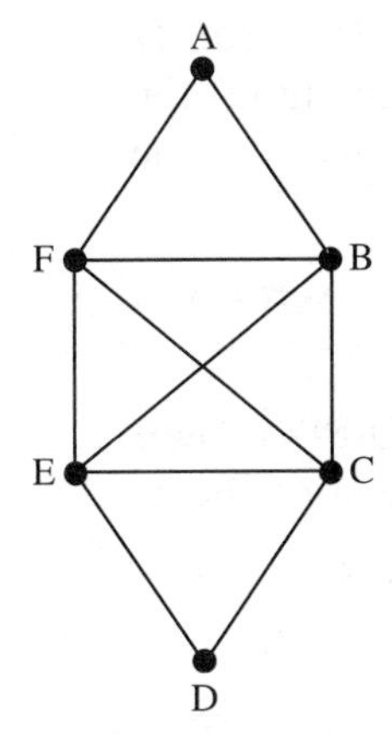

The graph above is a **Eulerian graph**. It can be drawn starting at A using the route ABCDEFCEBFA. This is of course a Eulerian cycle, as defined in Chapter 2.

Leonhard Euler (1707–1783) translated a practical problem into a graph theory problem, that of deciding if a graph was traversable or not.

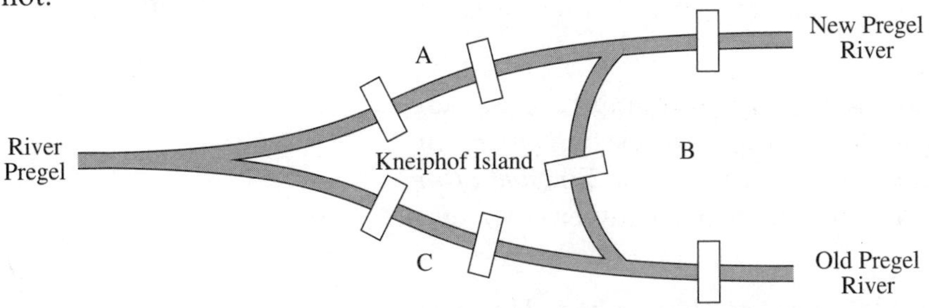

The sketch above shows the city of Konigsberg (now Kalingrad), which has seven bridges. The problem was to find a route for a walk crossing each of the seven bridges once and only once and returning to the starting point. If we represent the areas A, B, C and Kneiphof Island as vertices and the bridges as edges we obtain the following graph:

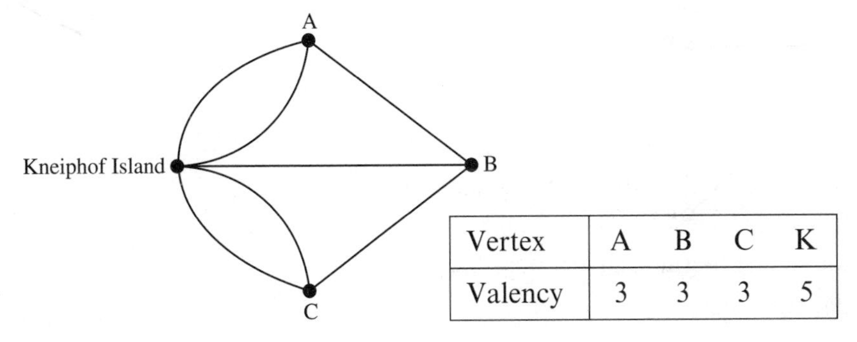

Vertex	A	B	C	K
Valency	3	3	3	5

All the vertices are **odd**. There is therefore no solution to the problem as the graph is not traversable.

It is easy to see from the above that the valencies of the vertices play a crucial part in deciding whether or not particular kinds of routes are possible. There is a very interesting result concerning the valencies of the vertices of a graph, which is known as the **handshaking theorem**:

- **The *sum* of the values of the valencies, taken over all the vertices of a graph *G*, is equal to *twice* the number of edges.**

Since each edge XY has two ends, X and Y, it contributes 1 to the valency of X and 1 to the valency of Y. So each edge contributes $(1 + 1) = 2$ to the sum of valencies:

$$\sum(\text{valencies}) = \underbrace{2+2+\ldots+2}_{\text{one for each edge}} = 2 \times (\text{number of edges})$$

This theorem has a very interesting and useful corollary:

- **In any graph, the number of odd vertices is even.**

To prove this, divide the graph into odd and even vertices. Then, from the handshaking theorem:

$$(\textit{sum of degrees of odd vertices}) + (\textit{sum of degrees of even vertices}) \\ = 2 \times (\textit{number of edges})$$

The right-hand side of this equation is an even number, so the left-hand side must also be an even number. However, (*sum of the degrees of the even vertices*) must be an even number. So (*sum of the degrees of the odd vertices*) is the difference between two even numbers, i.e. an even number.

(*Sum of degrees of odd vertices*) is a sum of odd numbers. However, in order for a sum of odd numbers to be even, it must have an even number of terms. Hence there are an even number of odd vertices.

Exercise 4A

1

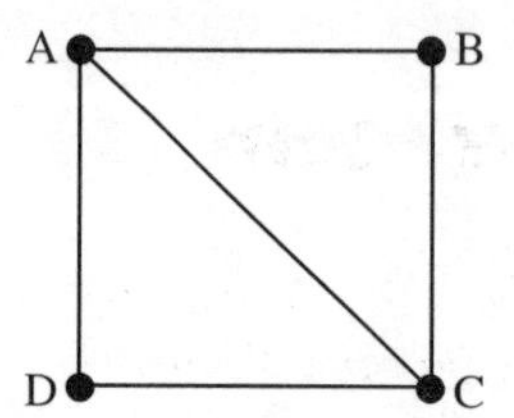

Show that the above graph is semi-Eulerian. Give a possible route that traverses each edge just once.

2

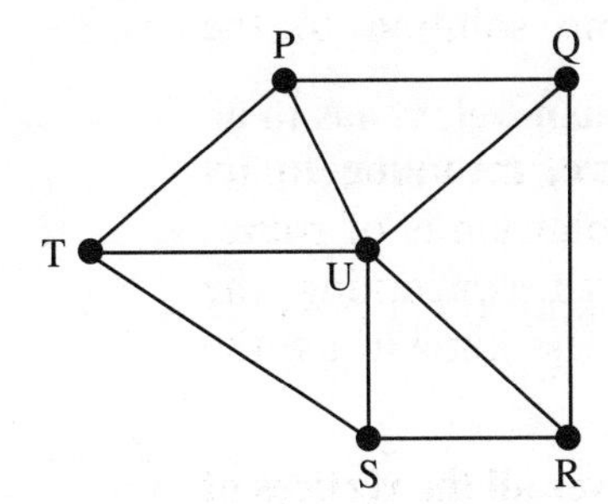

Show that the above graph is not semi-Eulerian.

3

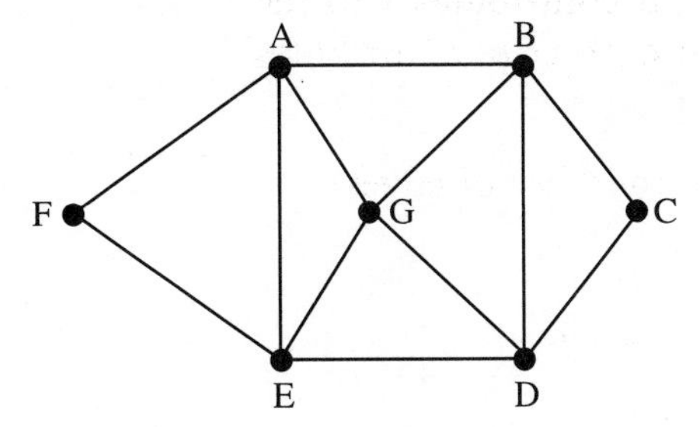

Show that the above graph is Eulerian. Give a Eulerian cycle for the graph.

4

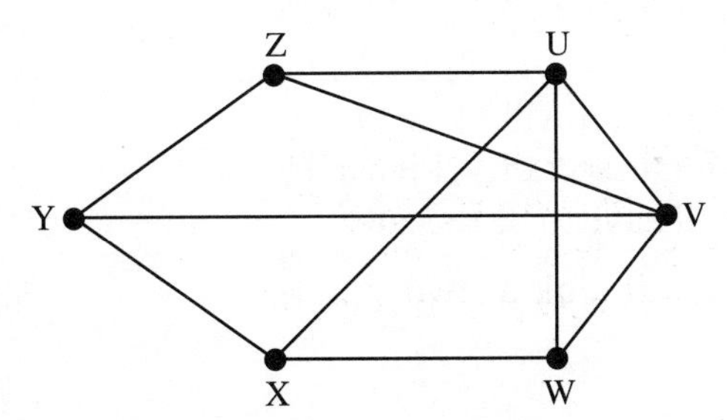

Show that the above graph is not Eulerian.

Show that it can be made Eulerian by adding two further edges.

5 For each of the graphs in questions 1 to 4 verify the handshaking theorem and its corollary.

4.2 The route inspection problem

The route inspection problem is also often called the Chinese postman problem. A Chinese mathematician, Mei-Ku Kwan, discussed the problem of a postman who wishes to deliver his letters by covering the shortest possible distance and returning to his starting point.

The problem can be stated as:

- **In a given undirected network a route of minimum weight has to be found that traverses every edge at least once, returning to its starting vertex.**

If the graph is Eulerian then we are simply looking for a possible route that starts and finishes at the same vertex.

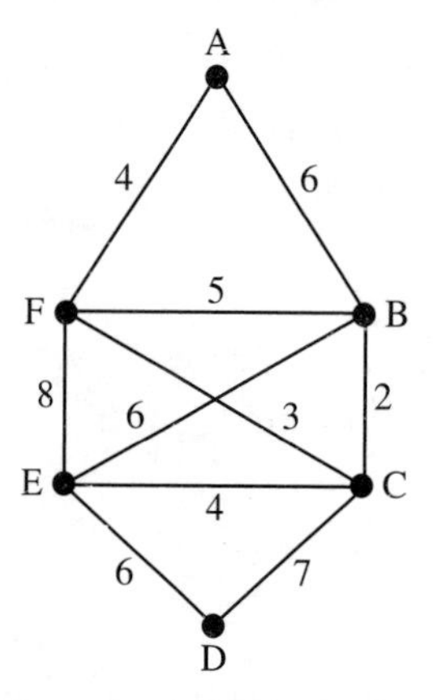

We saw in section 4.1 that the above graph is Eulerian. A possible route starting and finishing at A is ABCDEFCEBFA. This has a total weight of 51. It is suggested you mark your route on the network to make sure all edges are covered.

In general, when a network has some odd vertices, it will be necessary to repeat some of the edges. From the discussion in section 4.1 it can be seen that **at least one edge at each odd vertex** will have to be repeated.

Before considering the general algorithm we will look at two simple examples.

Example 1

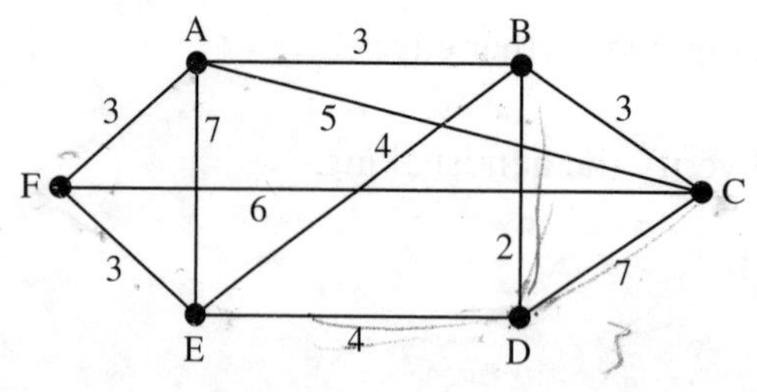

For the above network find a minimum weight route starting and finishing at A, and traversing every edge of the network at least once.

The first thing to do is find the valencies of the vertices:

Vertex	A	B	C	D	E	F
Valency	4	4	4	3	4	3

The only odd vertices are D and F. We need to find the shortest route between D and F. Dijkstra's algorithm may be used to find this but in this simple case we can see by inspection that it is DEF, of length 7. The edges to be repeated are DE and EF and this is shown on the modified network below.

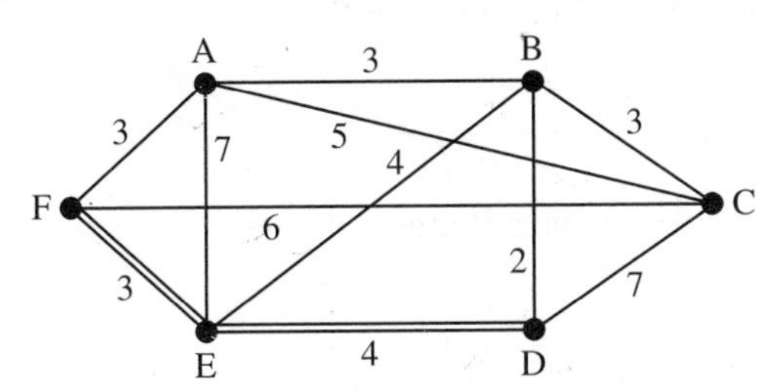

This modified network is now Eulerian as all vertices are even. A route starting and finishing at A is ABCDEBDEACFEFA. Its total length is:

(sum of weights of edges of original network) + (weight of repeated edges)

$$= (3 + 3 + 7 + 6 + 3 + 5 + 4 + 4 + 2 + 3 + 7) + 7$$
$$= (47) + 7 = 54$$

Alternative ways of pairing vertices

When there are more than two odd vertices, we know from section 4.1 that there will be an **even** number of odd vertices. Our considerations will be restricted to networks with at most four odd vertices. When there are four odd vertices, W, X, Y and Z (say), there are three possible pairings:

(i) (W with X) and (Y with Z)
(ii) (W with Y) and (X with Z)
(iii) (W with Z) and (X with Y)

Example 2

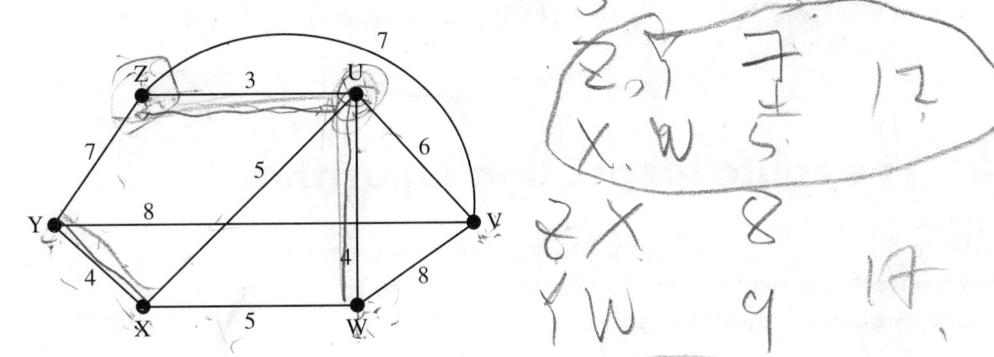

For the above network find a minimum weight route starting and finishing at V and traversing every edge of the network at least once.

The valencies of the vertices are U(4), V(4), W(3), X(3), Y(3) and Z(3).

The three possible pairings of the odd vertices W, X, Y and Z are those given above. For each possible pairing the shortest route between the two vertices must now be found. The edges on these shortest routes have to be repeated.

Pairing	Shortest routes	Length of repeated edges
(i) (W with X), (Y with Z)	WX and YZ	$5 + 7 = 12$
(ii) (W with Y), (X with Z)	WXY and XUZ	$(5 + 4) + (5 + 3)$ $= 9 + 8 = 17$
(iii) (W with Z), (X with Y)	WUZ and XY	$(4 + 3) + 4$ $= 7 + 4 = 11$

Since pairing (iii) gives the minimum sum this is the best pairing. The edges to be repeated are then WU, UZ and XY. This is shown in the modified network below.

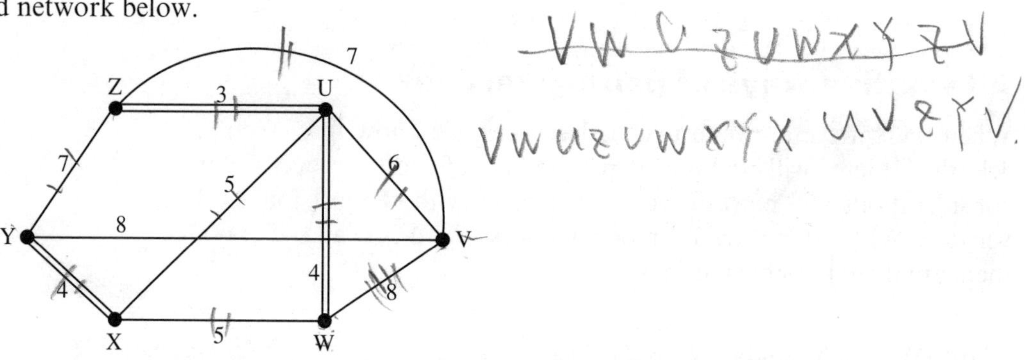

This is now Eulerian as all the vertices are even. A route starting and finishing at V is VWUWXYZVYXUZUV. Its total length is

$$(\textit{sum of weights of all edges}) + (\textit{lengths of repeated edges}) = 57 + 11 = 68$$

4.3 The route inspection algorithm

In order to find a minimum weight route that traverses every edge of a given connected network at least once and returns to the starting vertex, carry out the following steps:

Step 1 List all odd vertices.

Step 2 Form all possible pairings of odd vertices.

Step 3 For each pairing find the edges that are best to repeat and find the sum of the lengths of these edges.

Step 4 Choose the pairing with the smallest sum. Construct a route that repeats these edges. (The graph with repeated edges will be Eulerian and so traversable.)

Example 3

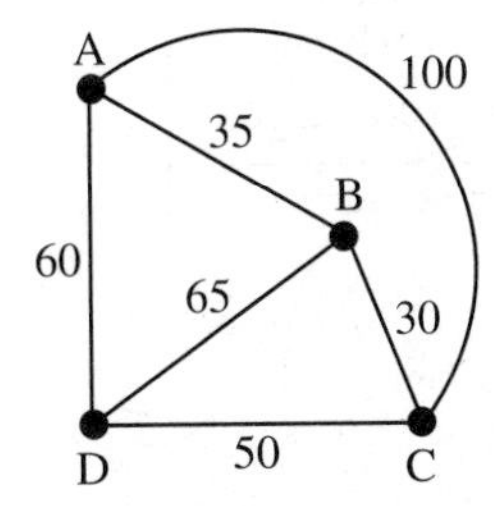

The diagram shows the layout of footpaths in a garden. The lengths of the paths are given in metres. The gardener wishes to inspect the paths for frost damage after the winter. Find a route, starting and finishing at A, that traverses each path and is of minimum length.

The valencies of the vertices are A(3), B(3), C(3) and D(3).

Step 1 The odd vertices are A, B, C and D.

Step 2 Possible pairings are
(i) (A with B), (C with D)
(ii) (A with C), (B with D)
(iii) (A with D), (B with C)

Step 3

Pairing	Shortest routes	Length of repeated edges
(i) (A with B), (C with D)	AB and CD	$35 + 50 = 85$
(ii) (A with C), (B with D)	ABC and BD	$(30 + 35) + 65$ $= 130$
(iii) (A with D), (B with C)	AD and BC	$60 + 30 = 90$

Step 4 The best pairing is (A with B), (C with D). The edges to be repeated are AB and CD. The modified network is now:

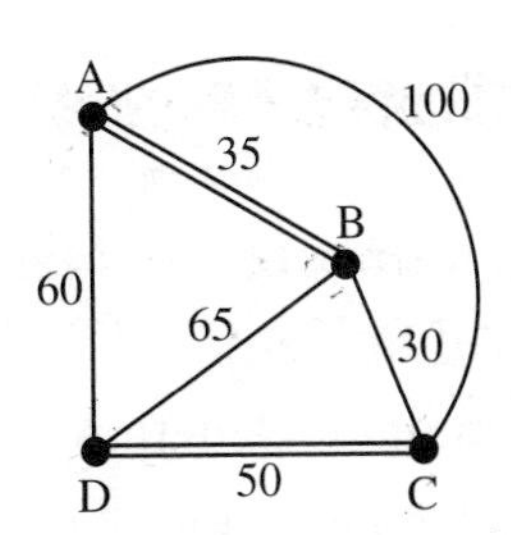

A possible route starting and finishing at A is ACDBCDABA of length $340 + 85 = 425$ m.

Special cases

As we saw in section 4.1, a graph with just two odd vertices X and Y is traversable, so it is possible to find a route from X to Y that traverses each edge once and only once.

In Example 3 there were four odd nodes. If we allow the gardener to start and finish at **two different nodes**, A and C say, this will reduce the number of pairings to be considered and the number of edges to be repeated.

As we do not need to consider vertices A and C, the only relevant odd vertices are B and D and so the edge to be repeated is the edge BD of weight 65. We then have:

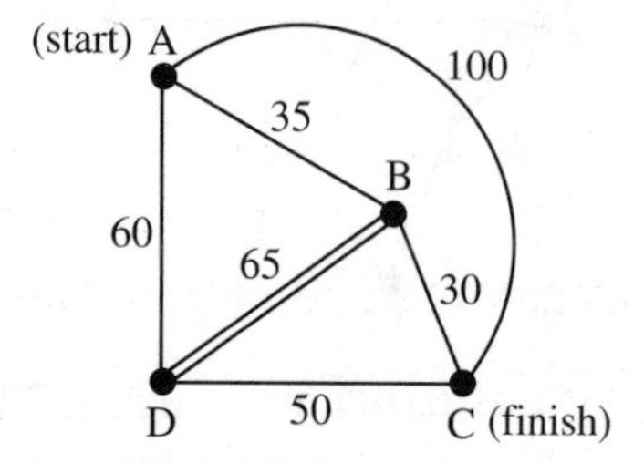

A possible route is ABDACDBC, of length $340 + 65 = 405$ m.

If instead the gardener chose to start at C and finish at D, then the repeated edge would be AB and the length of the route, for example CBACDBAD, would be $340 + 35 = 375$ m.

Exercise 4B

1

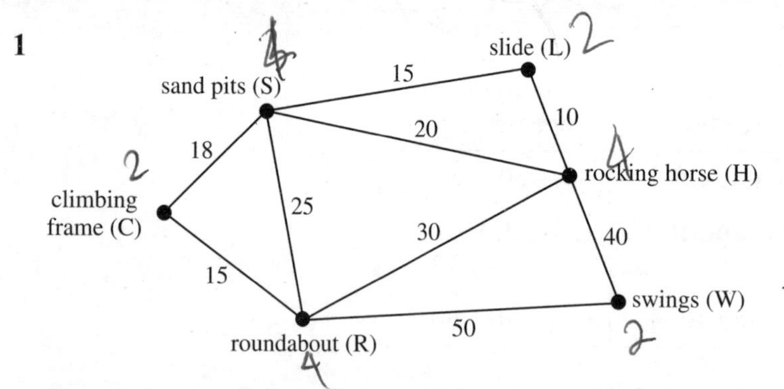

The above network models a children's playground. The arcs represent the footpaths between the different sites. The numbers on the arcs are the lengths, in metres, of the paths.

(a) Write down the valencies of the vertices.

(b) Hence obtain a possible minimum length route that starts and finishes at C and covers each path at least once. Give the length of this route.

2

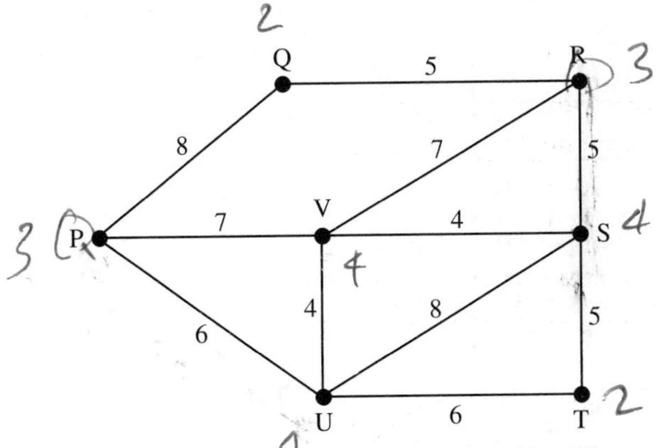

The network above models the area covered by a district council. The vertices represent towns and the arcs represent roads. The lengths of the roads are given in km. After a heavy snowfall the council wishes to use a snow plough to clear all the roads. The plough must drive along each road at least once to clear it. The snow plough is based at T, where the route must start and finish. Find a route of minimum length for the snow plough.

3 For the network shown above find a minimum weight route starting and finishing at A and traversing every edge at least once. Give the weight of this route.

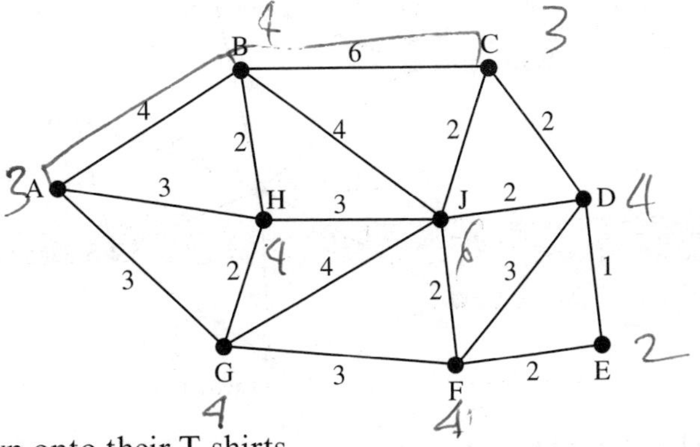

4 The Venturers wish to have this logo sewn onto their T-shirts. The vertices have been labelled A, B, C, D, E and F for convenience. The numbers indicate the lengths in centimetres of the edges.

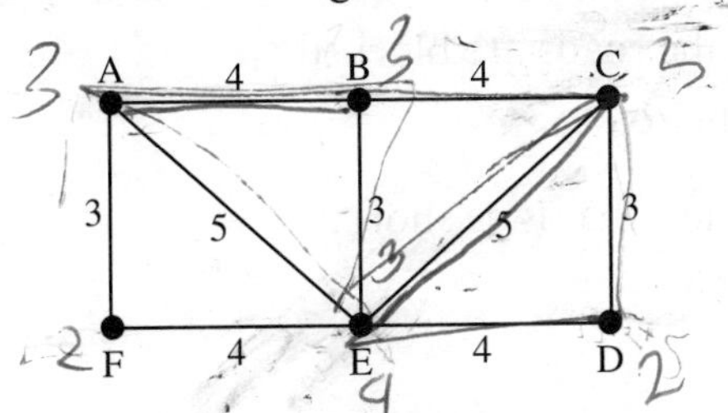

The logo is to be produced with a minimum amount of stitching. A machinist will start and finish at A with no breaks. Use the route inspection algorithm to find the minimum length of stitching required. Give a possible route with this length.

5

	A	B	C	D	E
A	—	20	—	—	25
B	20	—	32	70	30
C	—	32	—	40	60
D	—	70	40	—	35
E	25	30	60	35	—

The table above shows the distances, in km, between five locations A, B, C, D and E.

(a) Draw a network representation of this information.

(b) Find a minimum distance route starting and finishing at A and traversing each road at least once.

6

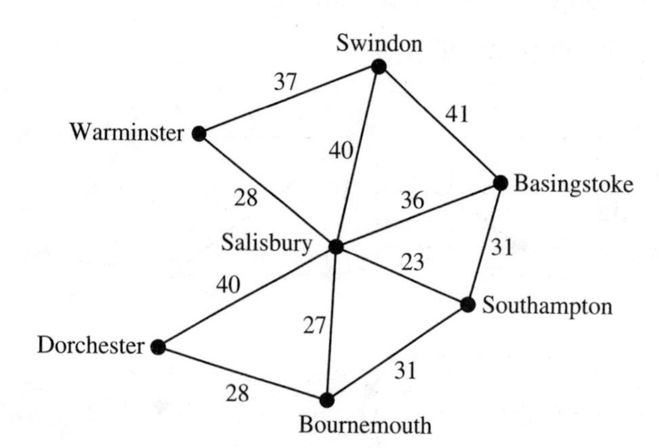

The above network shows some places in the south of England and the roads between them. The numbers indicate the distances, in miles, between towns. After a storm the highways authority wishes to make sure that none of the roads are blocked by fallen trees and so wishes to inspect all the roads.

Find a route starting and finishing at Salisbury that is as short as possible and traverses each road at least once.

7

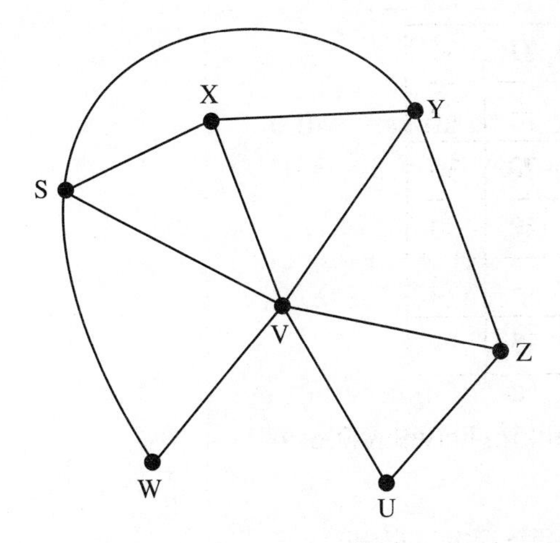

Joan and John are staying at town S. The above diagram shows the network of bus routes in the area. The table below gives the bus fares between the various towns in pence.

	S	X	Y	Z	U	V	W
S	—	90	150	—	—	85	78
X	90	—	50	—	—	60	—
Y	150	50	—	45	—	70	—
Z	—	—	45	—	88	83	—
U	—	—	—	88	—	72	—
V	85	60	70	83	72	—	56
W	78	—	—	—	—	56	—

Joan and John wish to start and finish at S, travel along each bus route and keep their cost to a minimum. Find a minimum cost route for them.

SUMMARY OF KEY POINTS

1 A **traversable graph** is one that can be drawn without removing your pen from the paper and without going over the same edge twice.

2 If a graph is not Eulerian but there is a route starting and finishing at different vertices and traversing every edge once and only once then the graph is **semi-Eulerian**.

3 If in a graph there is a route that starts and finishes at the same vertex and traverses every edge once and only once then the graph is **Eulerian**.

4 **The handshaking theorem**
The **sum** of the valencies, taken over all the vertices of a graph G, is equal to **twice** the number of edges.

5 **The route inspection problem**
In a given undirected network a route of **minimum weight** has to be found that traverses every edge at least once, returning to the starting vertex.

Critical path analysis

5

In this chapter we consider the modelling of complex projects by a network and the analysis of the resulting model. This is sometimes called network analysis or network planning but is more usually referred to as the **critical path method** (CPM) or **critical path analysis** (CPA).

Serious application of the method was first made in the mid-1950s. In Britain a team working for the Central Electricity Generating Board developed a method for scheduling the work of overhauling a generating plant. By 1957 they had devised a technique for identifying 'the longest irreducible sequence of events'. Using this technique, in 1958 they carried out an overhaul of a power station that reduced the overall time to 42% of the previous average time for the same work.

At about the same time a group of researchers working in the US for the E.I. du Pont Chemical Company used the critical path method to schedule the construction of a $10 million chemical plant. It was credited with saving the company $1 million.

Since that time the CPM or CPA has been used in a large number of areas, including overhaul, construction, civil engineering, town planning, marketing, ship building and design.

5.1 Precedence tables

The first step in scheduling a complex project is to break it down into a set of sub-projects known as **activities**. For example, if the project were building an extension to a house then this could be broken down into the following activities:

A Prepare the foundations
B Have foundations passed by inspector
C Obtain bricks
D Erect walls
E Construct roof
F Install plumbing
G Install wiring
H Plaster walls
I Decorate
J Landscape garden

It is clear that not all these activities are independent. Some of them are related in the sense that certain activities cannot start until others have been completed. The next step in the scheduling process is to identify which activities depend on which others being completed first. Some activities must precede (come before) others. The way in which the activities depend on each other can be summarised in a **precedence table** like this:

Activity	Depends on
A	—
B	A
C	—
D	B, C
E	D
F	D
G	E
H	F, G
I	H
J	E

A precedence table is sometimes called a **dependence table**.

The heading 'depends on' is shorthand for 'the activities that must be completed before it can be started'. A dash — indicates that there are no such activities. Notice that the 'depends on' column only shows activities *immediately* preceding each entry in the activity column. For example, we have written E depends only on D. The fact that E depends on B and C is therefore implied.

The relationships between the activities may not be given in such a direct form. When this is the case it is suggested that you first produce a dependence table from the given information.

Example 1

A project has been broken down into the activities A, B, C, D, E, F and G. After a committee meeting the information below was produced. Draw up a precedence table that summarises this information:

- activities A, C and D do not depend on the completion of any other activity
- activity A must be completed before activity B can start
- both activities B and C must be completed before activity E can begin
- activity F can only start when A, B, C, D and E are completed
- the project is completed when G is finished. G requires all other activities to be completed before it can start.

The table below is the precedence table. The reasons for the entries are given immediately following the table.

Activity	Depends on
A	—
B	A
C	—
D	—
E	B, C
F	D, E
G	F

The entries for A, C and D follow from their independence of other activities.

The entry for B follows from the given statement directly.

The entry for E follows as *both* B and C must be complete before it can start.

The entry for F requires some explanation. Since E depends on B and C, and B also depends on A, we need only include E here together with activity D.

As F depends on D and E, and E depends on B and C, and B depends on A the given statement implies G depends on all the others as required.

Exercise 5A

1 The project 'write and post a letter' may be broken down into the following activities:

A Purchase a pad of paper
B Purchase a packet of envelopes
C Purchase a stamp
D Write the letter
E Address the envelope
F Stick the stamp on the envelope
G Place the letter in the envelope
H Seal the envelope
I Post the letter

Draw up a precedence table for this project.

2 Mrs Brown decides that the lounge needs a total change of paint, wallpaper and curtains. She identifies the following activities:

A Buy new curtains
B Buy the paint
C Buy the wallpaper
D Take up the carpet
E Remove the curtains
F Paint the woodwork
G Hang the wallpaper
H Hang the new curtains
I Replace the carpet

Draw up a precedence table for this project.

3 A project consists of activities A, B, C, D and E. These activities must satisfy the following conditions:

- activities A and B are independent of the others
- only when activity A is completed can activity C start
- activity D requires activity B to be completed before it can start
- activity E can only start when all other activities are finished.

Draw up a precedence table for this project.

4 A project consists of activities A, B, C, ..., J. The following information has been obtained about these activities:

- activities A, B and C do not depend on the completion of any other activity
- activities A and B must be completed before activity D can begin
- before activities E and F can start, activity D must be completed
- activity H can only start when A, D and F are completed, whilst activity G can only start when activity E has also been completed
- one must wait until activities C, D, E, F and G are complete before activity I can begin
- the final activity J requires the completion of all the others.

Draw up a precedence table for this project.

5.2 Activity networks

Having produced a precedence table the next step in the critical path method is to produce an **activity network** (or project network) to model the situation.

- **The nodes (vertices) in this network represent *events*. Each event is the completion of one or more activities.**
- **The arcs in this network represent *activities* and the weight on each arc represents the *duration* of the corresponding activity.**
- **The *source node* represents the beginning of the project and the *sink node* represents the end of the project.**

Here is a typical activity network:

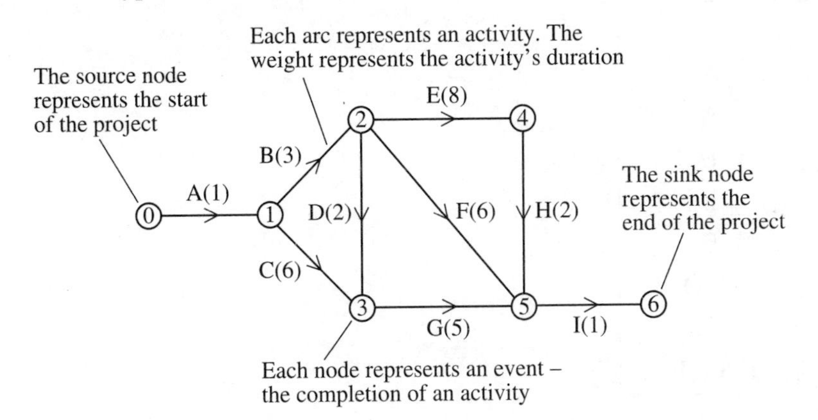

Notice that the events represented by the nodes are numbered. The lengths and orientation of the arcs are not significant.

- **It is conventional to assume that time flows from left to right when possible, but arrows are used to make the directionality clear.**

Consider node ④:

E(8) ④ H(2)

with the arc representing E *entering* and the arc representing H *leaving*. The numbers in brackets are the durations of these activities. This models the situation 'activity E must be completed before activity H can begin'. Event ④ denotes the completion of activity E.

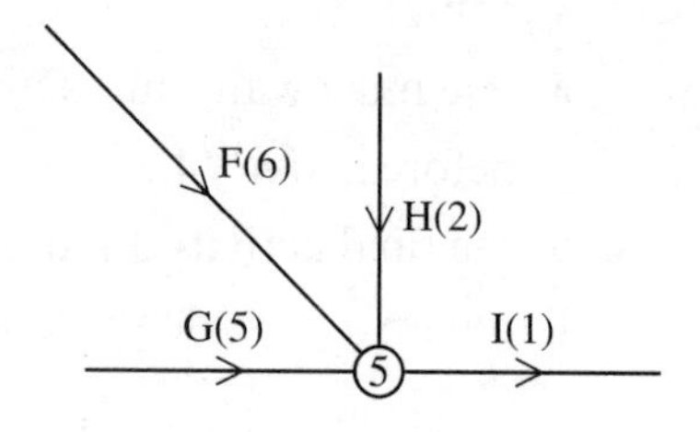

Node ⑤ represents a more complicated situation. This models the situation 'activity I cannot start until activities F, G and H are all complete'. Event ⑤ represents the completion of all of the activities F, G and H.
Consider now a typical activity, say G:

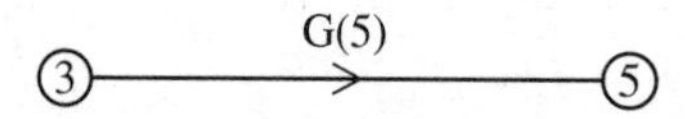

This activity has to be performed between the two events indicated by the numbers at the end of the arcs. The event at the beginning of the arc, in this case ③, is called the **tail event**; that at the end of the arc, in this case ⑤, is called the **head event**. Any activity can be specified by giving the numbers of its tail and head events. That is, arc (3, 5) represents G. In constructing a network the head event is always given a number greater than the corresponding tail event.

Drawing an activity network

The network on page 105 models this precedence table:

Activity	Depends on
A	—
B	A
C	A
D	B
E	B
F	B
G	C, D
H	E
I	F, G, H

The network was constructed like this:

1 Activity A has no predecessor and so may begin at any time. Label source node ⓪ and insert activity A.

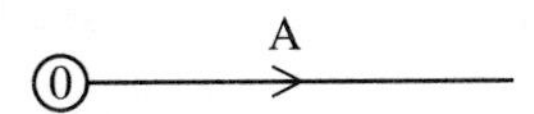

2 Activities B and C depend only on A therefore insert node ①, indicating the completion of activity A, and add activities B and C.

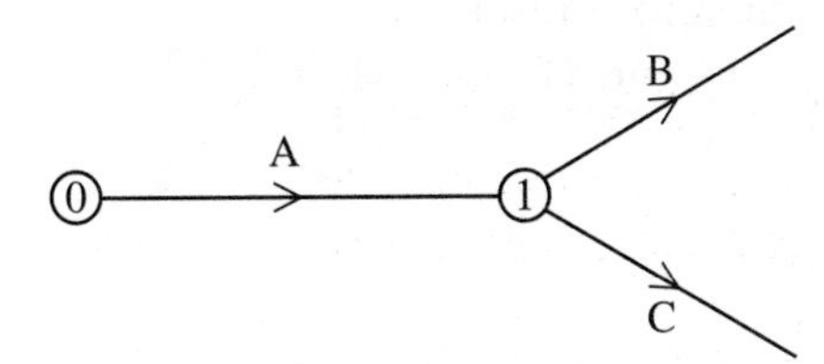

Arcs representing B and C are any two arcs leaving ①; their orientation is irrelevant but the convention is that they go from left to right when possible.

3 Activities D, E and F depend only on B therefore insert node ②, indicating the completion of activity B, and add activities D, E and F.

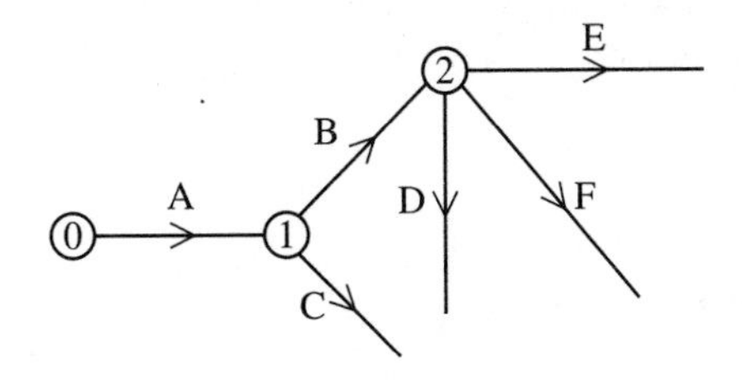

4 Activity G depends on C and D therefore insert node ③, indicating the completion of both activity C and activity D, and add activity G.

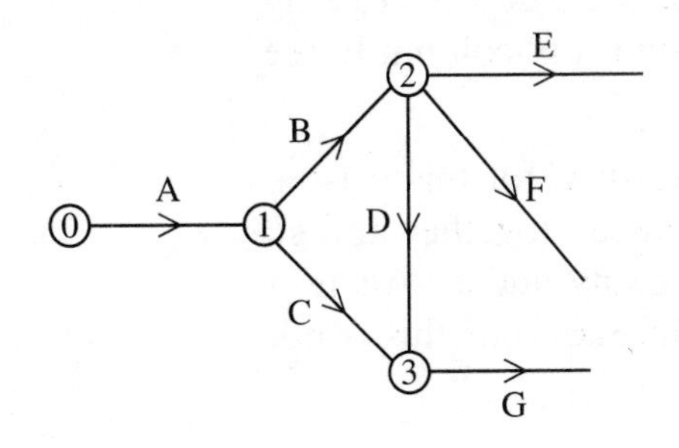

5 Activity H depends on E therefore insert node ④, indicating the completion of activity E, and add activity H.

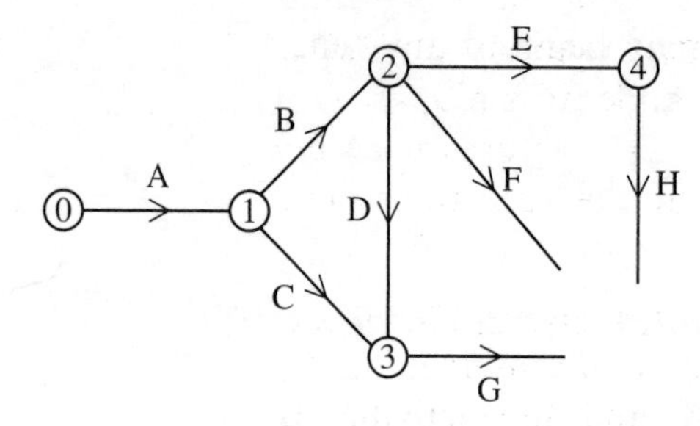

6 Finally, activity I depends on F, G and H therefore insert node ⑤, indicating the completion of activities F, G and H, and add activity I.

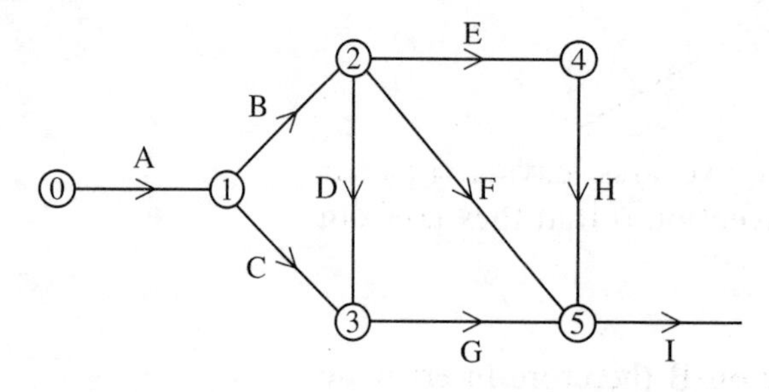

7 The project is complete when all activities are completed. So add a sink node ⑥.

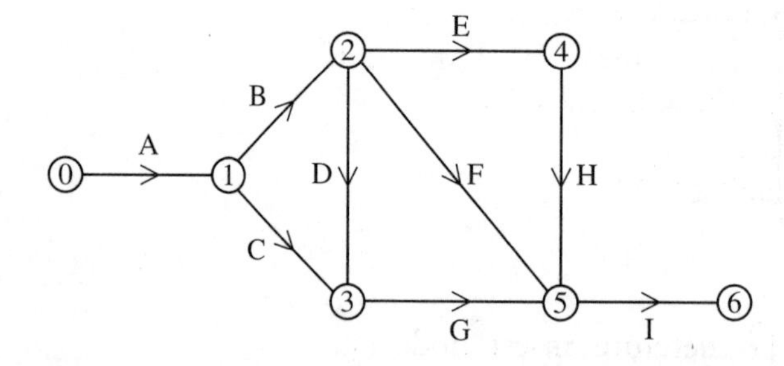

Your first attempt to draw a network may not look as neat as this. If possible you should try to produce a network without any crossing arcs. It is useful to look ahead down the 'depends on' column to see which arcs are likely to meet at a node.

In this example D was drawn downwards towards C for this reason. Further, it was noticed that F, G and H occurred together and so these arcs were drawn so that node ⑤ was easily added. Again a neat diagram results if all the arcs are straight lines, but this is not necessary.

Using dummies

In the above examples it was possible to represent the dependence table without too much difficulty. However, situations can occur when, in order to represent the dependence table, it is necessary to use a **dummy activity** or logical restraint. Such a situation occurs when two chains have a common event but they are themselves wholly or partly independent of each other.

- **A dummy activity is usually shown as a dotted line.** ***Its direction is important*** **but it has zero duration.**

Here is an example in which a dummy activity is necessary:

Activity	Depends on
C	A, B
D	B

This situation can be represented by:

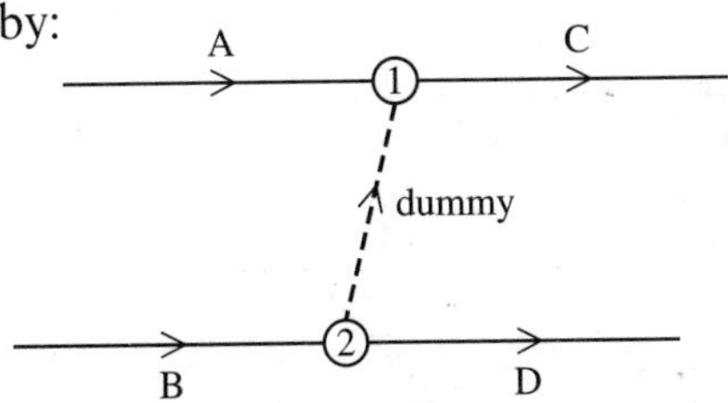

It is not possible to represent this situation without a dummy. Notice carefully the direction of the dummy.

It is a convention in the critical path method that we allow *at most one activity between any two events*. In order to meet this convention a dummy activity may be necessary. For example, consider

Activity	Depends on
A	—
B	A
C	A
D	B, C

This may be represented by

Example 2

Draw the activity network for the precedence table below.

Activity	Depends on
A	—
B	—
C	—
D	A
E	A
F	C
G	B, C, D
H	G
I	E, G
J	F

1 Activities A, B, C

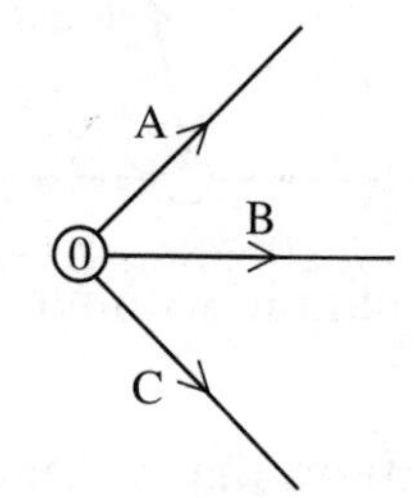

2 Activities D, E

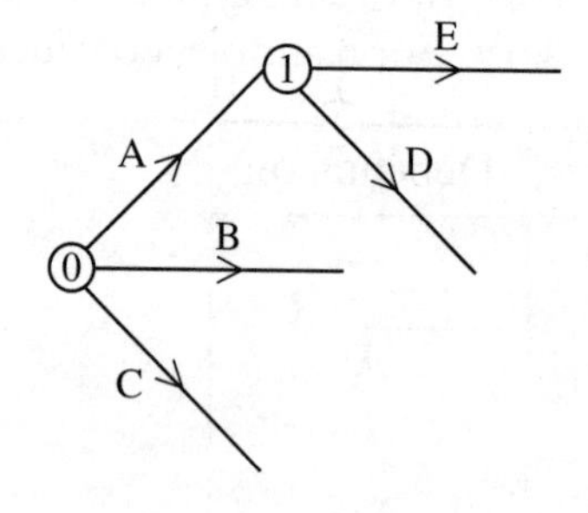

3 Activity F

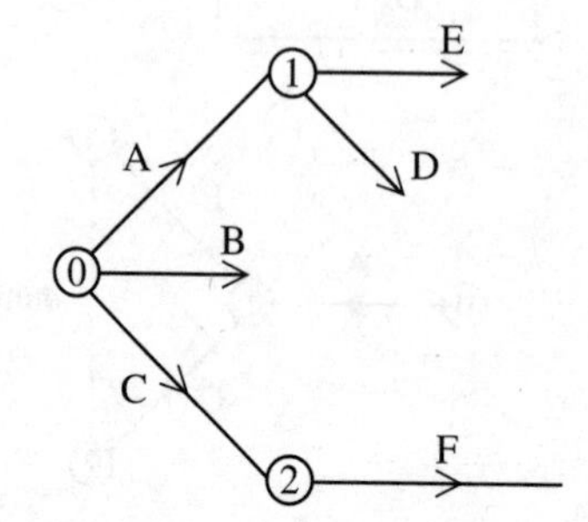

4 Activity G – you need a dummy here to represent this dependence:

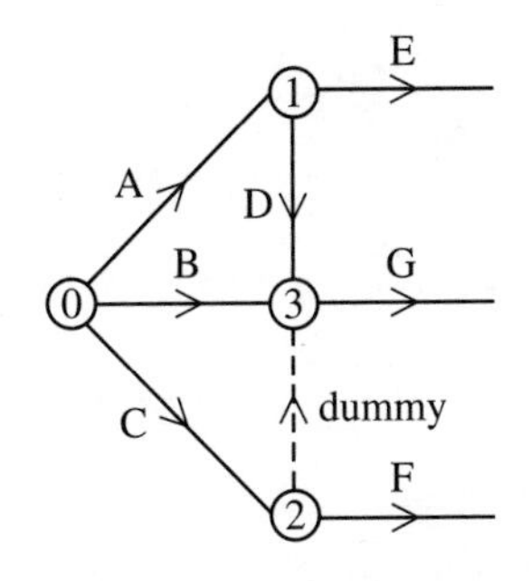

Notice the direction of the dummy. F does not depend on B.

5 Activity H

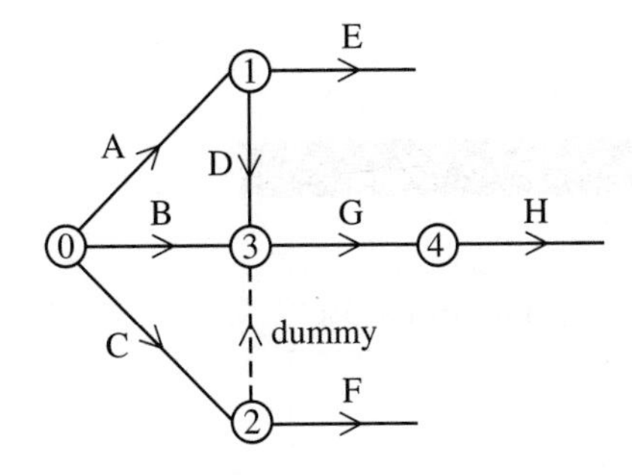

6 Activity I – as I depends on E and G, but H depends only on G, you need a dummy to represent this.

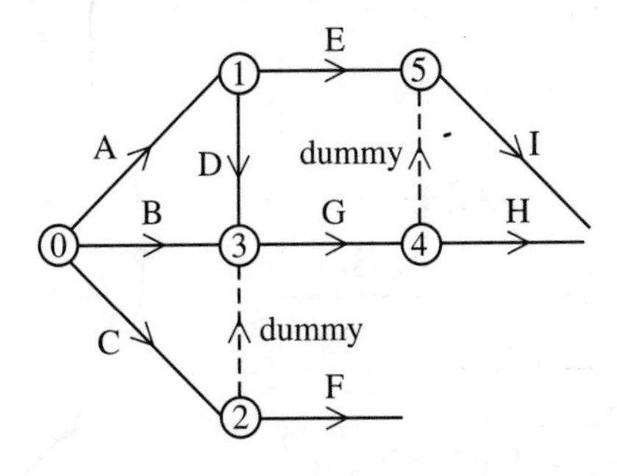

7 Activity J

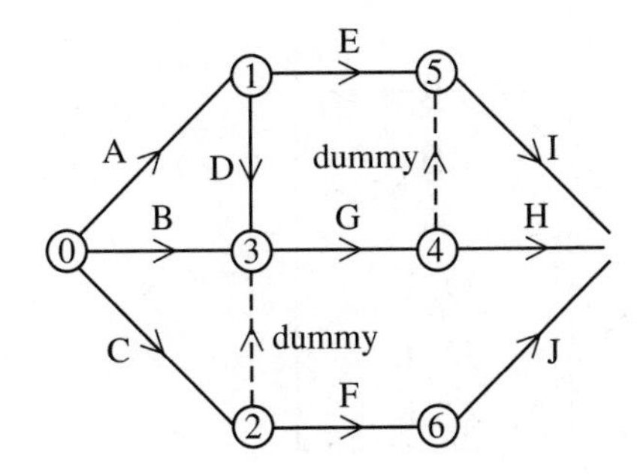

8 End of project when H, I and J are complete.

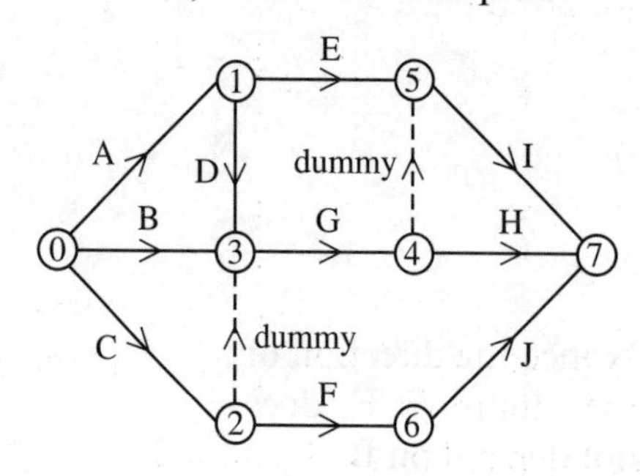

It is possible to draw different diagrams to represent the same dependencies.

Exercise 5B

1 Draw an activity network to represent the dependence table opposite, given that the project is complete when all activities are complete.

Activity	Depends on
A	—
B	—
C	—
D	A
E	A
F	B
G	C
H	D
I	E, F, G

2 Draw an activity network to represent the dependence table opposite, given that the project is complete when all activities are complete.

Activity	Depends on
A	—
B	—
C	B
D	B
E	A
F	D, E
G	D, E
H	C, F

3 Draw an activity network to represent the dependence table opposite, given that the project is complete when all activities are complete.

Activity	Depends on
A	—
B	—
C	—
D	A
E	B
F	B, C
G	B, D
H	G
I	E, F, H

4 The following information has been obtained concerning the activities A, B, C, . . . , N involved in a project:

Activity A precedes B, C, D, E and F
Activity J follows B
Activity K follows C
Activity G follows E
Activity H follows D and G and precedes L
Activity I follows D and G
Activities F and I precede M
Activity L follows J and K
Activity N follows L and M

(a) Draw up a precedence table for these activities.
(b) Draw an activity network to represent your precedence table.

5 The dependence table for a particular project is given in this table.

Draw an activity network for this project. Your network should include a minimum number of dummy activities.

Activity	Depends on
A	—
B	—
C	—
D	A
E	B, C
F	B, C
G	C
H	D, F, G
I	D, E
J	D

5.3 Analysing the project (the critical path algorithm)

The **critical path algorithm** identifies the longest path or paths from the source node to the sink node. This is the **critical path** (or paths). The algorithm proceeds by finding two times associated with each event: the **earliest event time** and the **latest event time**.

Earliest event time e_i (forward scan)

- **The earliest time e_i for node i is the earliest time that you can arrive at event i with all the incoming activities completed. We work from source node to sink node, that is forwards, or from left to right in the diagram.**

Initially you set the earliest time for the source node as zero. We will illustrate the algorithm by carrying out the calculations for the network on page 105, which for ease of reference we repeat here.

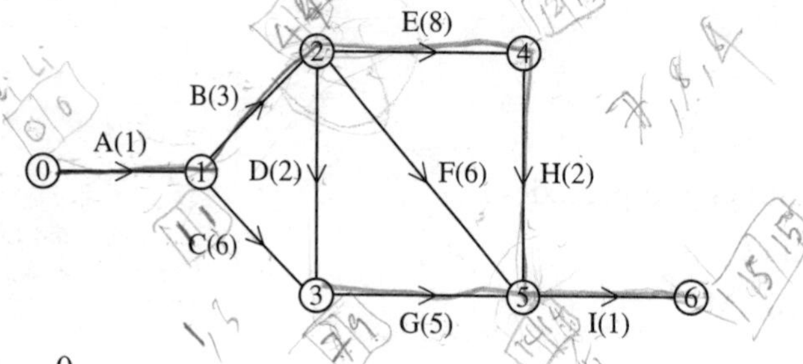

Node 0 $e_0 = 0$

Node 1 There is only one arc (0, 1) leading into node ① and so:

$$\begin{aligned} e_1 &= e_0 + (\text{duration of A}) \\ &= 0 + 1 \\ &= 1 \end{aligned}$$

Node 2 There is only one arc (1, 2) leading into node ② and so:

$$\begin{aligned} e_2 &= e_1 + (\text{duration of B}) \\ &= 1 + 3 \\ &= 4 \end{aligned}$$

Node 3 There are *two* arcs (1, 3) and (2, 3) leading into node ③ and so, since both activities C and D must be completed before event ③ can take place:

$$\begin{aligned} e_3 &= \text{larger of } [e_1 + (\text{duration of C}), e_2 + (\text{duration of D})] \\ &= \text{larger of } [1 + 6, 4 + 2] \\ &= \text{larger of } [7, 6] \\ &= 7 \end{aligned}$$

Node 4 There is only one arc (2, 4) leading into node ④ and so:

$$e_4 = e_2 + (\text{duration of E}) = 4 + 8 = 12$$

Node 5 There are *three* arcs (2, 5), (3, 5) and (4, 5) leading into node ⑤ and so activities F, G and H must all be completed before event ⑤ can take place.

$$e_5 = \text{largest of } [e_2 + (\text{duration of F}), e_3 + (\text{duration of G}), e_4 + (\text{duration of H}),]$$
$$= \text{largest of } [4+6, 7+5, 12+2]$$
$$= \text{largest of } [10, 12, 14] = 14$$

Node 6 There is only one arc (5, 6) leading into node ⑥ and so:

$$e_6 = e_5 + (\text{duration of I}) = 14 + 1 = 15$$

- **In general:**

  $$e_i = \underset{k}{\text{maximum}}\,[e_k + \text{duration}\,(k, i)]$$

  where the maximum is taken over all arcs (k, i) leading into vertex i.

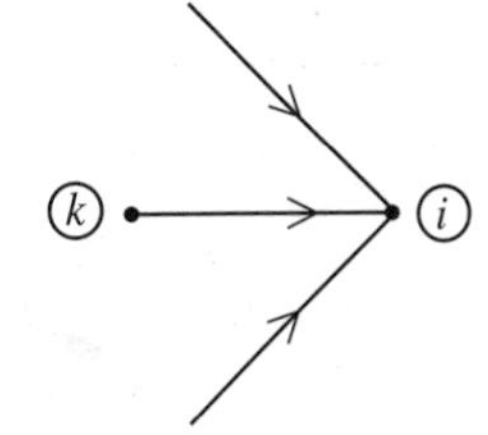

$\underset{k}{\text{maximum}}[e_k + \text{duration}\,(k, i)]$ means find the values of $[e_k + \text{duration}\,(k, i)]$ for all possible values of k, keeping i fixed, and then take the largest of these values.

The earliest time calculated for the sink node is the total amount of time required to complete the project. It is the length of the longest (critical) path (or paths).

For our project, therefore, the length of the critical path is 15.

Latest event time l_i (backward scan)

- **The latest time l_i for node i is the latest time you can leave event i without extending the length of the critical path. We work from sink node to source node, that is backwards, or from right to left in the diagram.**

Clearly for the sink node (finish event) the latest and earliest terms are the same so you have $l_6 = e_6 =$ the total project time $= 15$.

Node 5 Since there is only one arc (5, 6) leaving node ⑤:

$$l_5 = l_6 - (\text{duration of I}) = 15 - 1 = 14$$

Node 4 Since there is only one arc (4, 5) leaving node ④:

$$l_4 = l_5 - (\text{duration of H}) = 14 - 2 = 12$$

Node 3 Since there is only one arc (3, 5) leaving node ③:

$$\begin{aligned} l_3 &= l_5 - (\text{duration of G}) \\ &= 14 - 5 = 9 \end{aligned}$$

Node 2 There are *three* arcs (2, 3), (2, 4) and (2, 5) leaving node ② and so:

$$\begin{aligned} l_2 &= \text{smallest of } [l_3 - (\text{duration of D}), \\ &\qquad l_4 - (\text{duration of E}), \\ &\qquad l_5 - (\text{duration of F})] \\ &= \text{smallest of } [9 - 2, 12 - 8, 14 - 6] \\ &= \text{smallest of } [7, 4, 8] = 4 \end{aligned}$$

Node 1 There are *two* arcs (1, 2) and (1, 3) leaving node ① and so:

$$\begin{aligned} l_1 &= \text{smaller of } [l_2 - (\text{duration of B}), \\ &\qquad l_3 - (\text{duration of C})] \\ &= \text{smaller of } [4 - 3, 9 - 6] \\ &= \text{smaller of } [1, 3] = 1 \end{aligned}$$

Node 0 Since there is only one arc (0, 1) leaving node ⓪:

$$\begin{aligned} l_0 &= l_1 - (\text{duration of A}) \\ &= 1 - 1 = 0 \end{aligned}$$

- **In general:**

  $$\boldsymbol{l_i = \underset{j}{\text{minimum}}\left[l_j - \text{duration } (i, j)\right]}$$

  where the minimum is taken over all arcs (i, j) leaving the vertex i.

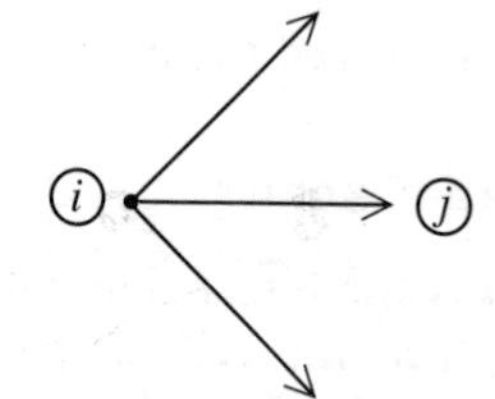

$\underset{j}{\text{minimum}}\left[l_j - \text{duration } (i, j)\right]$ means find the values of $\left[l_j - \text{duration } (i, j)\right]$ for all possible values of j, keeping i fixed, and then take the smallest of these values.

The information we have obtained so far is summarised in the table below:

Event, i	0	1	2	3	4	5	6
Earliest time, e_i	0	1	4	7	12	14	15
Latest time, l_i	0	1	4	9	12	14	15

This information may also be added to the network, as it is obtained, by placing at each node a box with e_i and l_i in the positions indicated:

e_i	l_i

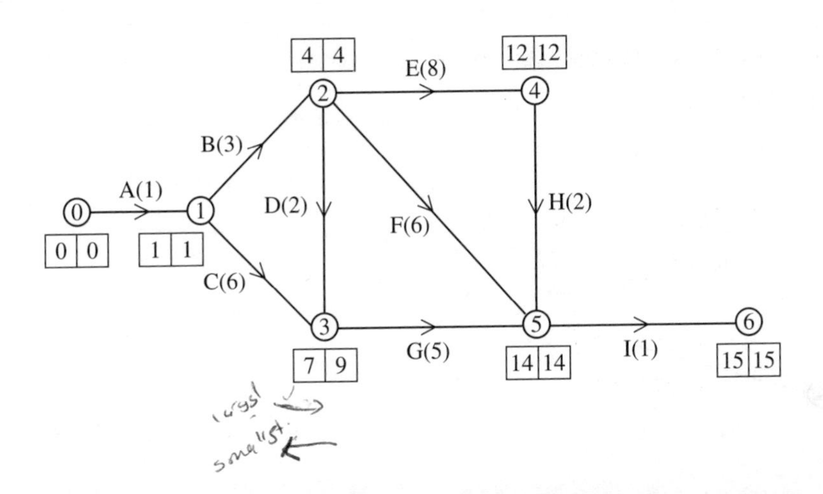

Exercise 5C

1

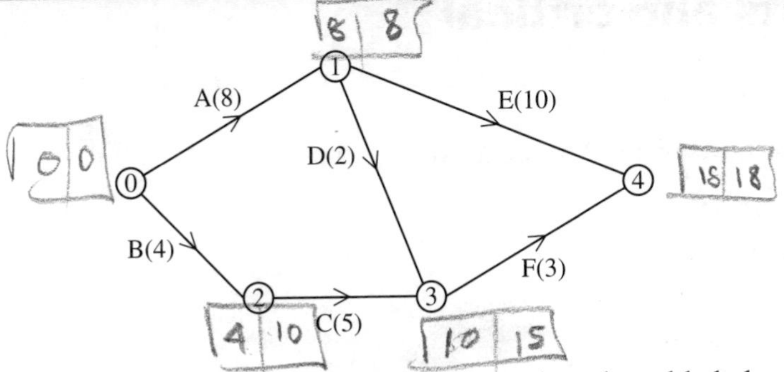

The above network models a project. Complete the table below by finding all the earliest times e_i and the latest times l_i.

Event, i	0	1	2	3	4
Earliest time, e_i	0	8	4		
Latest time, l_i				15	18

2 Complete the boxes $\boxed{e_i \mid l_i}$ for each event in the following network model.

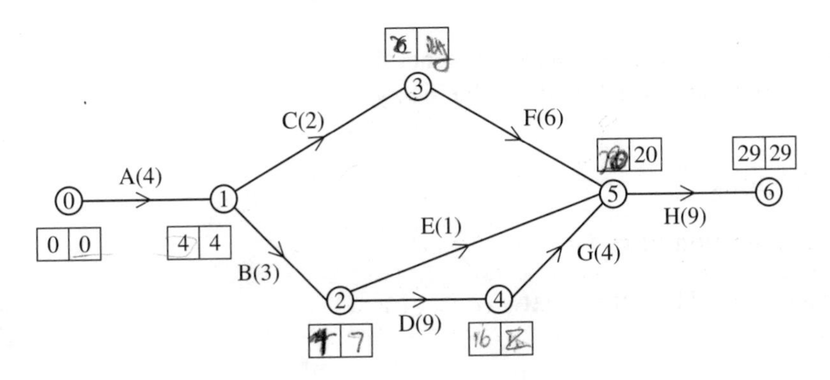

3 For the network model shown below find the earliest time e_i and the latest time l_i for each event.

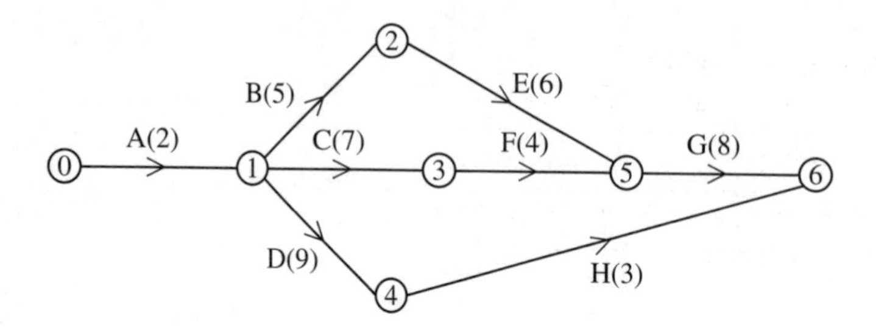

5.4 Critical events and critical activities

On a critical path(s) there can be no leeway. Any delay delays the completion of the entire project.

- **For all events on a critical path $e_i = l_i$.**

Such events are known as **critical events**.

In our example, events 0, 1, 2, 4, 5 and 6 are critical; only 3 is non-critical.

Any delay of a critical activity will increase the length of the critical path and so for any activity (i, j) on a critical path

$$[\text{duration of } (i, j)] = l_j - e_i$$

- **Activities for which $\left[l_j - e_i - \text{duration of } (i, j)\right] = 0$ are called critical activities.**

You should not assume that an activity joining critical events is a critical activity.

For any event the difference $s_i = l_i - e_i$ is called the **slack**. In general $s_i \geqslant 0$ and is **equal to zero for critical events**. It is the allowable delay at event i.

For any non-critical activity the quantity

$$F(i, j) = \left[l_j - e_i - \text{duration of } (i, j)\right]$$

is called the **total float of activity (i, j)**. The total float $F(i, j)$ of a critical activity is of course zero.

The total float represents the maximum possible delay that can be incurred in the processing of this activity without increasing the critical path length, providing there are no delays elsewhere. In our case we have

	A	B	C	D	E	F	G	H	I
Activity (i, j)	(0, 1)	(1, 2)	(1, 3)	(2, 3)	(2, 4)	(2, 5)	(3, 5)	(4, 5)	(5, 6)
Total float	0	0	2	3	0	4	2	0	0

Hence the critical activities are

$$\underset{A}{(0, 1)},\quad \underset{B}{(1, 2)},\quad \underset{E}{(2, 4)},\quad \underset{H}{(4, 5)},\quad \underset{I}{(5, 6)}$$

So ABEHI is the critical path of length 15. Notice that (2, 5) is *not* a critical activity although events 2 and 5 are critical events.

Exercise 5D

1 For the project in question 1 of Exercise 5C:

(a) obtain the slack s_i for each event i and so identify the critical events

(b) obtain the total float $F(i, j)$ for each activity (i, j) and so identify the critical activities

(c) hence state the critical path and give its length.

2 For the project in question 2 of Exercise 5C obtain the same information as requested in question 1.

3 For the project in question 3 of Exercise 5C obtain the same information as requested in question 1.

5.5 Time analysis of a network

It is useful in analysing a project to calculate four times associated with each *activity*: the earliest start time, the earliest finish time, the latest finish time and the latest start time. Let a typical activity be represented by the arc (i, j) and let the duration of this activity be a_{ij}.

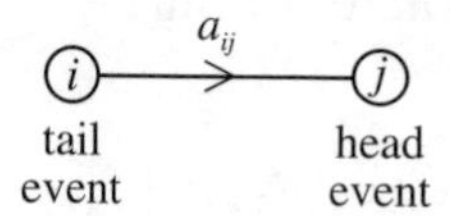

(i) Earliest start time – this is the *earliest possible time* at which activity (i, j) can *start*. By definition this is the earliest time for the tail event i and so is e_i.

(ii) Earliest finish time – this is the *earliest possible time* at which activity (i, j) can *finish*. It is obtained by adding the duration a_{ij} of the activity (i, j) to the earliest start time and so is

$$e_i + a_{ij}$$

(iii) Latest finish time – this is the *latest time* at which activity (i, j) can *finish* if the project is to be completed in the project time. It is therefore the latest time for the head event j and so is l_j.

(iv) Latest start time – this is the *latest time* at which activity (i, j) can *start* if the project is to be completed on time. This is therefore obtained by subtracting the duration a_{ij} of activity (i, j) from the latest finish time and so is

$$l_j - a_{ij}$$

For example, for our project consider activity (2, 5), that is activity F:

(i) earliest start time $= e_2 = 4$
(ii) earliest finish time $= e_2 + a_{25} = 4 + 6 = 10$
(iii) latest finish time $= l_5 = 14$
(iv) latest start time $= l_5 - a_{25} = 14 - 6 = 8$

Notice the order in which these times are obtained. If you calculate them in the order shown you will avoid errors.

The relationship between these times is often displayed in a diagram.

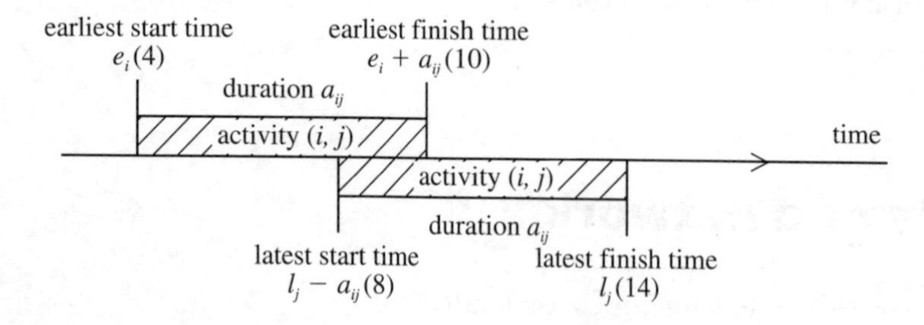

The figures in brackets are the values for the above example.

All the information obtained for the activities can be conveniently summarised in a table:

		Start		Finish		
Activity	Duration	Earliest	Latest	Earliest	Latest	Float
A (0, 1)	1	0	0	1	1	0
B (1, 2)	3	1	1	4	4	0
C (1, 3)	6	1	3	7	9	2
D (2, 3)	2	4	7	6	9	3
E (2, 4)	8	4	4	12	12	0
F (2, 5)	6	4	8	10	14	4
G (3, 5)	5	7	9	12	14	2
H (4, 5)	2	12	12	14	14	0
I (5, 6)	1	14	14	15	15	0
		(i)	(iv)	(ii)	(iii)	(v)

Notice that alternative definitions for the total float of activity (i, j), that is $(l_j - e_i - a_{ij})$, are:

(a) (latest finish time) – (earliest finish time)
(b) (latest start time) – (earliest start time).

The float (v) may therefore easily be calculated once columns (i), (ii), (iii) and (iv) have been obtained, in that order. Calculating the float using both (a) and (b) does provide a check on the other calculations.

Meaning of the total float

Let us look again at the activity F and illustrate the meaning of its entries in the table on the previous page.

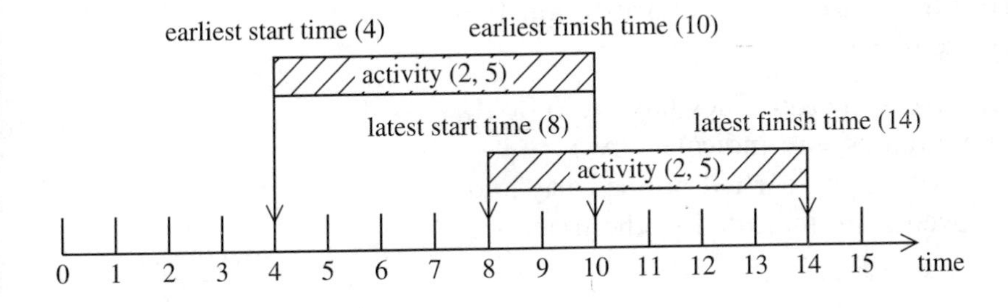

The activity F may start at any time between the earliest start time 4 and the latest start time 8 and it may finish at any time between the earliest finish time 10 and the latest finish time 14. The activity F may therefore float between the boundaries 4 and 14. The difference between the length of this interval (14 – 4 = 10) and the duration of the activity (6) is the total float 4. As we shall see in the next section there may be particular reasons for not starting this activity as early as possible.

Cascade (Gantt) charts

We may illustrate the information in the table in a graphical way.

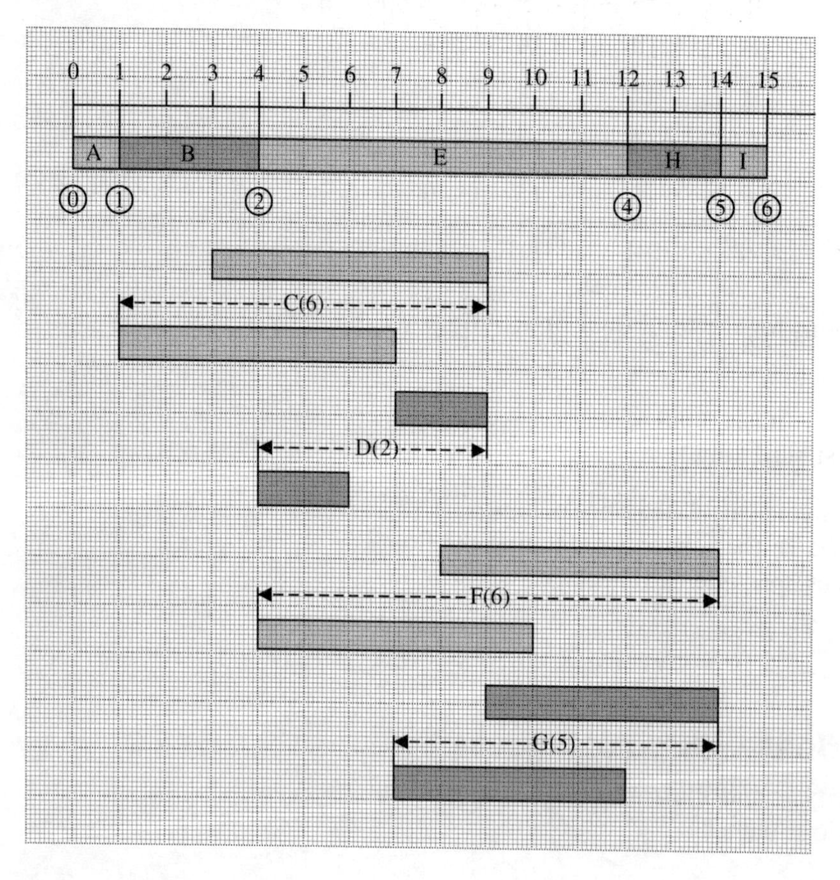

This is called a **cascade (Gantt) chart**. It summarises all the information obtained in a very useful way.

The critical activities are drawn across the top. They have zero float. For each other activity the 'boundaries' are indicated by ⊢-----⊣. The shaded bars indicate the activity being carried out as early and as late as possible. The critical events are indicated at the junctions of the critical activities.

Exercise 5E

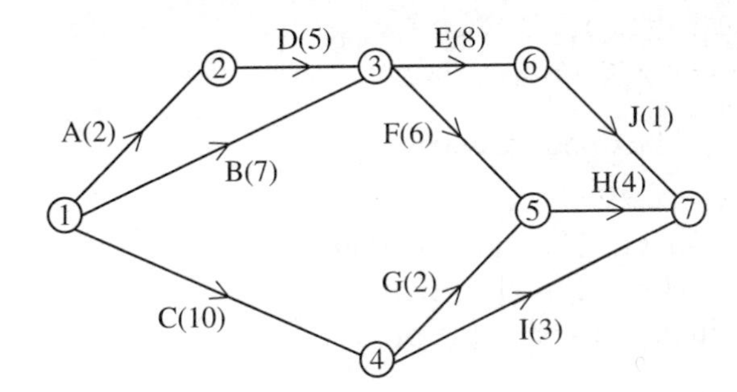

Analyse the above network and produce a table, giving for each activity the earliest and latest start times, the earliest and latest finish times and the total float. Illustrate the information in the table in a cascade (Gantt) chart.

5.6 Scheduling

For the present purpose we will assume that each activity of the project analysed in the previous section requires one worker. We now wish to consider such questions as:

(i) How many workers are required to complete the project in the critical time?
(ii) If only a limited number of workers is available, what is the minimum completion time for the project?

The following operating rules will be assumed:

(i) No worker may remain idle if there is an activity that can be started.
(ii) Once a worker starts an activity he must continue with that activity until it is finished.

The objective is, of course, to complete the project with as few workers as possible or to complete the project with the available workers in a minimum time. There is no algorithm that guarantees an optimal solution.

The procedure we shall adopt is:

(a) When a worker completes an activity consider all the activities that have not been started but that can now be started.
(b) Assign the worker to the activity whose latest start time is the smallest – this is in a sense the 'most critical' activity.
(c) If there are no activities that can be started at this time the worker will have to wait until an activity can be assigned.

With reference to our example we will try and answer three questions:

(i) Can the project be completed by two workers in the project time? (For the present purpose we will refer to this as 15 days.)

The critical activities A, B, E, H and I can all be done by one worker in the project time of 15 days.

The non-critical activities are C, D, F and G. The sum of their durations is $(6 + 2 + 6 + 5)$ days $= 19$ days. Hence it is clear that not all these activities can be completed by a single worker in the project time.

(ii) If only two workers are available what is the minimum time required to complete the project?

We will use the procedure described above and the table on page 121, which for ease of reference we repeat here.

		Start		Finish		
Activity	Duration	Earliest	Latest	Earliest	Latest	Float
A (0, 1)	1	0	0	1	1	0
B (1, 2)	3	1	1	4	4	0
C (1, 3)	6	1	3	7	9	2
D (2, 3)	2	4	7	6	9	3
E (2, 4)	8	4	4	12	12	0
F (2, 5)	6	4	8	10	14	4
G (3, 5)	5	7	9	12	14	2
H (4, 5)	2	12	12	14	14	0
I (5, 6)	1	14	14	15	15	0
		(i)	(iv)	(ii)	(iii)	(v)

Begin by allocating worker 1 to activity A. Worker 2 cannot be assigned until A(1) is complete. So at $t = 1$ allocate worker 1 to activity B(3) and worker 2 to activity C(6). You have then, using a time line,

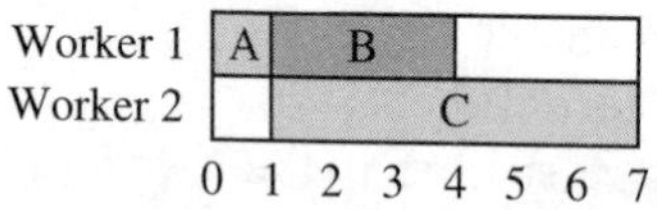

Worker 1 A B
Worker 2 C
0 1 2 3 4 5 6 7

When $t = 4$ worker 1 becomes available. The activity remaining that has the smallest of the latest start times is E.

So allocate worker 1 to E(8) and obtain

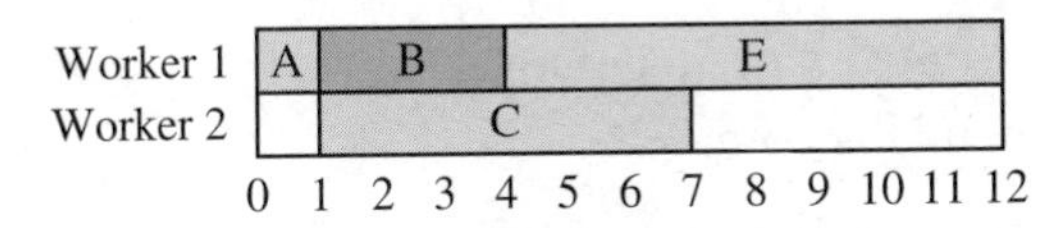

When $t = 7$ worker 2 becomes available. The activity remaining that has the smallest of the latest start times is D. So allocate worker 2 to D. This activity will be finished at $t = 9$, as the duration of D is 2 days. From the table you can see that worker 2 can then start activity F(6). You have then:

Worker 1 A B E
Worker 2 C D F
0 1 2 3 4 5 6 7 8 9 10 11 12 13 14 15

When $t = 12$ worker 1 becomes available. Using the criterion you can allocate worker 1 to activity G(5). When $t = 15$ worker 2 becomes available and will then start activity H. You have then:

Worker 1 A B E G
Worker 2 C D F H
0 1 2 3 4 5 6 7 8 9 10 11 12 13 14 15 16 17

The final activity I(1) may be done by either worker 1 or worker 2 as both G and H will be finished at $t = 17$. In either case the minimum time for the project when only two workers are available is 18 days.

(iii) Can the project be completed by three workers in the project time of 15 days?

Consider first the critical activities. Activity B depends only on A, activity E depends only on B and activity H depends only on E. But activity I depends also on non-critical activities. It is therefore clear that critical activities A, B, E and H can be completed by worker 1 in 14 days.

Of the remaining activities, activity C has the smallest of the latest start times therefore allocate worker 2 to activity C starting as early as possible, that is at $t = 1$.

The next activity to consider, according to (b) on page 123, is D. Allocate worker 3 to D starting as early as possible, that is at $t = 4$. We have then

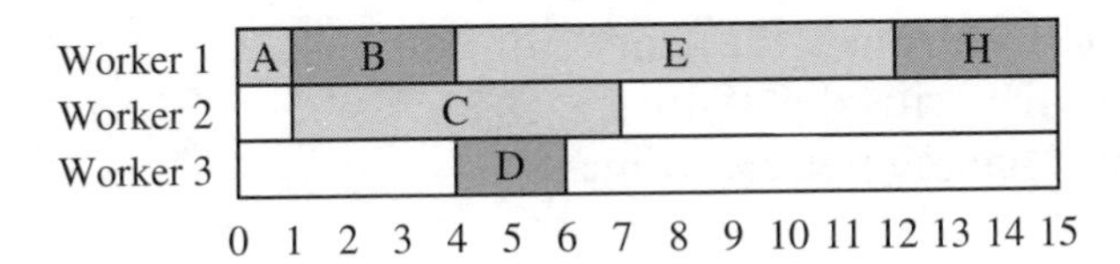

Activity F may be done by worker 3 starting at $t = 6$ and finishing at $t = 12$. Activity G, which depends on C and D, can be done by worker 2 starting at $t = 7$ and finishing at $t = 12$. The final activity I can be done by worker 1 starting at $t = 14$ when H is complete.

So the project can be completed by three workers in the project time of 15 days. The schedule below shows the allocation of jobs to workers.

Worker 1	A	B	E	H	I
Worker 2		C	G		
Worker 3		D	F		

0 1 2 3 4 5 6 7 8 9 10 11 12 13 14 15

Exercise 5F

In the project in Exercise 5E each activity requires one worker. Determine the minimum numbers of workers required if the project is to be completed in the project time. Draw a diagram to show how the workers should be allocated.

5.7 A further example involving a dummy

The activity network for an industrial process is shown below with each arc labelled with the time, in days, required for the corresponding activity. S is the source node and T is the sink node.

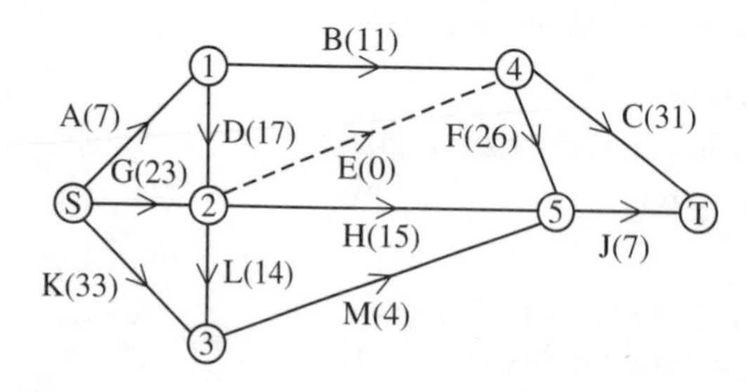

(a) Find the earliest time and the latest time for each event and the total float for each activity. Hence determine the length of the critical path, the critical events and the critical activities.
(b) What are the critical activities and the minimum project time if the dummy E is removed?

(a) The calculation of the earliest event times is summarised below:

$e_S = 0$, $e_1 = 7$

$e_2 = \max(7 + 17, 23) = \max(24, 23) = 24$

$e_3 = \max(33, 24 + 14) = \max(33, 38) = 38$

$e_4 = \max(7 + 11, 24 + 0) = \max(18, 24) = 24$

$e_5 = \max(24 + 15, 38 + 4, 24 + 26) = \max(39, 42, 50) = 50$

$e_T = \max(24 + 31, 50 + 7) = \max(55, 57) = 57$

The calculation of the latest event times for the events is summarised below:

$l_T = 57$, $l_5 = 50$

$l_4 = \min(57 - 31, 50 - 26) = \min(26, 24) = 24$

$l_3 = 50 - 4 = 46$

$l_2 = \min(24 - 0, 50 - 15, 46 - 14) = \min(24, 35, 32) = 24$

$l_1 = \min(24 - 11, 24 - 17) = \min(13, 7) = 7$

$l_S = \min(46 - 33, 24 - 23, 7 - 7) = \min(13, 1, 0) = 0$

The total floats are

A (S, 1) $= 7 - 0 - 7 = 0$
B (1, 4) $= 24 - 7 - 11 = 6$
C (4, T) $= 57 - 24 - 31 = 2$
D (1, 2) $= 24 - 7 - 17 = 0$
E (2, 4) $= 24 - 24 - 0 = 0$
F (4, 5) $= 50 - 24 - 26 = 0$
G (S, 2) $= 24 - 0 - 23 = 1$
H (2, 5) $= 50 - 24 - 15 = 11$
J (5, T) $= 57 - 50 - 7 = 0$
K (S, 3) $= 46 - 0 - 33 = 13$
L (2, 3) $= 46 - 24 - 14 = 8$
M (3, 5) $= 50 - 38 - 4 = 8$

The length of the critical path is obtained from e_T and so is 57 days.

The critical events are those for which $e_i = l_i$ and so are Ⓢ, ①, ②, ④, ⑤, Ⓣ.

The critical activities are those for which the total float is zero and so are activities A, D, E, F and J.

(b) When the dummy E is removed you need to consider the changes that are necessary in the above calculation.

Let us recalculate the earliest event times:

$$e_S = 0 \text{ (unchanged)}$$
$$e_1 = 7 \text{ (unchanged)}$$
$$e_2 = 24 \text{ (unchanged)}$$
$$e_3 = 38 \text{ (unchanged)}$$
$$e_4 = 18 \text{ (changed)}$$
$$e_5 = \max(24 + 15, 38 + 4, 18 + 26)$$
$$= \max(39, 42, 44) = 44$$
$$e_T = \max(44 + 7, 18 + 31) = 51$$

The length of the critical path is now 51 days. You can obtain the critical path in this case by using an alternative method. For each event, in addition to recording the earliest time e_i you can record the vertex $p(i)$ from which the maximum is achieved. It is then possible to determine the critical path by using these and working backwards from Ⓣ.

In this case:

$$e_S = 0$$
$$e_1 = 7 \qquad p(1) = Ⓢ$$
$$e_2 = 24 \qquad p(2) = ①$$
$$e_3 = 38 \qquad p(3) = ②$$
$$e_4 = 18 \qquad p(4) = ①$$
$$e_5 = 44 \qquad p(5) = ④$$
$$e_T = 51 \qquad p(T) = ⑤$$

Since $p(T) = ⑤$ you go to ⑤
and $p(5) = ④$ you go to ④
and $p(4) = ①$ you go to ①
and $p(1) = Ⓢ$

Then from the activity network the critical path is (S, 1) (1, 4) (4, 5) (5, T), that is A B F J.

Exercise 5G

1

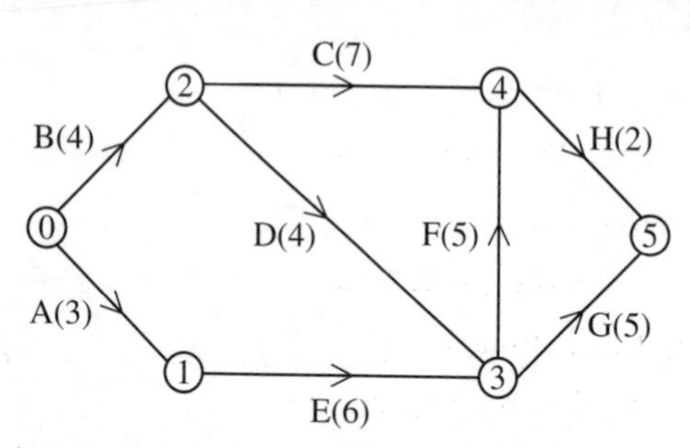

For the activity network shown above calculate the earliest time and the latest time for each event and the total float for each activity. Hence determine the length of the critical path, the critical activities and the critical events.

2

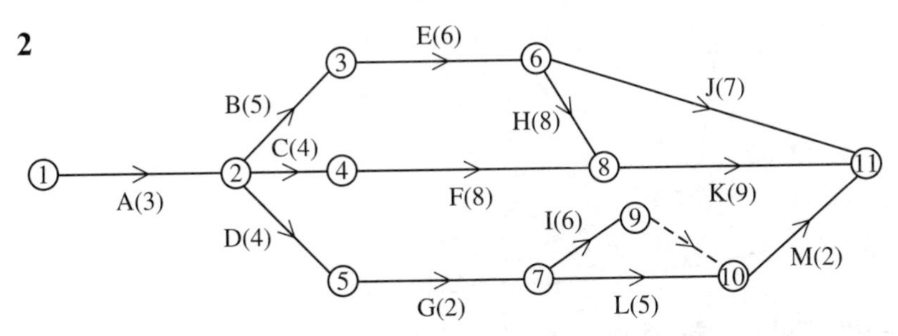

For the activity network shown above calculate the earliest and latest time for each event and record them in the form $\boxed{e_i \mid l_i}$ on a copy of the network. Indicate also the critical path by marking critical activities by a double dash, that is —‖— .

3 As a result of the breakdown of a piece of equipment the activity F in the previous example takes 16 days.

(a) Carry out further calculations to decide if the length of the project is changed.

(b) If the length of the project is changed, determine the new project time and the critical path.

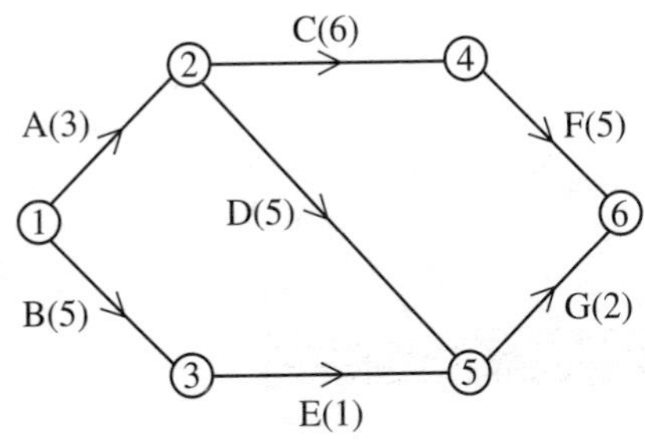

For the above activity network

(a) determine the earliest time and latest time for each event, and hence the project time and critical path

(b) produce a table including for each activity the earliest and latest start times and finish times and the total float.

(c) Given that each activity requires one worker determine the minimum number of workers required if the project is to be completed in the project time.

5

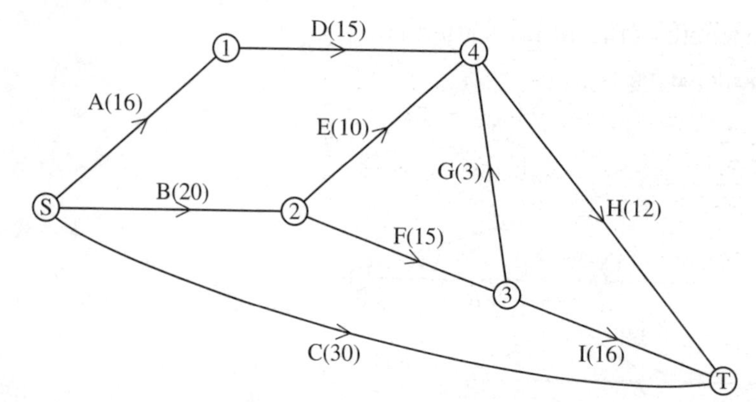

The activity network for an industrial project is shown above. The arcs are labelled with the time, in days, required to complete the corresponding activity. The start vertex is S and the terminal vertex is T.

(a) Find the earliest and latest time for each event and hence determine the length of the critical path.

(b) Determine the critical events, the critical activities and the critical path.

(c) Draw up time schedules for the project

(i) with non-critical activities as early as possible

(ii) with non-critical activities as late as possible.

(d) Given that each activity requires one worker determine the minimum number of workers required to complete the project in the critical time.

SUMMARY OF KEY POINTS

In an activity network:

1. The **nodes** represent events.
2. The **arcs** represent activities.
3. The **weights** represent the duration of the corresponding activity.
4. The **source node** represents the beginning of the project.
5. The **sink node** represents the end of the project.
6. The **earliest time** e_i for node i is the earliest time we can arrive at event i.

7 The **latest time** l_i for node i is the latest time we can leave event i without extending the length of the critical path.

8 The **critical path** is the longest path through the network from the source node to the sink node.

9 Events i for which $e_i = l_i$ are **critical events**.

10 Activities (i, j) for which $l_j - e_i - [\text{duration } (i, j)] = 0$ are **critical activities**.

11 For activity (i, j) of duration a_{ij}:

the earliest start time is e_i
the earliest finish time is $e_i + a_{ij}$
the latest finish time is l_j
the latest start time is $l_j - a_{ij}$
the total float on activity (i, j) is $l_j - e_i - a_{ij}$

6 Linear programming

Decision making is a process that has to be carried out in many areas of life. Usually there is a particular aim in making one decision rather than another. Two particular aims that are often considered in commerce are maximising profit and minimising costs.

During, and after, the Second World War a group of American mathematicians developed some mathematical methods to help with decision making. They sought to produce mathematical models of situations in which all the requirements, constraints and objectives were expressed as algebraic equations. They then developed methods for obtaining the **optimal solution** – the maximum or minimum value of a required function.

In this chapter you will study problems for which all the algebraic expressions are linear, that is of the form

$$(a\ number)\,x + (a\ number)\,y + (a\ number)\,z$$

For example

$$Profit = 4x + 3y + 2z$$

is a linear equation in x, y and z and

$$16x + 18y + 9z \leqslant 25$$

is a linear inequation or inequality in x, y and z. This area of mathematics is called **linear programming**.

Linear programming methods are some of the most widely used methods employed to solve management and economic problems. They have been applied to a variety of contexts, some of which will be discussed later in this chapter, with enormous savings in money and resources.

6.1 Formulating linear programming problems

The first step in formulating a linear programming problem is to determine which quantities you need to know to solve the problem. These are called the **decision variables**.

The second step is to decide what the **constraints** are in the problem. For example, there may be a limit on resources, or a maximum or minimum value a decision variable may take, or there could be a relationship between two decision variables.

The third step is to determine the objective to be achieved. This is the quantity to be maximised or minimised, that is optimised. The function of the decision variables that is to be optimised is called the **objective function**.

The examples which follow illustrate the varied nature of problems that can be modelled by a linear programming model. We will not at this stage attempt to solve these problems but instead concentrate on producing the objective function and the constraints, writing these in terms of the decision variables. As an aid to this it is often useful to summarise all the given information in the form of a table as is illustrated in example 1.

Example 1

A manufacturer makes two kinds of chairs, A and B, each of which has to be processed in two departments, I and II. Chair A has to be processed in department I for 3 hours and in department II for 2 hours. Chair B has to be processed in department I for 3 hours and in department II for 4 hours.

The time available in department I in a given month is 120 hours and the time available in department II, in the same month, is 150 hours.

Chair A has a selling price of £10 and chair B has a selling price of £12.

The manufacturer wishes to maximise his income. How many of each chair should be made in order to achieve this objective? You may assume that all chairs made can be sold.

The information given in the problem can be summarised by constructing the following table.

Type of chair	Time in dept I (hours)	Time in dept II (hours)	Selling price
A	3	2	10
B	3	4	12
Total time available	120	150	

Step 1 Which quantities do you need to know to solve the problem, that is what are the decision variables?
It is clear that in this case you wish to know how many type A chairs are to be made and how many type B chairs are to be made.
The decision variables are therefore:

x = number of type A chairs made
y = number of type B chairs made

Step 2 What are the constraints?
Consider what happens in department I, that is concentrate on the column in the table labelled time in dept I.

Since: the production of 1 type A chair uses 3 hours
then: the production of x type A chairs uses $3x$ hours.
Similarly: the production of 1 type B chair uses 3 hours
so: the production of y type B chairs uses $3y$ hours.

The total time used is therefore:

$$(3x + 3y) \text{ hours}$$

Since only 120 hours are available in department I, one constraint is:

$$(3x + 3y) \text{ hours} \leqslant 120 \text{ hours}$$

or

$$3x + 3y \leqslant 120$$

Considering department II in a similar way produces the second constraint:

$$2x + 4y \leqslant 150$$

In addition to these two constraints you also require that x and y be non-negative, that is:

$$x \geqslant 0, \quad y \geqslant 0$$

These are called the **non-negativity constraints**.

Step 3 What is the objective?
The objective is to maximise the income. If x chairs of type A are made the income is £$10x$ and if y chairs of type B are made the income is £$12y$.
The total income is then £Z = £$(10x + 12y)$.
The aim therefore is to maximise

$$Z = 10x + 12y$$

The problem can be modelled by the linear programming problem:

Maximise $Z = 10x + 12y$ subject to the constraints:

$$3x + 3y \leqslant 120$$
$$2x + 4y \leqslant 150$$
$$x \geqslant 0, \quad y \geqslant 0$$

Notice that all the constraints are dimensionless, that is they only involve numbers. Z also is a number. It is advisable, as in this case, to write the income as £Z so that all the variables in the problem have purely numerical values.

Example 2

A book publisher is planning to produce a book in three different bindings: paperback, book club and library. Each book goes through a sewing process and a gluing process. The table gives the time required, in minutes, for each process and for each of the bindings.

	Paperback	Book club	Library
Sewing (min)	2	2	3
Gluing (min)	4	6	10

The sewing process is available for 7 hours per day and the gluing process for 11 hours per day. The profits are 25p on a paperback edition, 40p on a book club edition and 60p on a library edition. How many books in each binding should be manufactured to maximise profits? (Assume that the publisher can sell as many of each type of book as is produced.)

For this problem it is a good idea to extend the table to include all the information given by adding the restrictions on time and the profits.

	Paperback	Book club	Library	Total time available
Sewing (min)	2	2	3	420
Gluing (min)	4	6	10	660
Profit (P)	25	40	60	

Notice that the time available has been converted to minutes to be consistent with the other times given.

Step 1 Decision variables
The decision variables are the numbers of books to be made in each binding.

Let x = number in paperback binding
y = number in book club binding
z = number in library binding.

Step 2 Constraints
The constraints are:

$$\text{sewing: } 2x + 2y + 3z \leqslant 420$$
$$\text{gluing: } 4x + 6y + 10z \leqslant 660$$

together with the non-negativity conditions
$x \geqslant 0$, $y \geqslant 0$, $z \geqslant 0$.

Step 3 Objective
The objective is to maximise the profit P pence. That is, to maximise $P = 25x + 40y + 60z$.

The linear programming problem to be solved is therefore:

Maximise $P = 25x + 40y + 60z$ subject to the constraints:

$$2x + 2y + 3z \leqslant 420$$
$$4x + 6y + 10z \leqslant 660$$
$$x \geqslant 0,\ y \geqslant 0,\ z \geqslant 0$$

Example 3

KJB Haulage receives an order to transport 1600 packages. They have large vans, which can take 200 packages each, and small vans, which can take 80 packages each.

The cost of running each large van on the required journey is £40 and the cost of running each small van on the same journey is £20.

There is a limited budget for the job which requires that not more than £340 be spent. It is additionally required that the number of small vans used must not exceed the number of large vans used.

How many of each kind of van should be used if costs are to be kept to a minimum?

The table summarises the given information:

	Capacity	Cost (in £)
Large van	200	40
Small van	80	20
Limits	1600	340

Step 1 Decision variables

The decision required is how many of each type of van to use to fulfil the order. The decision variables are:

l = number of large vans used
s = number of small vans used

Step 2 Constraints

If all the packages are to be transported then one constraint is:

$$200l + 80s \geqslant 1600$$

You can simplify this inequality as 40 is a factor of each of the numbers in it, giving:

$$5l + 2s \geqslant 40$$

The restriction on total cost means that another constraint is:

$$40l + 20s \leqslant 340$$

This inequality may also be simplified by dividing by 20, giving:

$$2l + s \leqslant 17$$

Also the number of small vans must not exceed the number of large, so a third constraint is:

$$s \leqslant l \quad \text{or} \quad s - l \leqslant 0$$

Finally you must have $l \geqslant 0$ and $s \geqslant 0$.

Step 3 Objective

The objective is to keep costs to a minimum, that is minimise the cost £C where $C = 40l + 20s$.

The linear programming problem to be solved is therefore:

Minimise $C = 40l + 20s$ subject to the constraints:

$$5l + 2s \geqslant 40$$
$$2l + s \leqslant 17$$
$$s - l \leqslant 0$$
$$l \geqslant 0, s \geqslant 0$$

Example 4

When a lecturer retires he will have a lump sum of £30 000 to invest. He decides to invest in stock AA and bond BBB. Stock AA yields 7% per annum and bond BBB yields 5% per annum. He decides that no more than £20 000 shall be invested in either option. How much should he invest in each option in order to maximise his yield?

Step 1 Decision variables

Suppose he invests £x in stock AA and £y in bond BBB.

Step 2 Constraints

The total amount he has to invest is £30 000 so £x + £y = £30 000

or $$x + y = 30\,000$$

Since he decides not to invest more than £20 000 in either option

$$x \leqslant 20\,000$$

$$y \leqslant 20\,000$$

We must also have here $x \geqslant 0$, $y \geqslant 0$.

Step 3 Objective

The yield £Y which he wishes to maximise is

$$\pounds Y = \tfrac{7}{100}\pounds x + \tfrac{5}{100}\pounds y$$

or $$Y = 0.07x + 0.05y$$

The linear programming problem to be solved is then:

Maximise $Y = 0.07x + 0.05y$ subject to the constraints:

$$x + y = 30\,000$$
$$x \leqslant 20\,000$$
$$y \leqslant 20\,000$$
$$x \geqslant 0,\ y \geqslant 0$$

Notice that one of the constraints is an equation, $x + y = 30\,000$, rather than an inequation.

Exercise 6A

The following problems will help you practise *formulating* linear programming problems that can be solved. You are not being asked to solve them at this stage.

1 Allwood PLC plans to make two kinds of table. For table A the cost of the materials is £20, the number of person-hours needed to complete it is 10 and the profit, when it is sold, is £15. For table B the cost of the materials is £12, the number of person-hours needed to complete it is 15 and the profit, when it is sold, is £17. The total money available for materials is £480 and the labour available is 330 person-hours. Find the maximum profit that can be made and the number of each type of table that should be made to produce it. Formulate this as a linear programming problem.

2 To ensure that her family has a healthy diet Mrs Brown decides that the family's daily intake of vitamins A, B and C should not fall below 25 units, 30 units and 15 units respectively. To provide these vitamins she relies on two fresh foods α and β. Food α provides 30 units of vitamin A, 20 units of vitamin B and 10 units of vitamin C per 100 g. Food β provides 10 units of vitamin A, 25 units of vitamin B and 40 units of vitamin C per 100 g. Food α costs 40p per 100 g and food β costs 30p per 100 g. How many grams of food α and food β should she purchase daily if the food bill is to be kept to a minimum? Formulate this as a linear programming problem.

3 A factory is to install two types of machines A and B. Type A requires one operator and occupies $3\,m^2$ of floor space, type B requires two operators and occupies $4\,m^2$ of floor space. The maximum number of operators available is 40 and the floor space available is $100\,m^2$. Given that the weekly profits on type A and type B machines are £75 and £120, find the number of each machine that should be bought to maximise the profit, and calculate the profit. Formulate this problem as a linear programming problem.

4 A sports club wishes to hire a minibus for a trip. Children under 16 can travel for £5 and adults for £10. The minibus company have the following conditions:

- the total number of passengers must not exceed 14. There must be no fewer than 10 passengers on each trip.
- there must be at least as many half-fare passengers as full-fare passengers.

Determine the maximum amount the minibus company can receive for the hire. Formulate this as a linear programming problem.

5 The Midlands Furniture Company manufactures bookshelves in three sizes: small, medium and large. Small bookshelves require 4 m of board, medium 8 m of board and large 16 m of board. The assembly times required are 2 hours, 4 hours and 6 hours

respectively. Only 500 m of board are available and assembly time is restricted to 400 hours. All bookshelves produced can be sold, the profits being £4, £6 and £12 respectively. How many of each should be made if profit is to be maximised? Formulate this as a linear programming problem.

6 Chair Supplies makes three types of wooden chairs. Each type is manufactured in a four-stage process. The company is able to obtain all the raw materials it needs. The available production capacity during the 60-hour production work week is as follows:

	Weekly capacity in number of chairs				
Process	Chair A		Chair B		Chair C
1	400	or	600	or	900
2	1800	or	400	or	300
3	200	or	900	or	600
4	600	or	400	or	450

It is assumed that there are 60 hours of labour available for each process. The profits on each of the three types of chair are £15, £20 and £25 respectively. Formulate this as a linear programming problem, given that the profit is to be maximised.

6.2 Graphical solutions for two-variable problems

Linear programming problems that involve only two decision variables, x and y, may be solved graphically, that is by drawing lines associated with the constraints, identifying a region of possible solutions and then locating a point in this region that has a particular property. To solve linear programming problems using graphical methods you need to be able to use the following basic mathematical ideas.

Sets of points defined by a linear inequality

Any equation of the form $ax + by = c$, where a, b and c are numbers, is called a **linear equation**. For example, $3x + 4y = 12$ is a linear equation. In the x, y plane this is the equation of a **straight line**. This line may be drawn by identifying any two points on it.

For example, when $x = 0$, $4y = 12 \Rightarrow y = 3$, therefore $x = 0$, $y = 3$, or (0, 3) is on the line.

Similarly, when $y = 0$, $3x = 12 \Rightarrow x = 4$, therefore $x = 4$, $y = 0$, or (4, 0) is also on the line. Plotting both points allows you to draw the line:

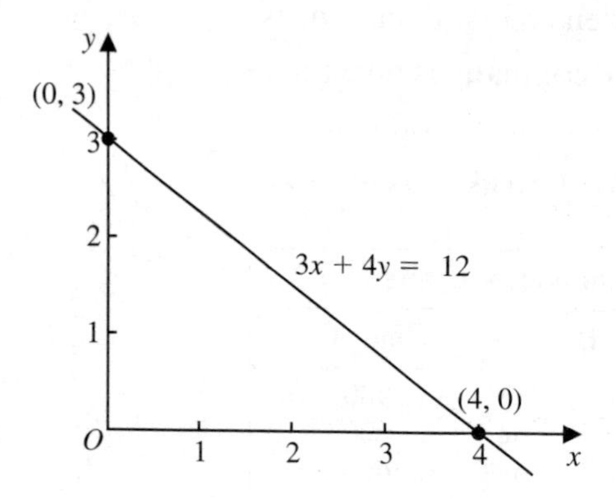

A special case of $ax + by = c$ is that in which $c = 0$, for example $2y - x = 0$. In such cases $x = 0$, $y = 0$, or (0, 0) is on the line. Taking $x = 1$ gives $y = \frac{1}{2}$, therefore a second point on the line is $x = 1$, $y = \frac{1}{2}$, or $(1, \frac{1}{2})$.

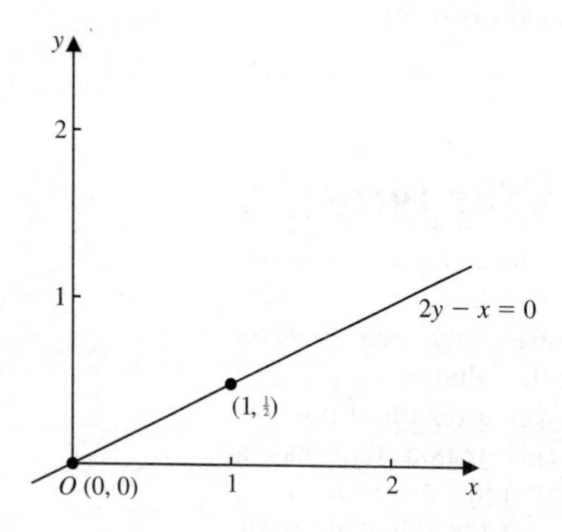

In linear programming problems we usually have the non-negativity conditions $x \geqslant 0$, $y \geqslant 0$, so we are only concerned with values of x and y *in the first quadrant*.

Any straight line divides the xy-plane into two half-planes. If the equation of the line is $ax + by = c$ then on one side of the line you have $ax + by < c$ and on the other side of the line you have $ax + by > c$.

For example:

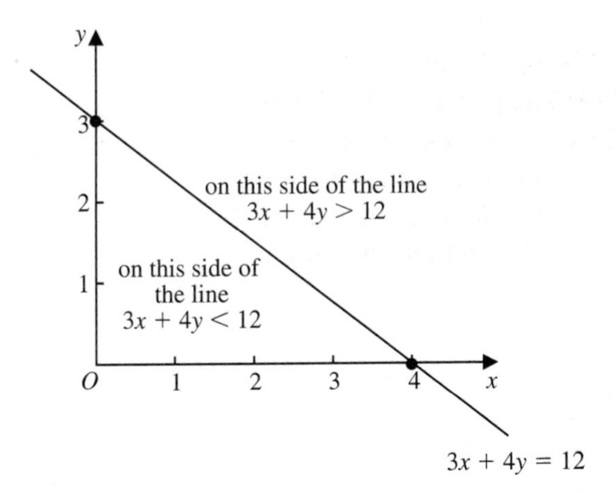

An inequality of the form $ax + by \leqslant c$ therefore defines a set of points (x, y) which *either lie on the line or else in the half-plane on one side of the line*. You can easily decide which side of the line is the required half-plane by inserting the point (0, 0) into the inequality. If $x = 0$, $y = 0$ satisfies the inequality then (0, 0), the origin, is in the required region.

Example 5

Show on a diagram the region for which $3x + 4y \leqslant 12$.

Draw the line $3x + 4y = 12$. Since $3(0) + 4(0) = 0$, which is less than 12, the origin lies in the required region, and the half-plane required contains the origin. So all points on the line or below it satisfy the inequality. We call this the **admissible set** defined by the inequality. You can show this by using arrows to indicate the required region and shading marks placed on the line to indicate the half-plane of **inadmissible** points.

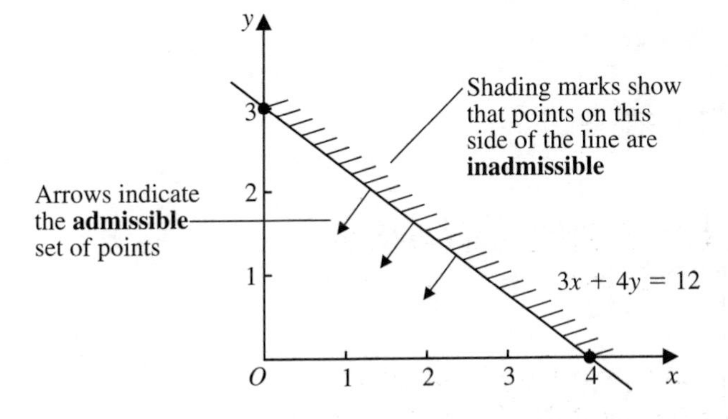

This method of indicating the admissible set is simple and produces clear solutions.

Sets of points defined by a collection of inequalities

Each inequality in a linear programming problem will produce an admissible set. To find a solution to the problem you need to find the set of points which satisfies all the inequalities **simultaneously**. This is obtained graphically by drawing a diagram like the one on page 143 but showing *all* the inequalities. **The required region is then the one which does not contain any shading marks and into which all arrows point**.

Example 6

Indicate on a diagram the region for which

$$3x + 4y \leqslant 12$$
$$3x + 2y \leqslant 9$$
$$x \geqslant 0,\ y \geqslant 0$$

You can find the region for which the second inequality is satisfied by following the procedure outlined in Example 5. The line with equation $3x + 2y = 9$ passes through the points with coordinates $(0, 4\frac{1}{2})$ and $(3, 0)$. The origin $(0, 0)$ is in the admissible set since $3(0) + 2(0) = 0$.

You then have:

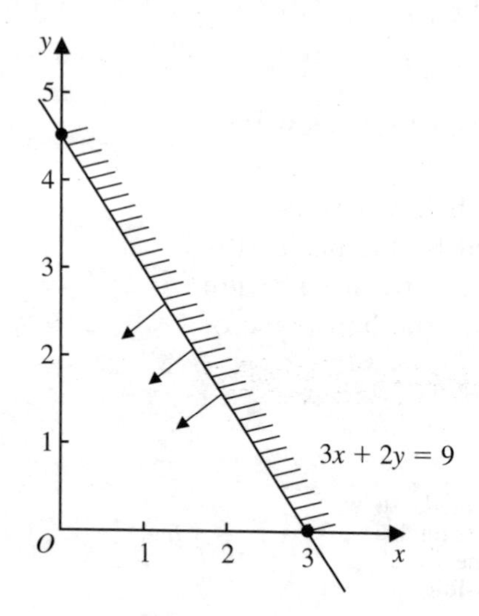

The inequality $x \geqslant 0$ gives:

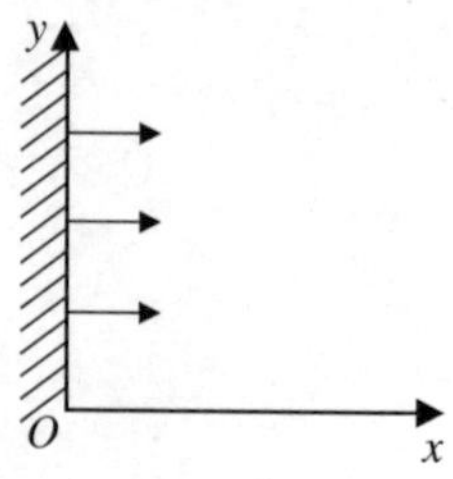

and the inequality $y \geqslant 0$ gives:

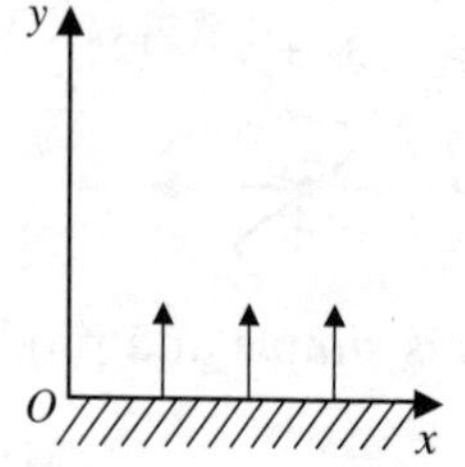

Combining these gives:

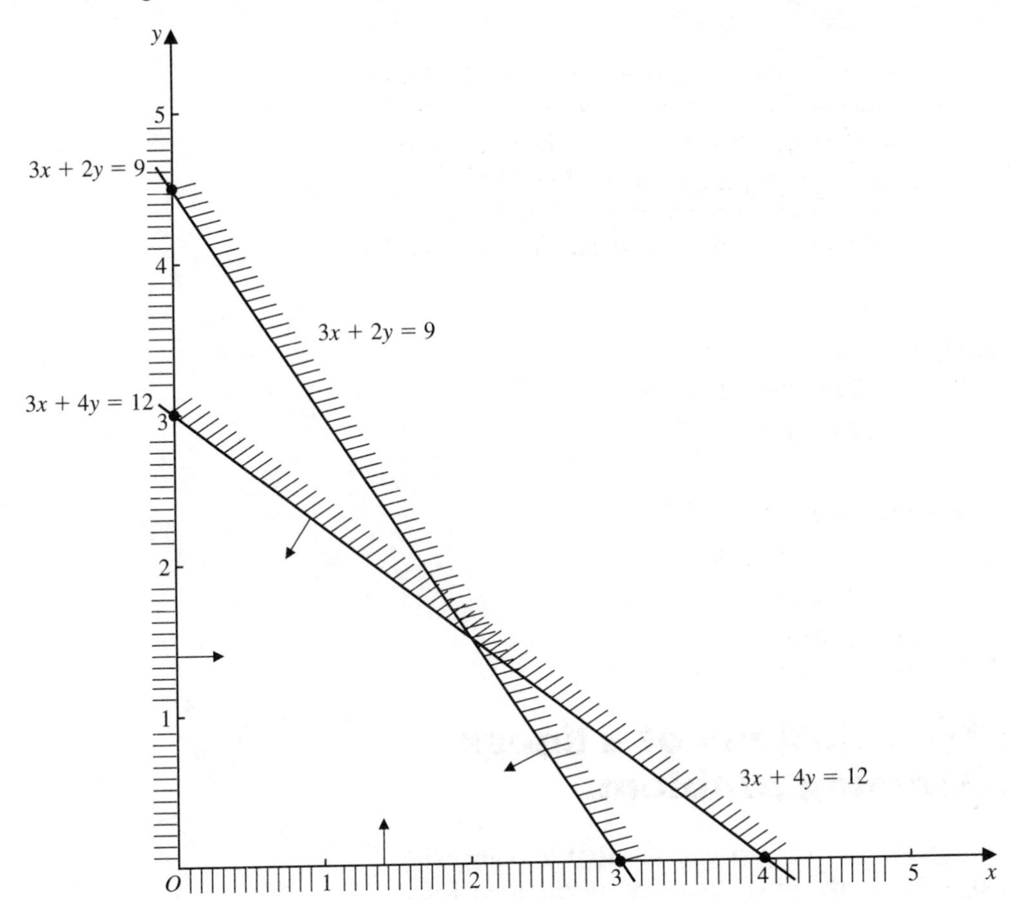

Exercise 6B

1 Indicate on a diagram the region for which

$$5x + 3y \leqslant 15$$
$$x \geqslant 0$$
$$y \geqslant 0$$

2 Indicate on a diagram the region for which

$$4x + 3y \leqslant 12$$
$$2x + 5y \leqslant 10$$
$$x \geqslant 0$$
$$y \geqslant 0$$

3 Indicate on a diagram the region for which

$$2x + y \leqslant 8$$
$$y \leqslant 7$$
$$x \leqslant 3$$
$$x \geqslant 0, y \geqslant 0$$

4 Indicate on a diagram the region for which

$$y + 2x \leqslant 12$$
$$x \geqslant 2$$
$$y \geqslant 4$$

5 Indicate on a diagram the region for which

$$3x + 2y \leqslant 12$$
$$3x + y \geqslant 6$$
$$x + y \geqslant 4$$

6 Indicate on a diagram the region for which

$$3x + 2y \leqslant 6$$
$$y \leqslant 2x$$
$$y \geqslant 0$$

6.3 Feasible solutions of a linear programming problem

From the discussion in section 6.2 you can see that a point with coordinates (x, y) that lies in the intersection of all the admissible sets, defined by the constraints, represents a **feasible solution**.

- **Any pair of values of x and y that satisfy all the constraints in a linear programming problem is called a feasible solution.**
  The region that contains all such points is called the *feasible region*.

In Example 6 all the constraints are satisfied in the unshaded region.

A linear programming problem is therefore solved by finding which member, or members, of the set of feasible solutions gives the optimal (maximum or minimum) value of the objective function.

6.4 Finding the optimal solution of a linear programming problem

Think about a linear programming problem in which you wish to maximise $P = \alpha x + \beta y$, where α and β are positive numbers. You have seen that $\alpha x + \beta y = P$ is the equation of a straight line. When P takes different values you get a family of parallel straight lines.

For example, the diagram below shows

$$x + 2y = P$$

for the cases $P = 2, 4, 6$.

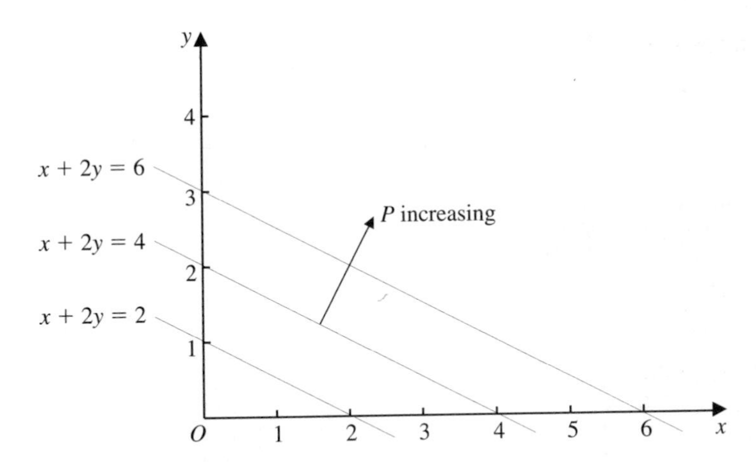

You can see that the points at which the line intercepts the x and y axes move further and further away from the origin O as P increases. To check each x intercept, find x when $y = 0$. To check each y intercept, find y when $x = 0$.

Now you can use the fact that the line with equation $\alpha x + \beta y = P$, α and β positive numbers, moves further from the origin as P increases to find the maximum value of P. The largest value of P will occur at the point in the feasible region that is on the line *furthest* from the origin. You can find this point by sliding a ruler over the feasible region so that it is always parallel to the family of straight lines given by $\alpha x + \beta y =$ constant. This will enable you to identify the point that is furthest from the origin:

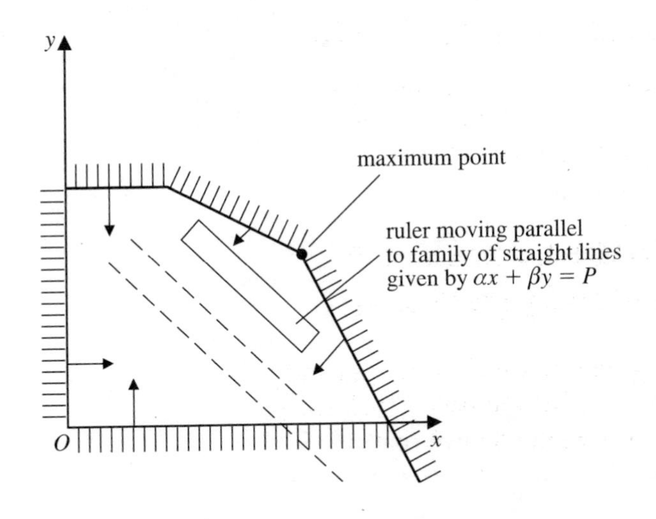

In the case of a *minimising* problem the point in the feasible region that is on the line *closest* to the origin is the one that is required:

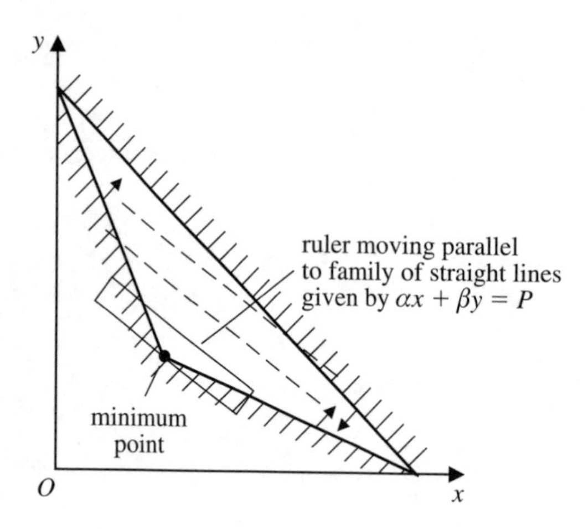

Suppose you have drawn the region on a graph in which the solution must lie. You can then start the process of maximising or minimising $\alpha x + \beta y$ by drawing an arbitrary member of the family of parallel lines, $\alpha x + \beta y = \gamma$, say.

Choose γ so that the intercepts on the x and y axes are integers, that is γ is a number that has both α and β as factors. For example, if $\alpha = 3$ and $\beta = 4$ draw $3x + 4y = 12$. Then when $y = 0$ the intercept is $x = 4$ and when $x = 0$ the intercept is $y = 3$. These ideas are illustrated in the following examples.

Example 7

Solve graphically the linear programming problem:

Maximise $Z = x + y$ subject to the constraints:

$$3x + 4y \leqslant 12$$
$$3x + 2y \leqslant 9$$
$$x \geqslant 0,\ y \geqslant 0$$

The feasible region for this problem was obtained above (page 145) and is shown on the next page.

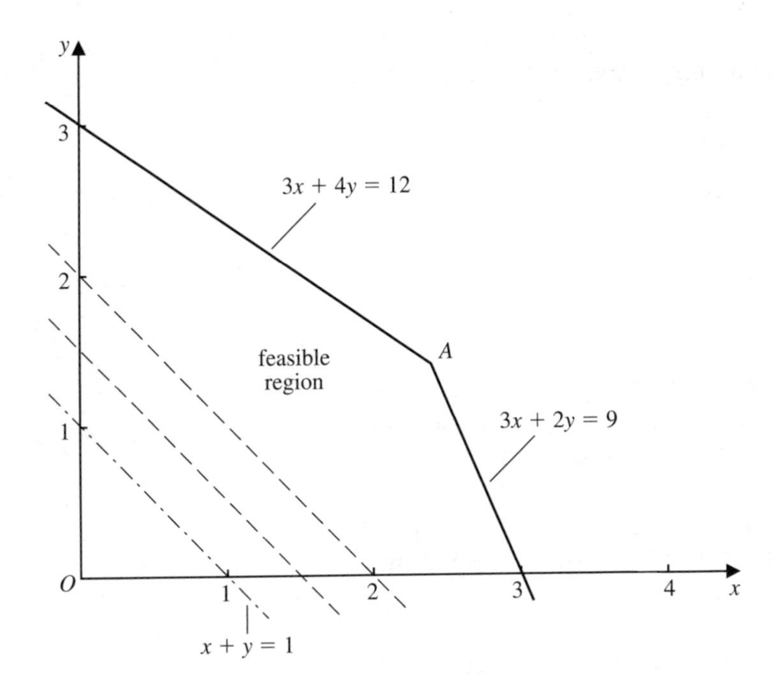

Start by drawing the line $x + y = 1$ (–·–· on the graph above). This passes through $(0, 1)$ and $(1, 0)$. Other members of the family $x + y = c$ are shown by – – –. The furthest you can move the ruler without losing contact with the feasible region is to the point A. The point A therefore gives the optimal value of Z.

The coordinates of the point A are obtained from the fact that A is the point of intersection of the two lines with equations

$$3x + 4y = 12$$

and $$3x + 2y = 9$$

Subtracting gives: $$2y = 3 \Rightarrow y = 1\tfrac{1}{2}$$

Substituting into the second equation:

$$3x + 2(1\tfrac{1}{2}) = 9$$

so: $$3x = 9 - 3 = 6 \Rightarrow x = 2$$

Do not try and obtain the coordinates of A from the graph. These should be obtained by solving a pair of simultaneous equations.

Now substitute these values in the equation $Z = x + y$.

When $x = 2$ and $y = 1\tfrac{1}{2}$ then $Z = 2 + 1\tfrac{1}{2} = 3\tfrac{1}{2}$.

The maximum value of Z is therefore $3\tfrac{1}{2}$ and occurs when $x = 2$ and $y = 1\tfrac{1}{2}$.

Example 8

Solve graphically the linear programming problem:

Minimise $Z = 3x + 4y$ subject to the constraints:

$$x + 4y \geqslant 20$$
$$x + y \geqslant 8$$
$$x + 2y \leqslant 16$$
$$x \geqslant 0,\ y \geqslant 0$$

(a) The line $x + 4y = 20$ passes through the points (0, 5) and (20, 0). Since $0 + 4(0) = 0$, which is not greater than 20, the origin does not lie in the admissible region for the inequality $x + 4y \geqslant 20$.

(b) The line $x + y = 8$ passes through the points $(0, 8)$ and $(8, 0)$. Since $0 + 0 = 0$, which is not greater than 8, the origin does not lie in the admissible region for the inequality $x + y \geqslant 8$.

(c) The line $x + 2y = 16$ passes through the points $(0, 8)$ and $(16, 0)$. Since $0 + 2(0) = 0$, which is less than 16, the origin does lie in the admissible region for the inequality $x + 2y \leqslant 16$.

The diagram below summarises the information in (a), (b) and (c) and shows the feasible region. The non-negativity conditions merely restrict you to the first quadrant.

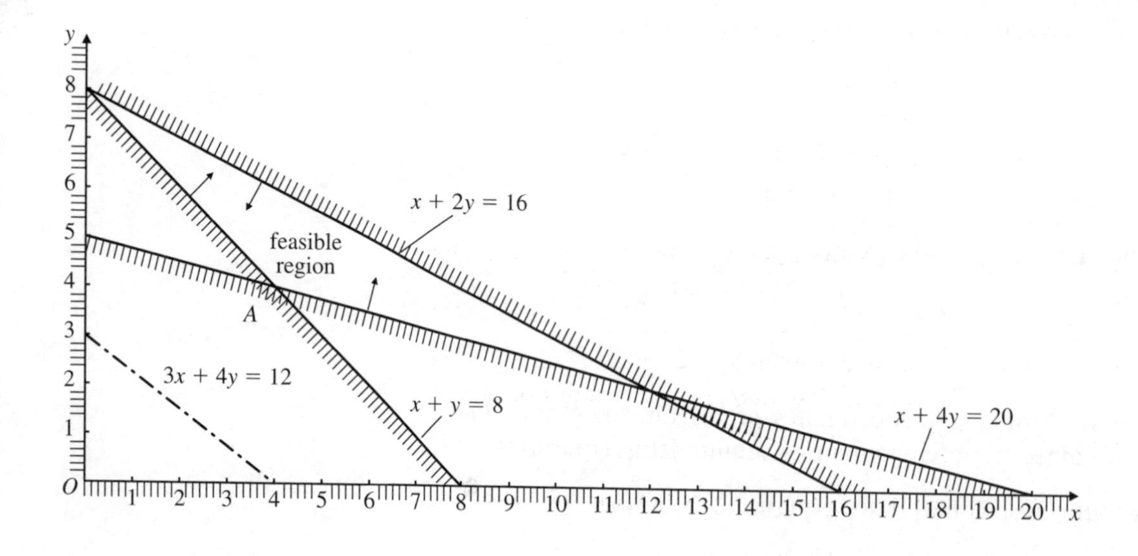

To locate the point where the objective function $Z = 3x + 4y$ has its minimum value first draw a line of the form $3x + 4y =$ constant.

The line $3x + 4y = 12$ is shown by $-\cdot-\cdot$ on the diagram. (The number 12 was chosen because 3 and 4 are factors of 12 and so the points where it intersects the axes have integer values.) This line passes through the points $(0, 3)$ and $(4, 0)$.

If you move a ruler parallel to this line then you see that the point of the feasible region *nearest* to the origin that is reached first is the point A. As A is the point of intersection of the lines $x + y = 8$ and $x + 4y = 20$ you can obtain the coordinates of A by solving the simultaneous equations

$$x + 4y = 20$$
$$x + y = 8$$

Subtracting gives: $$3y = 12 \Rightarrow y = 4$$

Substituting this value of y into the second equation gives

$$x + 4 = 8$$

so $$x = 8 - 4 = 4$$

The minimum value of Z is therefore $3(4) + 4(4) = 28$ and this occurs when $x = 4$ and $y = 4$.

Example 9

Solve graphically the linear programming problem:

Maximise $Z = 16x + 24y$ subject to the constraints:

$$2x + 3y \leqslant 24$$
$$2x + y \leqslant 16$$
$$y \leqslant 6$$
$$x \geqslant 0,\ y \geqslant 0$$

(a) The line $2x + 3y = 24$ passes through the points $(0, 8)$ and $(12, 0)$. Since $2(0) + 3(0) = 0$, which is less than 24, the origin does lie in the admissible region for the inequality $2x + 3y \leqslant 24$.

(b) The line $2x + y = 16$ passes through the points $(0, 16)$ and $(8, 0)$. Since $2(0) + 0 = 0$, which is less than 16, the origin does lie in the admissible region for the inequality $2x + y \leqslant 16$.

(c) The admissible region for the inequality $y \leqslant 6$ is the region on and below the line $y = 6$.

The diagram on the next page summarises the information in (a), (b) and (c), together with the non-negativity conditions, and shows the feasible region.

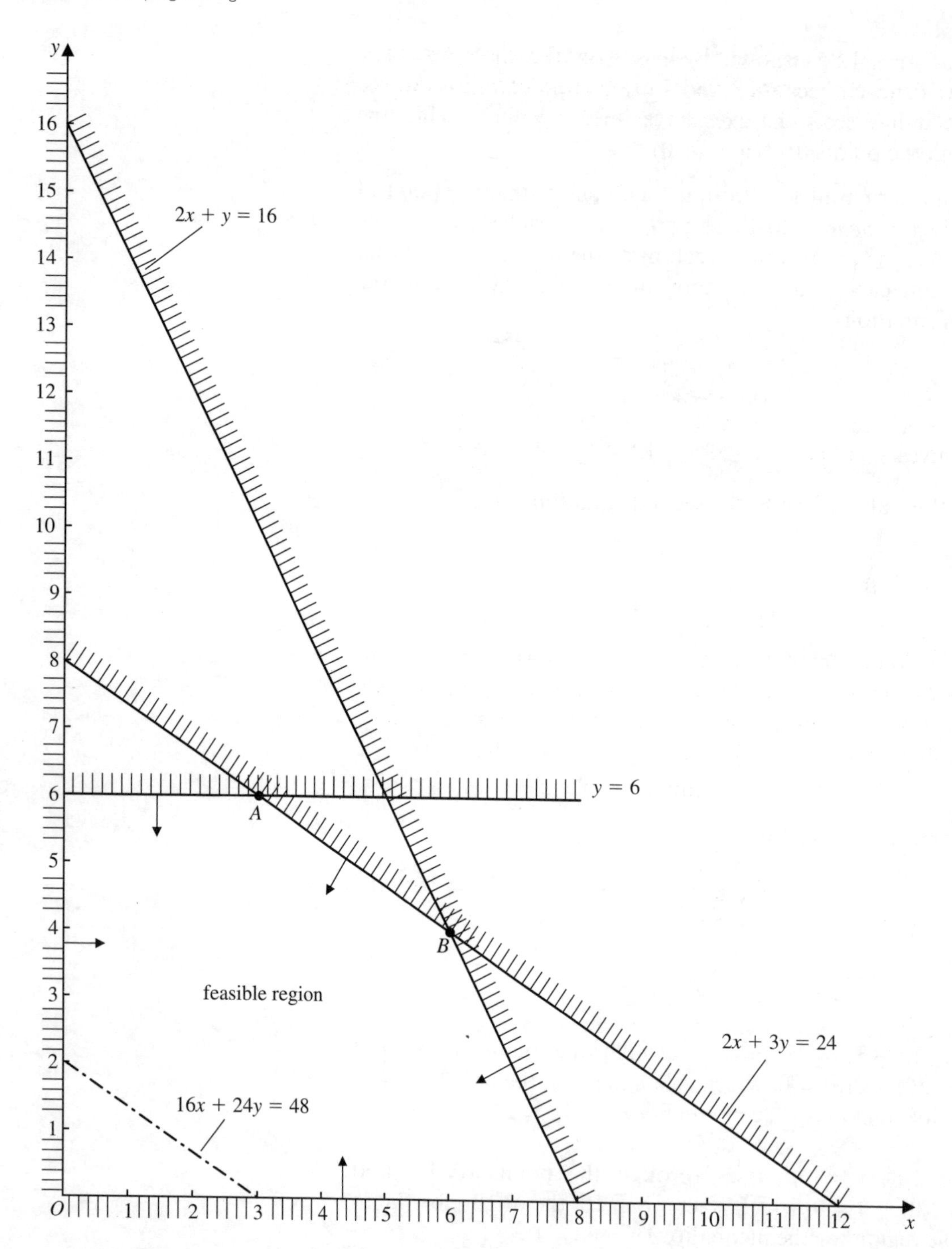

As in the previous examples you need first to draw a member of the family of parallel lines $16x + 24y =$ constant. In the diagram the line $16x + 24y = 48$ has been drawn. This line passes through the points (0, 2) and (3, 0). If you move the ruler parallel to this line as far as possible, so it is still in contact with the feasible region, you ultimately reach the boundary AB of the feasible region.

The coordinates of the point A are easily obtained. Its y-coordinate is 6 and its x-coordinate can be found by using the fact that A lies on the line with equation $2x + 3y = 24$, therefore

$$2x + 18 = 24$$
$$2x = 24 - 18 = 6$$
$$\Rightarrow \quad x = 3$$

A has coordinates (3, 6).

The point B is the intersection of the lines

$$2x + y = 16$$

and
$$2x + 3y = 24$$

Subtracting gives: $2y = 8 \Rightarrow y = 4$

Substituting in the first equation gives: $2x + 4 = 16$

$$2x = 16 - 4 = 12$$

so
$$x = 6$$

B has coordinates (6, 4).

The value of Z at A is $Z_A = 16(3) + 24(6) = 192$ and the value of Z at B is $Z_B = 16(6) + 24(4) = 192$.

In fact the value of Z is 192 for every point on the line AB. The maximum value of Z for this problem occurs at every point of the line segment AB rather than for a single point, as in the previous cases.

Example 10

Solve Example 9 graphically when the objective function is changed to

(a) $Z = 24x + 16y$ (b) $Z = 12x + 24y$

(a)

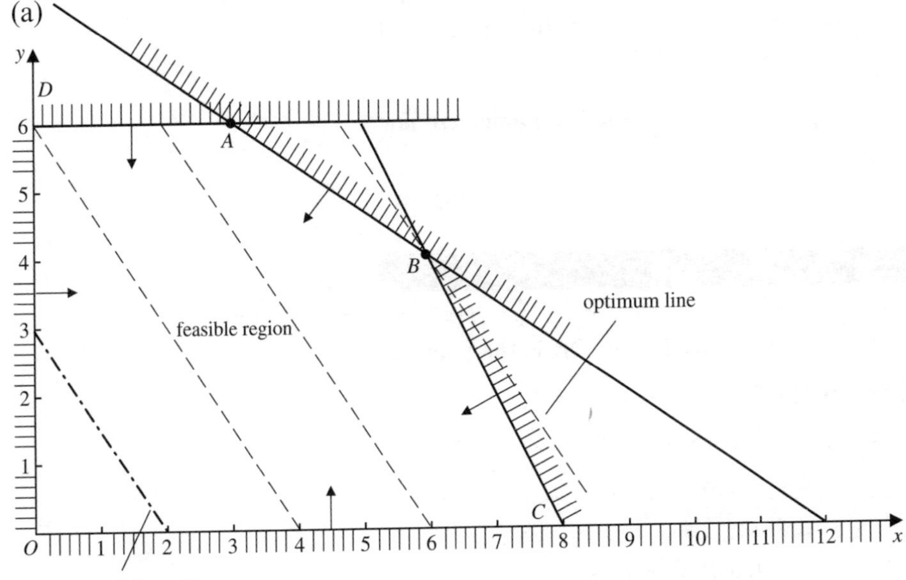

One member of the family of lines $24x + 16y = c$ is $24x + 16y = 48$, which passes through (0, 3) and (2, 0) (shown by $-\cdot-\cdot$). Other members of the family are shown by $- - -$. The maximum value of Z occurs at the point $B(6, 4)$ and $Z_{\max} = 24(6) + 16(4) = 208$. The optimum line passes through point B.

(b)

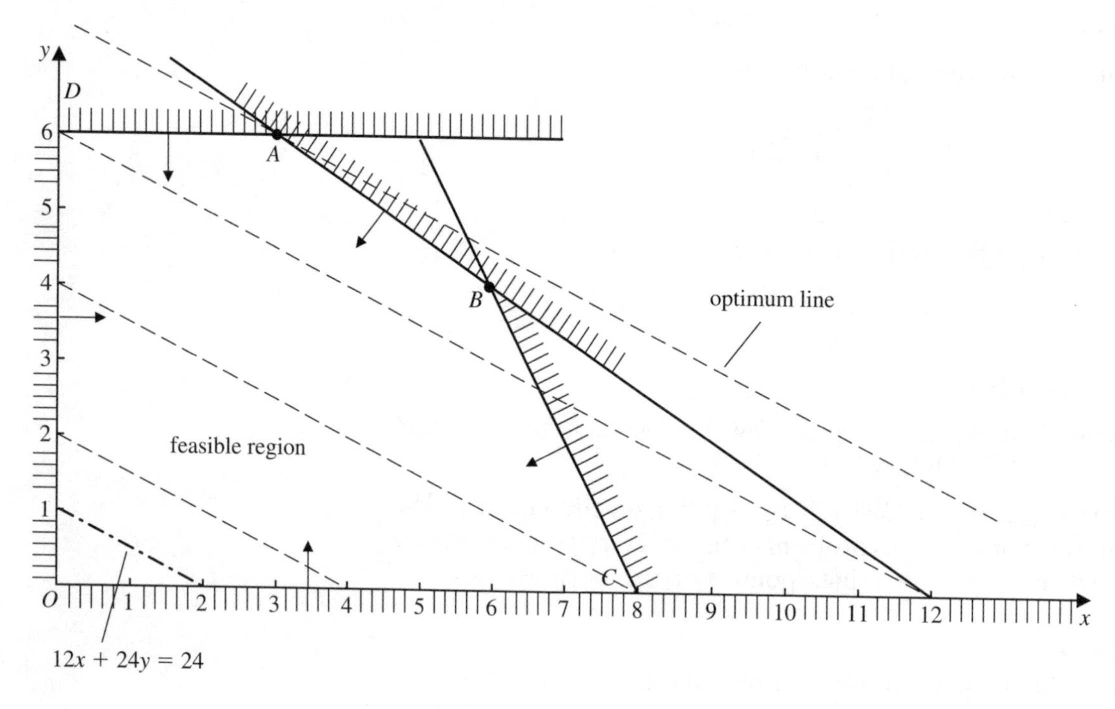

One member of the family of lines $12x + 24y = c$ is $12x + 24y = 24$, which passes through (0, 1) and (2, 0) (shown by $-\cdot-\cdot$). Other members of the family are shown by $- - -$. The maximum value of Z occurs at $A\,(3, 6)$ and $Z_{\max} = 12(3) + 24(6) = 180$.

The method used in these examples is usually called the **objective line method** or **ruler method**.

Exercise 6C

Solve the problems in questions 1 to 5 graphically. Draw the feasible region and use the objective line method.

1 Maximise $Z = 2x + y$ subject to the constraints:

$$x + y \leqslant 6$$
$$x \leqslant 5$$
$$y \leqslant 4, x \geqslant 0, y \geqslant 0$$

2 Maximise $Z = x + 5y$ subject to the constraints:

$$4x + 3y \leqslant 12$$
$$2x + 5y \leqslant 10$$
$$x \geqslant 0$$
$$y \geqslant 0$$

3 Maximise $Z = 3x + 2y$ subject to the constraints:

$$y + 2x \leqslant 12$$
$$x \geqslant 2$$
$$y \geqslant 4$$

4 Minimise $Z = 4x + y$ subject to the constraints:

$$3x + y \geqslant 6$$
$$x + y \geqslant 4$$
$$x \leqslant 4,\ y \leqslant 6$$

5 Minimise $Z = 2x + y$ subject to the constraints:

$$3x + y \geqslant 6$$
$$x + y \geqslant 4$$
$$x \leqslant 3$$
$$y \leqslant 4$$

6.5 Extreme points and optimality

In each of the above examples *the optimal value occurred at a corner or vertex (extreme point) of the feasible region.* In Example 9 it occurred at two corners and at every point on the line joining them. This is not a coincidence. It can be proved mathematically that:

- **The optimal solution of a linear programming problem, if it exists, will occur at one or more of the extreme points (vertices) of the feasible region.**

This provides us with an alternative way of finding the optimal value of Z and the values of x and y for which it occurs. This is usually called the **vertex method**. For a maximising problem:

1 Having determined the feasible region obtain the coordinates of the vertices of this region.

2 Evaluate the objective function at each of these vertices.
3 The maximum of the values found in step 2 gives the optimal value of Z and the coordinates of the corresponding vertex give the values of x and y for which this occurs.

Example 11

Use the vertex method to obtain the maximum values of
(a) $Z = 24x + 16y$
(b) $Z = 12x + 24y$

subject to the constraints:

$$2x + 3y \leqslant 24$$
$$2x + y \leqslant 16$$
$$y \leqslant 6$$
$$x \geqslant 0, y \geqslant 0$$

The feasible region has been determined in Example 9 and is shown below. The vertices are labelled A, B, C and D as shown.

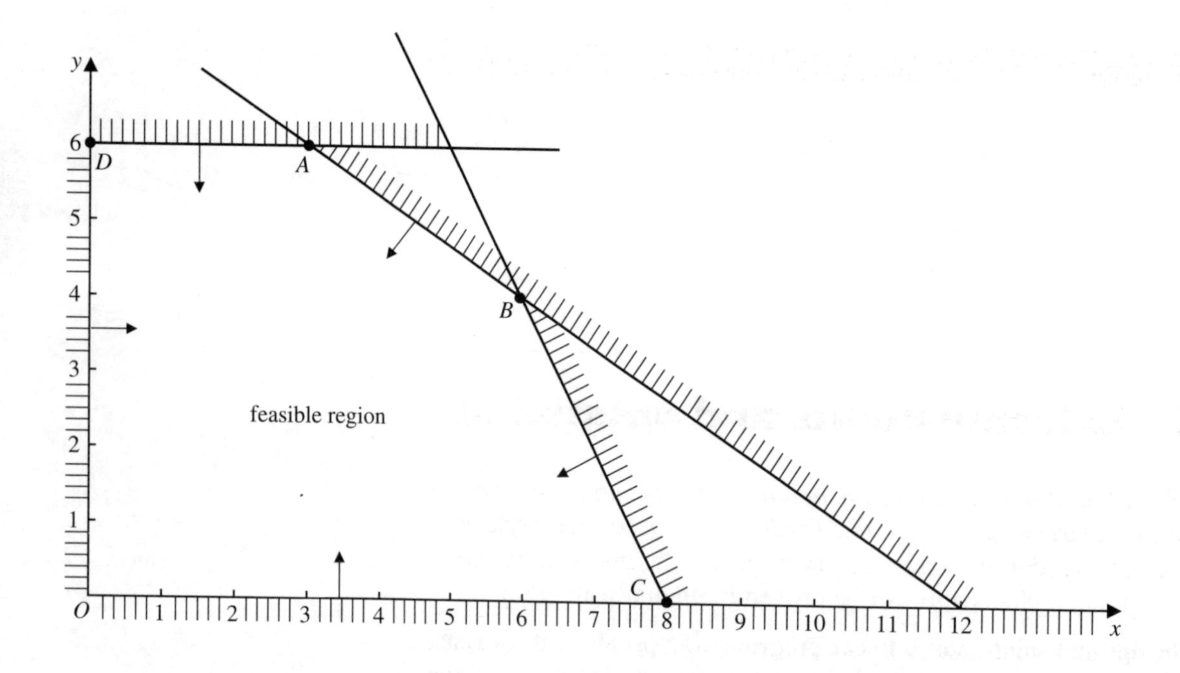

We have previously shown that A has coordinates (3, 6) and B has coordinates (6, 4).

It is clear from the construction of the feasible region that C has coordinates (8, 0) and D has coordinates (0, 6). The origin O has coordinates (0, 0).

(a) Which of these vertices gives the maximum value of $Z = 24x + 16y$? Substituting the coordinates of each point in turn in the equation gives:

$$Z_O = 0,\ Z_A = 24(3) + 16(6) = 168$$
$$Z_B = 24(6) + 16(4) = 208$$
$$Z_C = 24(8) + 16(0) = 192$$
$$Z_D = 24(0) + 16(6) = 96$$

The maximum value occurs at B where $x = 6$, $y = 4$ and $Z = 208$, as found previously.

(b) Now substitute the coordinates of each vertex in turn in the equation $Z = 12x + 24y$.

$$Z_O = 0,\ Z_A = 12(3) + 24(6) = 180$$
$$Z_B = 12(6) + 24(4) = 168$$
$$Z_C = 12(8) + 24(0) = 96$$
$$Z_D = 12(0) + 24(6) = 144$$

The maximum value occurs at A where $x = 3$, $y = 6$ and $Z = 180$, as found previously.

Exercise 6D

For each of the problems given in Exercise 6C:

(a) determine the coordinates of the vertices of the feasible region

(b) evaluate the objective function at each of these vertices

(c) hence solve the linear programming problem and confirm your previous answers.

6.6 Integer-valued solutions

All the examples considered so far have produced solutions for which the values of the decision variables have been integers. This is not true of all linear programming problems, in fact it is rare.

For some contexts solutions with non-integer values are acceptable. For example, in an investment problem, if £x and £y are the investments in two options a solution of $x = \frac{4}{5}$ and $y = 1\frac{4}{5}$ is acceptable since £s are subdivisible. However in other contexts non-integer solutions are not acceptable. In Example 3 (page 137), involving large and small vans, the solution must clearly involve integer numbers of large and small vans.

In the next example we will show how to proceed when the additional constraint *decision variables must be integers* is included.

Example 12

A manufacturer makes two kinds of toys, A and B. The machine time and the craftsman's time required for each are shown in the table, together with the limitations on time per week and the profit on each.

	Machine time (h)	Craftsman's time (h)	Profit (£)
Toy A	3	2	10
Toy B	3	4	12
Time available (h)	40	50	

The manufacturer wishes to maximise profit. In this context the numbers of A and B made must be integers.

Let x be the number of toy A made and y be the number of toy B made.

Then you have:

machine time:	$3x+3y \leqslant 40$	constraint (i)
craftsman's time:	$2x+4y \leqslant 50$	constraint (ii)
profit:	$Z=10x+12y$	
non-negativity condition:	$x \geqslant 0, y \geqslant 0$	constraint (iii)
integer condition:	x and y must be integers	constraint (iv)

The problem is to maximise the profit $Z = 10x + 12y$ subject to the constraints.

First draw the feasible region:

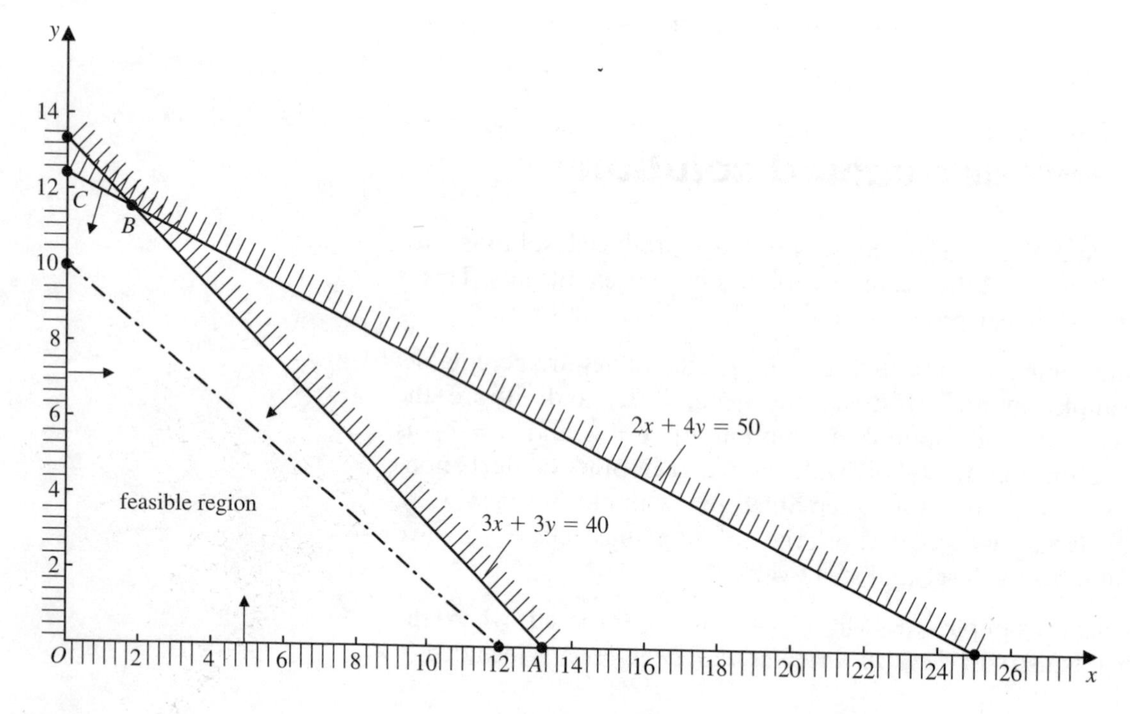

(i) $3x + 3y = 40$ passes through $(13\frac{1}{3}, 0)$, $(0, 13\frac{1}{3})$. The origin lies in the admissible set for the inequality $3x + 3y \leqslant 40$.
(ii) $2x + 4y = 50$ passes through $(0, 12\frac{1}{2})$, $(25, 0)$. The origin lies in the admissible set for the inequality $2x + 4y \leqslant 50$.

The problem can now be solved using either the ruler method or the vertex method.

Method 1 – using a ruler Draw an arbitrary member of the family of lines $10x + 12y =$ constant.

The line $10x + 12y = 120$ is drawn in the diagram as –·–·. If you move a ruler parallel to this you will find the optimal value at the point B. The coordinates of B are obtained by solving the simultaneous equations

$$3x + 3y = 40$$
$$2x + 4y = 50$$

Multiplying the first equation by 2 gives $6x + 6y = 80$.

Multiplying the second equation by 3 gives $6x + 12y = 150$.

Subtracting gives: $6y = 70$

$$\Rightarrow \quad y = 11\tfrac{2}{3}$$

Substituting into the first equation:

$$3x + 3(11\tfrac{2}{3}) = 40$$
$$3x = 40 - 35 = 5$$
$$\Rightarrow \quad x = 1\tfrac{2}{3} \quad \text{and} \quad Z = 10(1\tfrac{2}{3}) + 12(11\tfrac{2}{3})$$

At $B(1\frac{2}{3}, 11\frac{2}{3})$ the value of Z is $156\frac{2}{3}$.

Method 2 – vertices The coordinates of the vertices (extreme points) of the feasible region are:

$B\ (1\frac{2}{3}, 11\frac{2}{3})$ – from above
$O\ (0, 0)$ – origin
$A\ (13\frac{1}{3}, 0)$ – from (i) above
$C\ (0, 12\frac{1}{2})$ – from (ii) above

The values of $Z = 10x + 12y$ at B, O, A and C are:

$$Z_B = 10(1\tfrac{2}{3}) + 12(11\tfrac{2}{3}) = 156\tfrac{2}{3}$$
$$Z_O = 10(0) + 12(0) = 0$$
$$Z_A = 10(13\tfrac{1}{3}) + 12(0) = 133\tfrac{1}{3}$$
$$Z_C = 10(0) + 12(12\tfrac{1}{2}) = 150$$

The maximum occurs at $B(1\frac{2}{3}, 11\frac{2}{3})$ and $Z_{max} = 156\frac{2}{3}$.

Now let us consider how to apply the **integer constraint**. The solution obtained above is clearly not an acceptable one as $1\frac{2}{3}$ of a toy is not a credible answer.

You may be tempted just to take the integer part of the solutions but, as you will see below, this does not give the optimal solution. If the feasible region is small then it is possible to find Z for each point (x, y) in the feasible region for which both x and y are integers. In our case there are many such points and this is therefore not a sensible approach.

Here is a systematic procedure which, for many situations, leads to the optimal value with integer values of the decision variables.

First find the points close to the optimal point that have integer coordinates and are in the feasible region. Then calculate z for each of these points and choose the maximum.

In this case, since $x = 1\frac{2}{3}$, $y = 11\frac{2}{3}$ at the optimal point, consider the lines $x = 1$ and $x = 2$ and the lines $y = 11$ and $y = 12$:

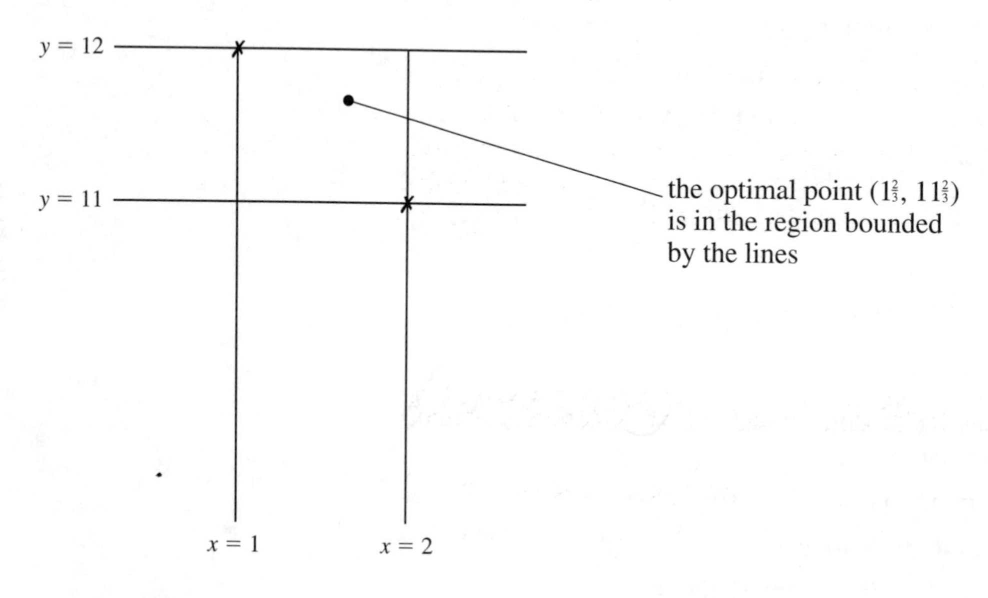

$x = 1$: When $x = 1$:

constraint (i) gives $3(1) + 3y \leqslant 40$, that is

$$3y \leqslant 37 \quad \text{or} \quad y \leqslant 12\tfrac{1}{3}$$

constraint (ii) gives $2(1) + 4y \leqslant 50$, that is

$$4y \leqslant 48 \quad \text{or} \quad y \leqslant 12$$

For integral values of y, you therefore require $y \leqslant 12$, so $(1, 12)$ does lie in the feasible region and $Z = 10(1) + 12(12) = 154$.

$x = 2$: When $x = 2$

constraint (i) gives $3(2) + 3y \leqslant 40$, that is

$$3y \leqslant 34 \quad \text{or} \quad y \leqslant 11\tfrac{1}{3}$$

constraint (ii) gives $2(2) + 4y \leqslant 50$, that is

$$4y \leqslant 46 \quad \text{or} \quad y \leqslant 11\tfrac{1}{2}$$

For integral values of y, you therefore require $y \leqslant 11$, so (2, 11) does lie in the feasible region and $Z = 10(2) + 12(11) = 152$.

The point (1, 11) also satisfies constraint (i) and the value of Z there is $Z = 10(1) + 12(11) = 142$.

From these calculations it is clear that as your ruler moves out from the origin it first reaches the point (1, 11), where Z is 142, then the point (2, 11), where Z is 152, and then the point (1, 12), where Z attains its maximum value for integer solutions of 154. Notice that (1, 11), obtained by rounding down each of the coordinates, *does not give the optimal solution.*

In this situation we found the optimal integer solution by looking at integer solutions close to the optimal point. However, in some optimisation problems the optimal integer solution may not lie close to the optimal point. Finding optimal integer solutions is, in general, a time-consuming and difficult problem. If you are asked to find integer solutions in your D1 exam it will be sufficient to look for integer solutions close to the optimal point.

Exercise 6E

Solve the problems in questions 1 to 3 graphically. Draw the feasible region using (a) the objective line method and (b) the vertex method.

1 Maximise $Z = 2x + 3y$ subject to the constraints:

$$3x + y \leqslant 30$$
$$x + 2y \leqslant 30$$
$$x \geqslant 0,\ y \geqslant 0$$

2 Minimise $W = 12u + 10v$ subject to the constraints:

$$2u + 3v \geqslant 16$$
$$4u + 2v \geqslant 24$$
$$u \geqslant 0,\ v \geqslant 0$$

3 Maximise $Z = 16x + 12y$ subject to the constraints:

$$3x + 4y \leqslant 84$$
$$4x + 3y \leqslant 84$$
$$x \leqslant 15$$
$$x \geqslant 0,\ y \geqslant 0$$

4 The variables x and y are subject to the constraints

$$3x + 4y \leqslant 84$$
$$4x + 3y \leqslant 84$$
$$x \leqslant 15$$
$$x \geqslant 0,\ y \geqslant 0$$

as in question 3. Use graphical methods, both the ruler and vertex methods, to determine the maximum value of

(a) $Z = 10x + 10y$

(b) $Z = 30x + 15y$.

5 Maximise $Z = x + y$ subject to the constraints:

$$3x + 4y \leqslant 12$$
$$2x + y \leqslant 4$$
$$x \geqslant 0,\ y \geqslant 0$$

(a) Solve this linear programming problem.

(b) Identify all the points in the feasible region that have integer coordinates. Which ones give the maximum value of Z?

(c) Proceeding from your answer to part (a) find the solution to the linear programming problem when the integer constraint is added.

6 A builder has a plot of land available on which he can build houses. He can build either luxury houses or standard houses. He decides to built at least five luxury houses and at least 10 standard houses. Planning regulations prevent him from building more than 30 houses altogether. Each luxury house requires 300 m^2 of land and each standard house requires 150 m^2 of land. The total area of the plot is 6000 m^2. The builder makes a profit of £12 000 on each luxury house and a profit of £8000 on each standard house. The builder wishes to make the maximum possible profit.

(a) Formulate the above problem as a linear programming problem.

(b) Determine how many of each type of house he should build to make the maximum profit and state what this profit is.

7 Example 3 on page 137 is about the KJB Haulage company. Solve the linear programming problem formulated there to obtain the number of large and small vans to be used to minimise the cost.

8 Solve the linear programming problem formulated in Example 4 (page 138) to determine how much the lecturer should invest in each option to maximise his yield.

9 A minibus operator is contracted to transport 50 workers to their place of employment. He has available three type A minibuses and four type B minibuses. A type A minibus carries 15 workers and a type B minibus carries 10 workers. Only five drivers are available. It costs £50 to operate a type A minibus and £40 to operate a type B minibus. Let x be the number of type A minibuses used and y the number of type B minibuses used.

(a) Write down four inequalities satisfied by x and y, other than $x \geqslant 0, y \geqslant 0$.

(b) Display these inequalities on a graph and label clearly the feasible region.

Given that x and y must be integers:

(c) determine the possible combinations of x and y that satisfy the constraints

(d) obtain the minimum cost of the operation and the number of type A and type B minibuses used in this case.

6.7 The algebraic method for solving linear programming problems

The graphical method for solving linear programming problems discussed above obviously has limitations since it requires the drawing of the feasible region. This is only possible when you have just two decision variables. For more than two decision variables you need an algebraic method. Such a method has been developed which relies on the fact stated earlier that 'the optimal solution may be found by examining the value of the objective function at the extreme points of the feasible region'.

Slack variables

Before we discuss this method let us consider another way of looking at the graphical methods discussed above. Example 1 (page 134) produced the following linear programming problem:

Maximise $Z = 10x + 12y$ subject to the constraints:

$$3x + 3y \leqslant 120 \quad \text{(i)}$$
$$2x + 4y \leqslant 150 \quad \text{(ii)}$$
$$x \geqslant 0,\ y \geqslant 0$$

You can easily construct the feasible region:

(i) the line $3x + 3y = 120$ passes through $(0, 40)$ and $(40, 0)$ and the origin does lie in the admissible set for the inequality $3x + 3y \leqslant 120$

(ii) the line $2x + 4y = 150$ passes through $\left(0, 37\frac{1}{2}\right)$ and $(75, 0)$ and the origin does lie in the admissible set for the inequality $2x + 4y \leqslant 150$.

Here is the feasible region:

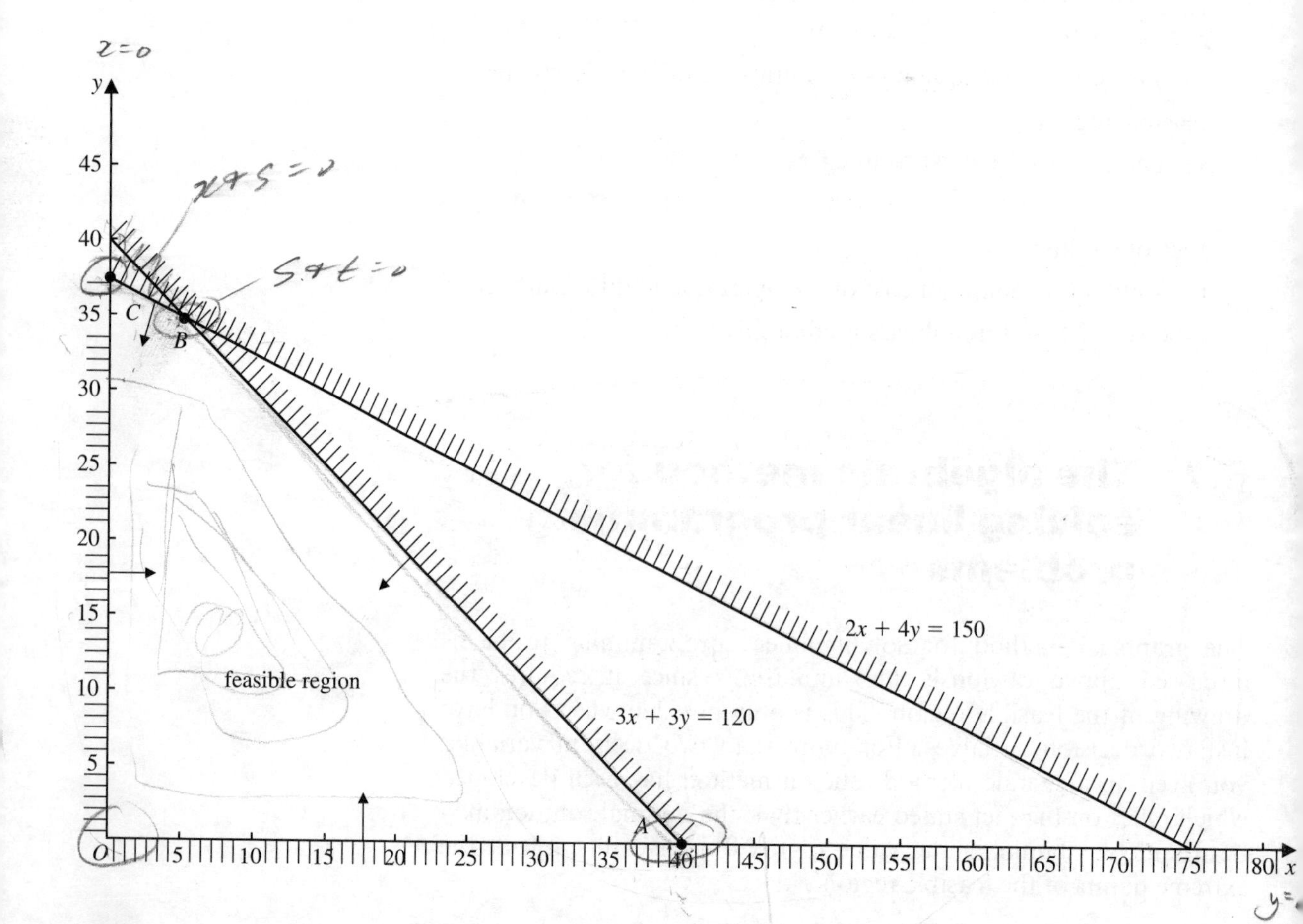

It is convenient for present purposes, and also for future work, to convert inequalities (i) and (ii) into equations. There are standard methods for dealing with systems of equations. If you replace inequality (i) by the equation

$$3x + 3y + s = 120$$

then

$$s = 120 - (3x + 3y)$$

and from inequality (i) this means $s \geqslant 0$.

Similarly, if the inequality (ii) is replaced by the equation

$$2x + 4y + t = 150$$

then

$$t = 150 - (2x + 4y)$$

and from inequality (ii) this means $t \geqslant 0$.

In terms of x, y, s and t you now have:

equation of OA is $y = 0$
equation of AB is $s = 0$
equation of BC is $t = 0$
equation of OC is $x = 0$

and therefore at O: $x = 0, \quad y = 0$
at A: $y = 0, \quad s = 0$
at B: $s = 0, \quad t = 0$
at C: $x = 0, \quad t = 0$

- **At a vertex of the feasible region two of the variables x, y, s and t are zero.**

The variable s is called the **slack variable** for inequality (i). It gives the difference between the amount of resource available (120) and the amount used $(3x + 3y)$.

Similarly the variable t is called the **slack variable** for inequality (ii). It gives the difference between the amount of resource available (150) and the amount used $(2x + 4y)$.

The coordinates of the vertices of the feasible region are now easy to handle. This makes the process of considering extreme points very straightforward. These ideas are easily generalised to problems involving more than two decision variables.

6.8 Basic solutions

When we converted the inequalities (i) and (ii) in the previous problem, we obtained the two equations

$$3x + 3y + s = 120$$
$$2x + 4y + t = 150$$

These two equations contain four unknowns: x, y, s and t. In this case

$$(\textit{number of variables}) - (\textit{number of equations}) = 4 - 2 = 2$$

If you take this number (2 in this case) of variables and set them to zero, then solve the equations, the solution you obtain is called a **basic solution**.

The two variables set to zero are called **non-basic variables**.

The variables solved for are called **basic variables**.

A basic solution that is also a feasible solution is called a **basic feasible solution**. The table below summarises the position for this problem. The equations have been solved after setting two variables at a time to zero.

Vertex	x	y	s	t	Type of solution	Value of $Z = 10x + 12y$	(x, y)
O	0	0	120	150	Basic feasible	0	(0, 0)
C	0	$37\frac{1}{2}$	$7\frac{1}{2}$	0	Basic feasible	450	$(0, 37\frac{1}{2})$
B	5	35	0	0	Basic feasible	470	(5, 35)
	0	40	0	−10	Basic non-feasible	—	(0, 40)
A	40	0	0	70	Basic feasible	400	(40, 0)
	75	0	−75	0	Basic non-feasible	—	(75, 0)

6.9 Simplex method

The algebraic method for solving linear programming problems is called the **simplex method** and was developed by George Dantzig in 1947. The **simplex algorithm** that is used consists of two steps:

1 A way of finding out whether a given solution, corresponding to an extreme point of the feasible region, is an optimal solution.
2 A way of obtaining an adjacent extreme point with a larger value for the objective function.

It is useful at this point to define a **standard form** for a linear programming problem:

- **A linear programming problem is in standard form if**
  (a) the objective function $\alpha x + \beta y + \gamma z$ is to be maximised
  (b) all the constraints, other than the non-negativity conditions, are of the form

$$ax + by + cz \leqslant d$$

A problem that is not in standard form may be rewritten in standard form using the following observation:

$$\text{minimise}\,(\alpha x + \beta y + \gamma z) = -\text{maximise}\,(-\alpha x - \beta y - \gamma z)$$

That is, to minimise the objective function you could maximise its negative instead.

The calculations involved in using the simplex algorithm are recorded in a sequence of tables that are known as **simplex tableaux**. The method is illustrated by solving Example 1 using the simplex algorithm.

To use the simplex method as we shall describe it, the problem must be written in standard form.

Example 13

Use the simplex method to solve Example 1.

The first step is to add slack variables to the inequalities, other than the non-negativity conditions, as described earlier.

For Example 1:

$$3x + 3y \leqslant 120 \Rightarrow 3x + 3y + s = 120$$

$$2x + 4y \leqslant 150 \Rightarrow 2x + 4y + t = 150$$

The second step is to rewrite the linear programming problem so that each equation contains all variables x, y, s and t.

The problem now becomes:
Maximise $Z = 10x + 12y + 0s + 0t$ subject to the constraints:

$$3x + 3y + 1s + 0t = 120$$

$$2x + 4y + 0s + 1t = 150$$

$$x \geqslant 0,\ y \geqslant 0,\ s \geqslant 0,\ t \geqslant 0$$

The third step is to put all this information in the form of a table, which is known as the **initial tableau**. In order to do this, write the objective function in the form

$$Z - 10x - 12y - 0s - 0t = 0$$

Now draw up this table:

Tableau 1

Basic variable	x	y	s	t	Value
s	3	3	1	0	120
t	2	4	0	1	150
Z	−10	−12	0	0	0

← objective row

↑ most-negative coefficient

The first row corresponds to the first constraint.
The second row corresponds to the second constraint.
The third row corresponds to the objective function.

In drawing up this table we have taken as the initial basic feasible solution $x = 0$, $y = 0$, $s = 120$ and $t = 150$.

s and t are basic variables and x and y are non-basic variables. Since s appears only in the first inequality and its value is obtained from this, it appears as the basic variable for that row. For a similar reason t appears at the beginning of row 2.

Notice that the basic variables s and t each appear in *only one row* and they appear there with a coefficient of 1.

The column that the basic variable labels has all zeros except for the 1 in the row which it labels.

To interpret the solution that corresponds to this tableau, look at the first and last columns:

$$s = 120,\ t = 150 \Rightarrow x = y = 0 \quad \text{and} \quad Z = 0$$

The **optimality condition** states:

- **If the objective row of a tableau has zero entries in the columns labelled by basic variables and no negative entries in the columns labelled by non-basic variables then the solution represented by the tableau is optimal.**

Clearly this condition is not satisfied by Tableau 1 and so we do not have an optimal solution. We must therefore construct a new tableau corresponding to an adjacent extreme point. We will describe the steps for doing this without going into the theoretical details.

The first step in the process is to choose the non-basic variable that is to become a basic variable. (This is often called **entering the basis**.) The most common rule for selecting this variable is to select the variable with the *most negative entry* in the objective function row. In our case this is (−12) and the corresponding variable is y. We usually indicate this with a vertical arrow as we have done in Tableau 1.

The new basic variable is called the **entering variable** and the column it is in is called the **pivotal column**.

The second step in the process is to choose *which variable is to leave the basis* (the **leaving variable**). In order to do this we calculate **θ-values**. These are calculated for each row, other than the objective row. You find the θ-value for a row by dividing the entry in the value column, which must always be positive, by the entry in the pivotal column. Only θ-values for rows in which the entries in the pivotal column are *positive* are used.

From Tableau 1 you get:

row (i) $\theta = \dfrac{120}{3} = 40$ row (ii) $\theta = \dfrac{150}{4} = 37\frac{1}{2}$

The row with the smallest θ-value is called the **pivotal row**. Here, the pivotal row is row (ii).

The entry at the intersection of the pivotal row and the pivotal column is called the **pivot**. It is usually ringed.

In practice all the above information is included in Tableau 1:

Basic variable	x	y	s	t	Value	
s	3	3	1	0	120	$\theta = \frac{120}{3} = 40$
t	2	(4)	0	1	150	$\theta = \frac{150}{4} = 37\frac{1}{2}$ ← pivotal row
Z	−10	−12	0	0	0	

pivot

pivotal column

Forming a new tableau

The first step in forming a new tableau is to divide the pivotal row by the pivot so that the pivot now becomes 1:

Basic variable	x	y	s	t	Value
s	3	3	1	0	120
t	$\frac{1}{2}$	(1)	0	$\frac{1}{4}$	$37\frac{1}{2}$
Z	−10	−12	0	0	0

The second step is to add suitable multiples of the new pivotal row to all other rows, including the objective row, so that all other elements in the pivotal column become zero.

1st row To get the required zero in the y column, take row (i) − 3 × row (ii):

x	y	s	t	Value
$3 - 1\frac{1}{2}$ $= 1\frac{1}{2}$	$3 - 3$ $= 0$	$1 - 0$ $= 1$	$0 - \frac{3}{4}$ $= -\frac{3}{4}$	$120 - 3 \times 37\frac{1}{2}$ $= 7\frac{1}{2}$

3rd row To get the required zero in the y column, take row (iii) + 12 × row (ii):

x	y	s	t	Value
$-10 + 6$ $= -4$	$-12 + 12$ $= 0$	$0 + 0$ $= 0$	$0 + 3$ $= 3$	$0 + 450$ $= 450$

To complete the new tableau, replace the label on the pivotal row by the entering variable (in this case y).

The second tableau is then:

Tableau 2

Basic variable	x	y	s	t	Value	
s	$(1\frac{1}{2})$	0	1	$-\frac{3}{4}$	$7\frac{1}{2}$	←
y	$\frac{1}{2}$	1	0	$\frac{1}{4}$	$37\frac{1}{2}$	
Z	-4	0	0	3	450	
	↑					

This tableau corresponds to the solution $s = 7\frac{1}{2}$, $y = 37\frac{1}{2}$, $x = 0$, $t = 0$ with $Z = 450$.

If you apply the optimality condition to this tableau you will see that it is not optimal – the objective row (Z row) contains a negative entry. You must therefore form a new tableau by pivoting.

Step 1 Choose the pivotal column. As there is only one negative entry (−4) in the Z row, choose the corresponding variable x as the entering variable. The pivotal column is indicated by the arrow ↑.

Step 2 Calculate θ-values:

row (i) $\dfrac{(7\frac{1}{2})}{(1\frac{1}{2})} = 5$ row (ii) $\dfrac{(37\frac{1}{2})}{(\frac{1}{2})} = 75$

The θ-value for row (i) is the smaller and the pivotal row is therefore row (i), shown by ←. The leaving variable is therefore s. The pivot is therefore $1\frac{1}{2}$ at the intersection of the pivotal row and pivotal column. It is shown ringed.

Step 3 Divide the pivotal row by the pivot:

Basic variable	x	y	s	t	Value
s	①	0	$\frac{2}{3}$	$-\frac{1}{2}$	5
y	$\frac{1}{2}$	1	0	$\frac{1}{4}$	$37\frac{1}{2}$
Z	-4	0	0	3	450

Step 4 Add multiples of the new pivotal row to all other rows so that all other elements in the pivotal column become zero:

Tableau 3

Basic variable	x	y	s	t	Value	
ⓧ	1	0	$\frac{2}{3}$	$-\frac{1}{2}$	5	
y	0	1	$-\frac{1}{3}$	$\frac{1}{2}$	35	← row (ii) $-\frac{1}{2}$ row (i)
Z	0	0	$\frac{8}{3}$	1	470	← row (iii) + 4 row (i)

This tableau corresponds to the solution $x = 5$, $y = 35$, $s = 0$, $t = 0$ and $Z = 470$.

Since there are no negative entries in the Z row this solution is optimal. This final tableau is called the **optimal tableau**.

The simplex method can be readily applied to problems when there are more than two decision variables, as is illustrated by the following example.

Example 14

Use the simplex method to solve the linear programming problem:

Maximise $P = 8x + 9y + 5z$ subject to the constraints:

$$2x + 3y + 4z \leqslant 3$$
$$6x + 6y + 2z \leqslant 8$$
$$x \geqslant 0,\ y \geqslant 0,\ z \geqslant 0$$

The problem is in standard form.
First introduce the slack variables s and t into the constraints to produce the required form. The problem then becomes

Maximise $P = 8x + 9y + 5z + 0s + 0t$ subject to the constraints:

$$2x + 3y + 4z + s = 3$$
$$6x + 6y + 2z + t = 8$$
$$x \geqslant 0,\ y \geqslant 0,\ z \geqslant 0,\ s \geqslant 0,\ t \geqslant 0$$

A basic solution in this case has $(5 - 2) = 3$ variables zero.

The initial tableau is then:

Tableau 1

Basic variable	x	y	z	s	t	Value	
s	2	③	4	1	0	3	$\theta = \frac{3}{3} = 1 \leftarrow$
t	6	6	2	0	1	8	$\theta = \frac{8}{6} = 1\frac{1}{3}$
P	−8	−9	−5	0	0	0	
		↑					

This corresponds to $s = 3$, $t = 8$, $x = 0$, $y = 0$, $z = 0$ and $P = 0$. The pivotal column, the θ-values giving the pivotal row and the pivot are indicated on this initial tableau. The entering variable is y and the leaving variable is s.

Dividing the pivotal row by the pivot gives:

Basic variable	x	y	z	s	t	Value
s	$\frac{2}{3}$	①	$\frac{4}{3}$	$\frac{1}{3}$	0	1
t	6	6	2	0	1	8
P	−8	−9	−5	0	0	0

The second tableau is then:

Tableau 2

Basic variable	x	y	z	s	t	Value	
ⓨ	$\frac{2}{3}$	1	$\frac{4}{3}$	$\frac{1}{3}$	0	1	
t	②	0	−6	−2	1	2	row (ii) − 6 row (i) ←
P	−2	0	7	3	0	9	row (iii) + 9 row (i)
	↑						

This corresponds to the solution $y = 1$, $t = 2$, $x = 0$, $z = 0$, $s = 0$ and $P = 9$.

It is not optimal as there is a negative entry in the objective row. We therefore form a new tableau.

The variable to enter the basis is now x. The θ-values are:

$$\text{row (i)} \ \frac{1}{\left(\frac{2}{3}\right)} = \frac{3}{2}$$

$$\text{row (ii)} \ \frac{2}{2} = 1$$

and so row (ii) is the pivotal row and the ringed ② is the pivot. The leaving variable is t.

Dividing the pivotal row by the pivot gives:

Basic variable	x	y	z	s	t	Value
y	$\frac{2}{3}$	1	$\frac{4}{3}$	$\frac{1}{3}$	0	1
t	①	0	-3	-1	$\frac{1}{2}$	1
P	-2	0	7	3	0	9

Adding multiples of the new pivotal row to all other rows so that all other elements in the pivotal column become zero gives the tableau:

Tableau 3

Basic variable	x	y	z	s	t	Value	
y	0	1	$\frac{10}{3}$	1	$-\frac{1}{3}$	$\frac{1}{3}$	row (i) $-\frac{2}{3}$ row (ii)
ⓧ	1	0	-3	-1	$\frac{1}{2}$	1	
P	0	0	1	1	1	11	row (iii) + 2 row (ii)

This corresponds to the solution $y = \frac{1}{3}$, $x = 1$, $z = 0$, $s = 0$, $t = 0$ and $P = 11$.

It is the optimal solution as there are no negative entries in the objective row.

Example 15

As a final example, and for comparison, apply the simplex method to Example 9:

Maximise $Z = 16x + 24y$ subject to the constraints:

$$2x + 3y \leqslant 24$$
$$2x + y \leqslant 16$$
$$y \leqslant 6$$
$$x \geqslant 0, y \geqslant 0$$

As there are three constraints, other than the non-negativity condition, we need three slack variables s, t, u. Introducing these gives:

Maximise $Z = 16x + 24y + 0s + 0t + 0u$ subject to the constraints:

$$2x + 3y + s = 24$$
$$2x + y + t = 16$$
$$y + u = 6$$
$$x \geqslant 0, y \geqslant 0, s \geqslant 0, t \geqslant 0, u \geqslant 0$$

The initial tableau is:

Basic variable	x	y	s	t	u	Value	
s	2	3	1	0	0	24	$\theta = \frac{24}{3} = 8$
t	2	1	0	1	0	16	$\theta = \frac{16}{1} = 16$
u	0	①	0	0	1	6	$\theta = \frac{6}{1} = 6 \leftarrow$
Z	−16	−24	0	0	0	0	
		↑					

The entering variable is y and the leaving variable is u. This corresponds to the solution $s = 24$, $t = 16$, $u = 6$, $x = 0$, $y = 0$ and $Z = 0$.

Since the pivot is 1 you can proceed immediately to produce the next tableau by adding suitable multiples of row (iii) to the others:

Basic variable	x	y	s	t	u	Value	
s	②	0	1	0	−3	6	row (i) − 3 row (iii) ←
t	2	0	0	1	−1	10	row (ii) − row (iii)
ⓨ	0	1	0	0	1	6	
Z	−16	0	0	0	24	144	row (iv) + 24 row (iii)
	↑						

The θ-values are:

$$\text{row (i) } \tfrac{6}{2} = 3$$

$$\text{row (ii) } \tfrac{10}{2} = 5$$

and so the entering variable is x and the leaving variable is s, the pivot being the ringed ②.

Dividing the pivotal row by the pivot gives:

Basic variable	x	y	s	t	u	Value
s	①	0	$\frac{1}{2}$	0	$-\frac{3}{2}$	3
t	2	0	0	1	−1	10
y	0	1	0	0	1	6
Z	−16	0	0	0	24	144

You can now produce the next tableau by proceeding in the usual way:

Basic variable	x	y	s	t	u	Value	
ⓧ	1	0	$\frac{1}{2}$	0	$-\frac{3}{2}$	3	
t	0	0	−1	1	2	4	row (ii) − 2 row (i)
y	0	1	0	0	1	6	row (iii) + 0 row (i)
Z	0	0	8	0	0	192	row (iv) + 16 row (i)

(Notice that row (iii) remains unchanged as there is already a zero in the pivot column in this row.)

This tableau corresponds to the solution $x = 3$, $t = 4$, $y = 6$ and $Z = 192$.

Compare this with the previous results (on page 153). Notice that in addition to telling us that the maximum value of Z occurs when $x = 3$ and $y = 6$ it also tells us that there is a non-zero slack of 4 in the second inequality. This corresponds to unused time or material, depending on the context.

Exercise 6F

Use the simplex algorithm to solve the linear programming problems in questions 1 to 4.

1 Maximise $Z = 2x + 3y$ subject to the constraints:

$$3x + y \leqslant 30$$
$$x + 2y \leqslant 30$$
$$x \geqslant 0,\ y \geqslant 0$$

2 Maximise $Z = 30x + 15y$ subject to the constraints:

$$3x + 4y \leqslant 84$$
$$4x + 3y \leqslant 84$$
$$x \leqslant 15$$
$$x \geqslant 0,\ y \geqslant 0$$

3 Maximise $P = 3x + 6y + 2z$ subject to the constraints:

$$3x + 4y + z \leqslant 20$$
$$x + 3y + 2z \leqslant 10$$
$$x \geqslant 0,\ y \geqslant 0,\ z \geqslant 0$$

4 Maximise $P = x - y + 2z$ subject to the constraints:

$$3x + 6y + z \leqslant 6$$
$$4x + 2y + z \leqslant 4$$
$$x - y + z \leqslant 3$$
$$x \geqslant 0,\ y \geqslant 0,\ z \geqslant 0$$

5 A company manufactures two kinds of cloth, A and B, and uses three different colours of wool. The material required to make a unit length of each type of cloth and the total amount of wool of each colour that is available are shown in the table.

Requirements for unit length of cloth of type		Colour of wool	Wool available (kg)
A (kg)	B (kg)		
4	1	Red	56
5	3	Green	105
1	2	Blue	56

The manufacturer makes a profit of £12 on a unit length of cloth A and a profit of £15 on a unit length of cloth B. He wants to find out how to use the available material to make the largest possible profit.

(a) Formulate this as a linear programming problem.

(b) Solve the resulting problem by using the simplex algorithm.

(c) Confirm your answer to (b) by solving the problem graphically.

6 A firm estimates that its profit £P is dependent on three variables x, y and z and is given by the equation

$$P = x + 4y + 10z$$

The variables x, y and z are all non-negative and also satisfy the constraints:

$$x + 4y + 2z \leqslant 40$$

$$x + 4z \leqslant 8$$

Use the simplex algorithm to determine the maximum value of P and the values of x, y and z for which it occurs.

SUMMARY OF KEY POINTS

1 Any pair of values of x and y that satisfy all the constraints in a linear programming problem is called a **feasible solution.**

2 The region that contains all feasible solutions is called the **feasible region.**

3 The **optimal solution** of a linear programming problem, if it exists, will occur at one or more of the extreme points (vertices) of the feasible region.

4 The **simplex method** is an algebraic method for solving linear programming problems.
   (i) The column that contains the entering variable is called the **pivotal column**.
   (ii) The row with the smallest θ-value is called the **pivotal row**.
   (iii) The entry at the intersection of the pivotal row and the pivotal column is called the **pivot**.

5 **Optimality condition**:
If the objective row of a tableau has zero entries in the columns labelled by basic variables and no negative entries in the columns labelled by non-basic variables then the solution represented by the tableau is optimal.

Review exercise 2

1

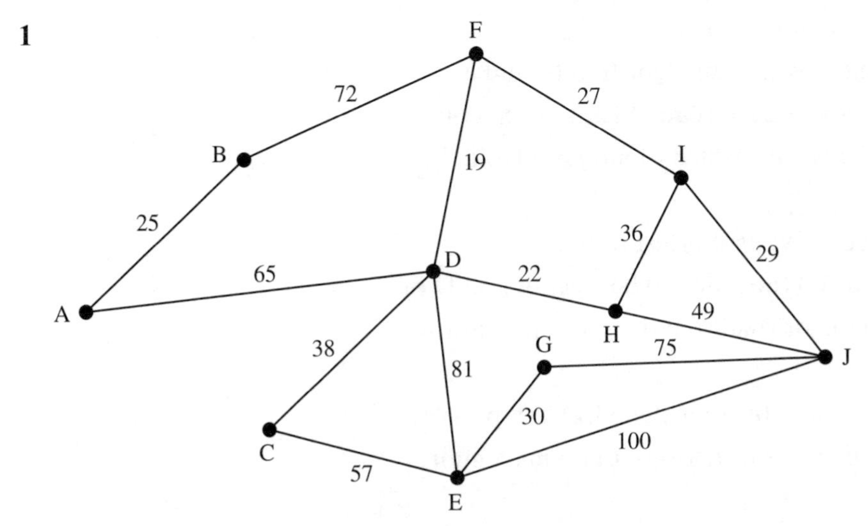

The network of paths in a garden is shown above. The numbers on the paths give their lengths in metres. The gardener wishes to inspect each of the paths to check for broken paving slabs so that they can be repaired before the garden is opened to the public. The gardener has to walk along each of the paths at least once.

(a) Write down the degree (valency) of each of the ten vertices A, B, ..., J.

(b) Hence find a route of minimum length. You should clearly state, with reasons, which, if any, paths will be covered twice.

(c) State the total length of your shortest route. [E]

2 The network at the top of page 180 shows the major roads that are to be gritted by a council in bad weather. The number on each arc is the length of the road in kilometres.

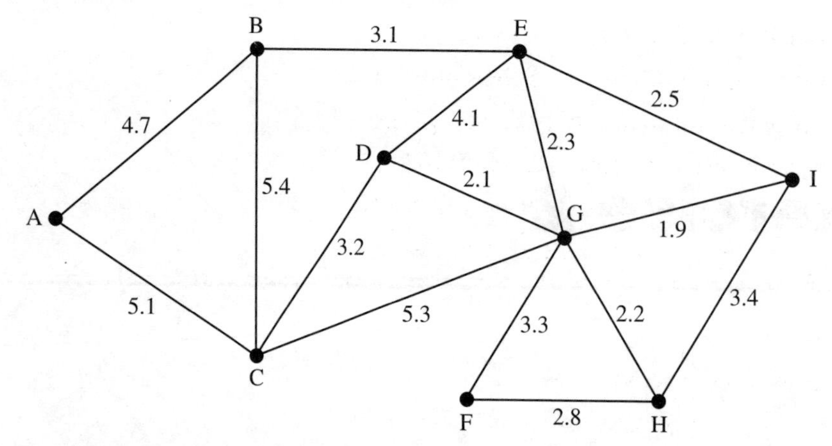

(a) List the valency of each of the vertices.

(b) Starting and finishing at A, use an algorithm to find a route of minimum length that covers each road at least once. You should clearly state, with reasons, which (if any) roads will be traversed twice.

(c) Obtain the total length of your shortest route.

(d) There is a minor road BD (not shown) between B and D of length 6.4 km. It is not a major road so it does not need gritting urgently.

Decide whether or not it is sensible to include BD as a part of the main gritting route, giving your reasons. (You may ignore the cost of the grit.)

[E]

3

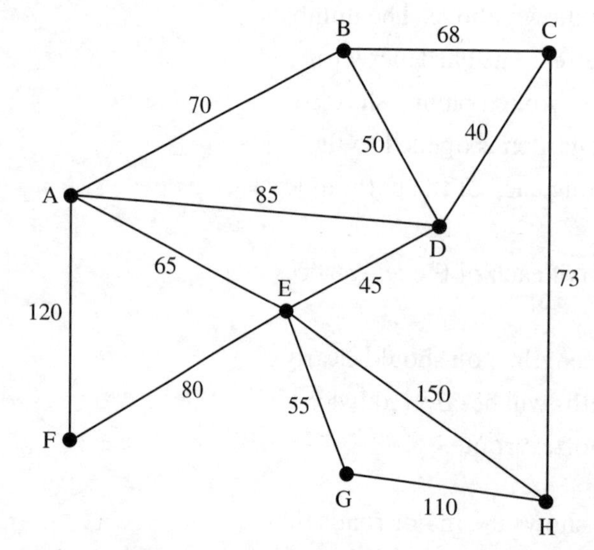

The network above represents the streets in a village. The number on each arc represents the length of the street in metres.

The junctions have been labelled A, B, C, D, E, F, G and H.
An aerial photographer has taken photographs of the houses in the village. A salesman visits each house to see if the occupants would like to buy a photograph of their house. He needs to travel along each street at least once. He parks his car at A and starts and finishes there. He wishes to minimise the total distance he has to walk.

(a) Describe an appropriate algorithm that can be used to find the minimum distance the salesman needs to walk.

(b) Apply the algorithm and hence find a route that the salesman could take, stating the total distance he has to walk.

(c) A friend offers to drive the salesman to B at the start of the day and collect him from C later in the day.
Explaining your reasoning, carefully determine whether this would increase or decrease the total distance the salesman has to walk. [E]

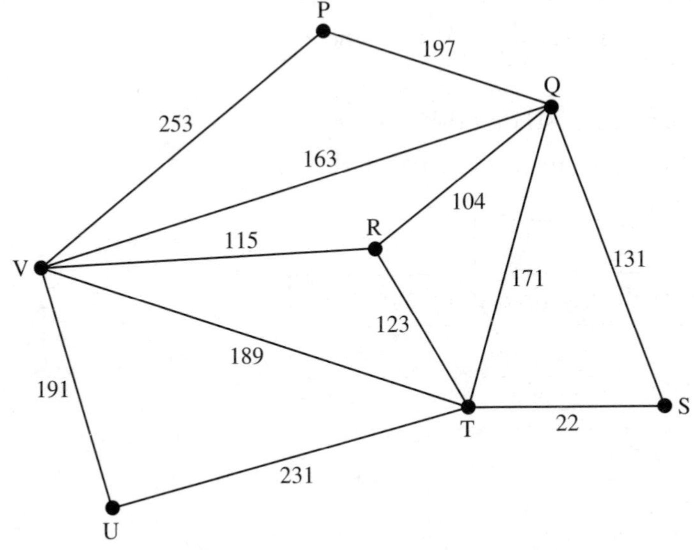

Starting and finishing at P, solve the route inspection (Chinese postman) problem for the network shown above. You must make your method and working clear.

State:

(i) your route, using vertices to describe the arcs

(ii) the total length of your route. [E]

5

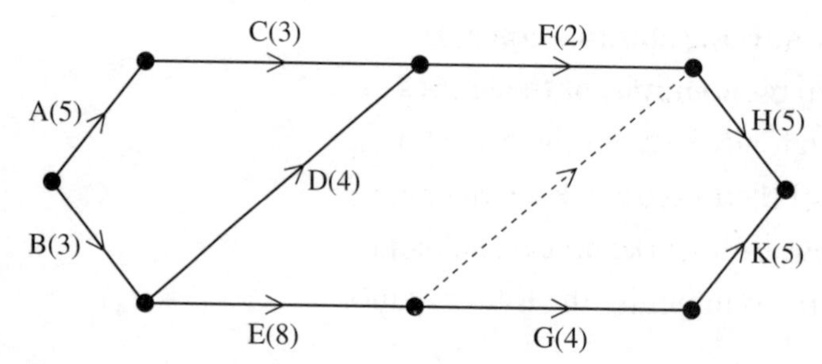

An activity network is shown above. The nodes represent events and the activities A, B, C, D, E, F, G, H and K are represented by arcs. The number in each bracket gives the time, in hours, needed to complete the activity.

(a) Explain what the dotted line means in practical terms.

(b) (i) Calculate the early time and the late time for each event.

(ii) Hence state the critical activities.

(c) Use a time line to schedule the activities. Each activity needs just one worker. Two workers are available. The process is to be completed in the shortest time. You must make clear the order in which the activities are to be completed.

(d) New safety regulations state that activities B, C, D, G and H each need two people working together. Activities A, E, F and K still need just one worker.

Schedule the activities using three workers so that the process is completed in the shortest time. You must make clear the start time and the finish time of each activity.

6

Activity	Time (hours)	Immediately preceding activities
A	10	—
B	2	A
C	1	—
D	4	A, C
E	3	B, D
F	2	B
G	1	E, F

The table above indicates the relationships between certain activities involved in the first stage of a manufacturing process.

(a) Draw an activity network to show this information.

(b) Explain why it is necessary to add some dummies when drawing this activity network.

(c) The second, independent stage of the process is represented by the activity network below. The numbers in brackets indicate the time in hours needed to complete each task.

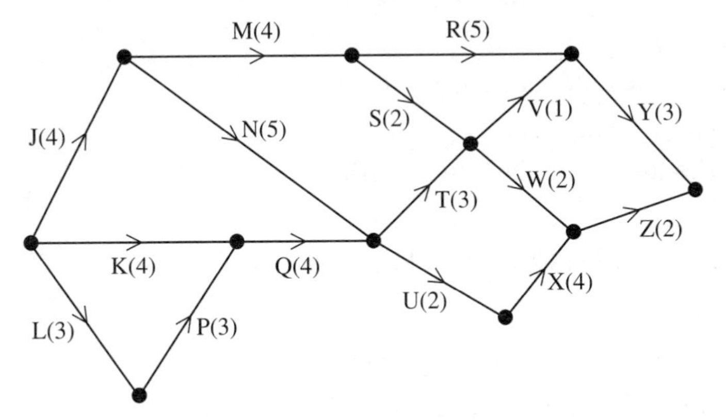

Use the critical path algorithm to determine the activities that lie on the critical path. You must make your method clear.

(d) State the shortest completion time.

(e) Each task requires only one person and the second stage of the process is to be completed in the shortest time by the minimum of workers.

Use a time line to schedule the activities in this second stage of the process. You must make clear the order in which the activities are to be completed by each worker. [E]

7

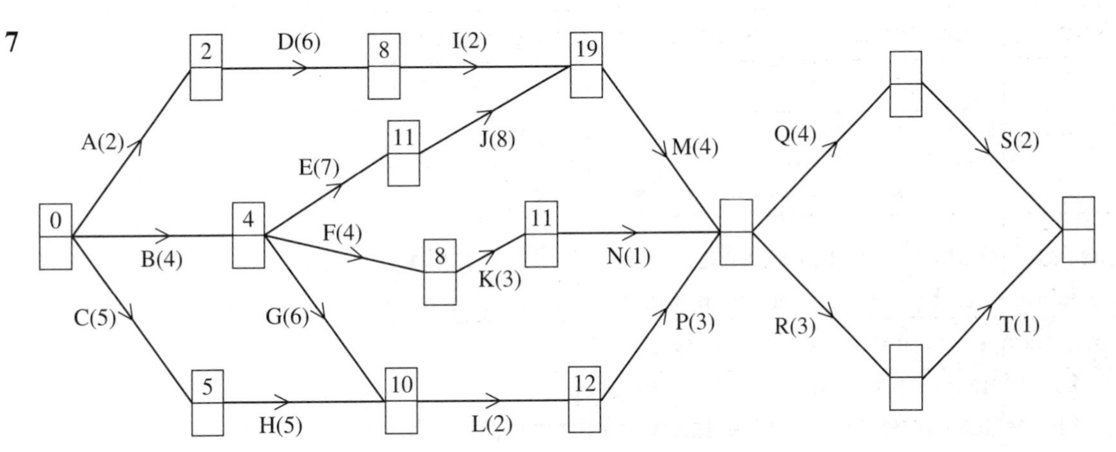

The activity network above shows the activities involved in the production of a magazine. The activities are represented by the

arcs and each number in brackets gives the time, in hours, taken to complete the activity.
At each vertex two boxes have been drawn. The upper box is for the early time and the lower box is for the late time.

(a) Calculate the values of the three remaining early times and the finish time for the project and enter them in the empty upper boxes.
(b) Use a backward scan to find the late time for each event and enter these in the lower box at each vertex.
(c) Hence determine those activities that lie on the critical path and list them in order.
(d) State the minimum length of time needed to produce the magazine. [E]

8

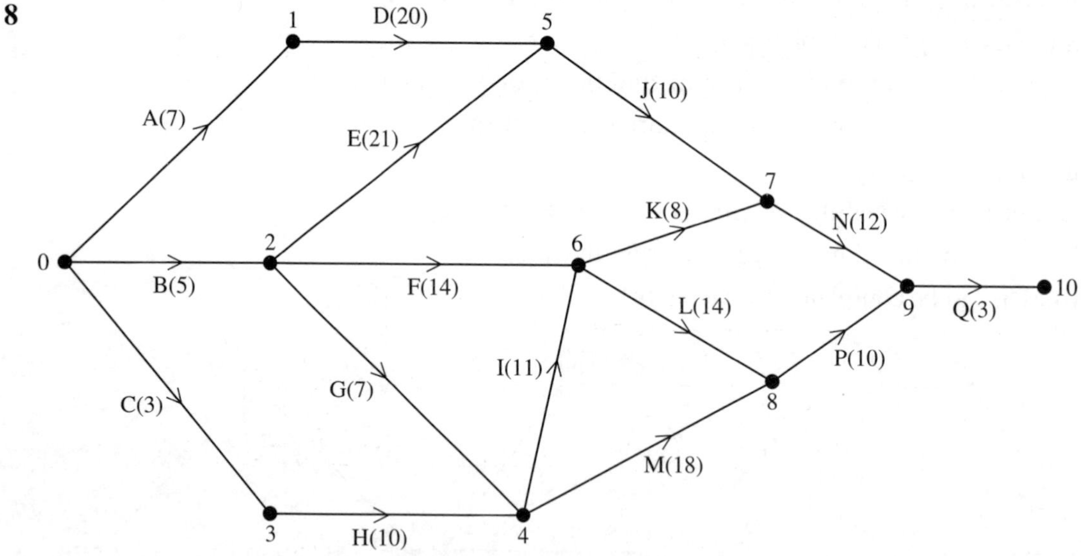

The network above is used to plan a school play. The activities are represented by the arcs and the numbers in brackets give the time, in days, to complete each activity.
(a) Calculate the early and late times for each event.
(b) Hence determine the activities that lie on the critical path.
(c) State the minimum number of days needed to produce the school play. [E]

9 A process is broken down into activities A, B, C, D, E, F, G, H, I, J, K and L. The table below gives the duration of each activity, in minutes, and the preceding activities that must be completed prior to its start.

Activity	A	B	C	D	E	F	G	H	I	J	K	L
Duration (minutes)	11	11	23	5	12	7	9	8	6	9	13	15
Preceding activities	—	A	A	A	B	E	D	G	FCH	I	FCH	JK

(a) Draw an activity network that represents this process. The activities should be represented by arcs, and the vertices, representing the events, should be numbered.

(b) Use critical path analysis, including a forward and backward pass, to determine the length of the critical path and those activities that lie on it.

(c) Calculate the float at C.

(d) As a result of a change in the process, activity I now must be preceded by only activity H, and activity J preceded by activities F, C and I.

Using a new diagram find the new critical path and its length. [E]

10

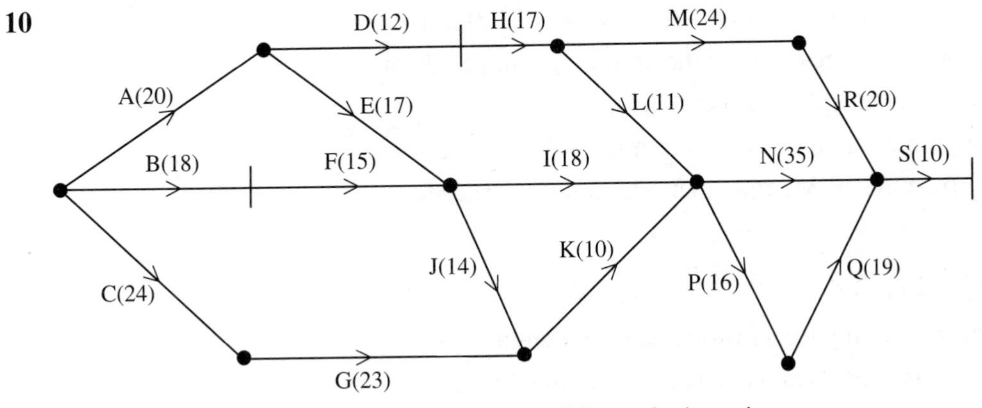

The activity network used to model the building of a boat is shown above. The activities are represented by the arcs. The number in brackets on each arc represents the time, in hours, taken to complete the activity.

(a) Calculate the early and late times for each event.

(b) Hence determine the critical activities.

(c) State the minimum time needed to build the boat.

(d) Verify that the boat cannot be built by only three people in the minimum time.

11 Mr Baker is making cakes and fruit loaves for sale at a charity cake stall. Each cake requires 200 g of flour and 125 g of fruit. Each fruit loaf requires 200 g of flour and 50 g of fruit. He has 2800 g of flour and 1000 g of fruit available.

Let the number of cakes that he makes be x and the number of fruit loaves he makes be y.

(a) Show that these constraints can be modelled by the inequalities

$$x + y \leqslant 14 \quad \text{and} \quad 5x + 2y \leqslant 40.$$

Each cake takes 50 minutes to cook and each fruit loaf takes 30 minutes to cook. There are 8 hours of cooking time available.

(b) Obtain a further inequality, other than $x \geqslant 0$ and $y \geqslant 0$, which models this time constraint.

(c) On graph paper illustrate these three inequalities, indicating clearly the feasible region.

(d) It is decided to sell the cakes for £3.50 each and the fruit loaves for £1.50 each. Assuming that Mr Baker sells all that he makes, write down an expression for the amount of money P, in pounds, raised by the sale of Mr Baker's products.

(e) Explaining your method clearly, determine how many cakes and how many fruit loaves Mr Baker should make in order to maximise P.

(f) Write down the greatest value of P. [E]

12 A junior librarian is setting up a music recording lending section to loan CDs and cassette tapes. He has a budget of £420 to spend on storage units to display these items.

Let x be the number of CD storage units and y the number of cassette storage units he plans to buy.

Each type of storage unit occupies $0.08\,\text{m}^3$, and there is a total area of $6.4\,\text{m}^3$ available for the display.

(a) Show that this information can be modelled by the inequality

$$x + y \leqslant 80$$

The CD storage units cost £6 each and the cassette storage units cost £4.80 each.

(b) Write down a second inequality, other than $x \geqslant 0$ and $y \geqslant 0$, to model this constraint.

The CD storage unit displays 30 CDs and the cassette storage unit displays 20 cassettes. The chief librarian advises the junior librarian that he should plan to display at least half as many cassettes as CDs.

(c) Show that this implies that $3x \leqslant 4y$.

(d) On graph paper, display your three inequalities, indicating clearly the feasible region.

The librarian wishes to maximise the total number of items, T, on display. Given that

$$T = 30x + 20y$$

(e) determine how many CD storage units and how many cassette storage units he should buy, briefly explaining your method. [E]

13 The headteacher of a school needs to hire coaches to transport all the year 7, 8 and 9 pupils to take part in the recording of a children's television programme. There are 408 pupils to be taken and 24 adults will accompany them on the coaches.

The headteacher can hire either 54 seater (large) or 24 seater (small) coaches. She needs at least two adults per coach.

The bus company has only seven large coaches but an ample supply of small coaches.

Let x and y be the number of large and small coaches hired respectively.

(a) Show that the situation can be modelled by the three inequalities:

(i) $9x + 4y \geqslant 72$

(ii) $x + y \leqslant 12$

(iii) $x \leqslant 7.$

(b) On graph paper display the three inequalities, indicating clearly the feasible region.

A large coach costs £336 and a small coach costs £252 to hire.

(c) Write down an expression, in terms of x and y, for the total cost of hiring the coaches.

(d) Explain how you would locate the best option for the headteacher, given that she wishes to minimise the total cost.

(e) Find the number of large and small coaches that the headteacher should hire in order to minimise the total cost and calculate this minimum total cost. [E]

14 The graph below was drawn to solve a linear programming problem. The feasible region, R, includes the points on its boundary.

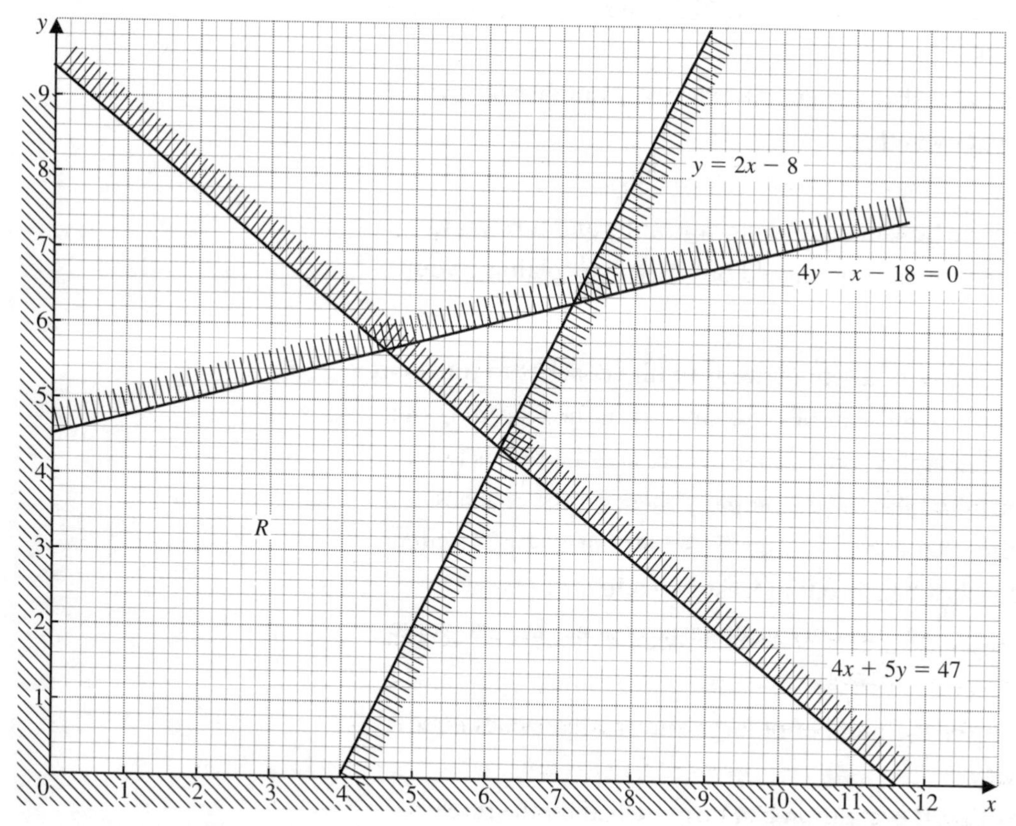

(a) Write down the inequalities that define the region R.

The objective function, P, is given by $P = 3x + 2y$.

(b) Find the value of x and the value of y that lead to the maximum value of P. Make your method clear.

(c) (i) Give an example of a practical linear programming problem in which it would be necessary for the variables to have integer values.
(ii) Given that the solution must have integer values of x and y, find the values that lead to the maximum value of P.

15 In a particular factory three types of product, A, B and C, are made. The numbers of each of the products made are x, y and z respectively and P is the profit in pounds. There are two machines involved in making the products and these have only a limited time available. These time limitations produce two constraints. In the process of using the simplex algorithm the following tableau is obtained, where r and s are slack variables.

P	x	y	z	r	s	
1	$\frac{3}{2}$	0	0	$\frac{3}{4}$	0	840
0	$\frac{1}{3}$	0	1	-8	1	75
0	$\frac{2}{11}$	1	0	$\frac{17}{11}$	0	56

(a) Give the reason why this tableau can be seen to be optimal (final).
(b) By writing out the profit equation, or otherwise, explain why a further increase in profit is not possible under these constraints.
(c) From this tableau deduce:
(i) the maximum profit
(ii) the optimum numbers of types A, B and C that should be produced to maximise the profit. [E]

16 A sweet manufacturer produces packets of orange and lemon flavoured sweets. The manufacturer can produce up to 25 000 orange sweets and up to 36 000 lemon sweets per day.
Small packets contain five orange and five lemon sweets.
Medium packets contain eight orange and six lemon sweets.
Large packets contain 10 orange and 15 lemon sweets.
The manufacturer makes a profit of 14p, 20p and 30p on each of the small, medium and large packets respectively. He wishes to maximise his total daily profit.

Use x, y and z to represent the number of small, medium and large packets respectively, produced each day.

(a) Formulate this information as a linear programming problem, making your objective function and constraints clear. Change any inequalities to equations using r and s as slack variables.

The tableau below is obtained after one complete iteration of the simplex algorithm.

P	x	y	z	r	s	
1	−4	−8	0	0	2	72 000
0	$1\frac{2}{3}$	4	0	1	$-\frac{2}{3}$	1 000
0	$\frac{1}{3}$	$\frac{2}{3}$	1	0	$\frac{1}{15}$	2 400

(b) Start from this tableau and continue the simplex algorithm by increasing y until you have either completed two complete iterations or found an optimal solution.

(c) From your final tableau:

(i) write down the numbers of small, medium and large packets indicated

(ii) write down the profit

(iii) state whether or not this is an optimal solution, giving your reason. [E]

17 Tables are to be bought for a new restaurant. The owners may buy small, medium and large tables that seat two, four and six people respectively.

The owners require at most 20% of the total number of tables to be medium sized. The tables cost £60, £100 and £160 respectively for small, medium and large. The owners have a budget of £2000 for buying tables.

Let the number of small, medium and large tables be x, y and z respectively.

(a) Write down five inequalities implied by the constraints. Simplify these where appropriate.

The owners wish to maximise the total seating capacity, S, of the restaurant.

(b) Write down the objective function for S in terms of x, y and z.

(c) Explain why it is not appropriate to use a graphical method to solve this problem.

(d) It is decided to use the simplex algorithm to solve this problem. Show that a possible initial tableau is:

S	x	y	z	r	t	
1	−2	−4	−6	0	0	0
0	−1	4	−1	1	0	0
0	3	5	8	0	1	100

It is decided to increase z first.

(e) Show that, after one complete iteration, the tableau becomes:

S	x	y	z	r	t	
1	$\frac{1}{4}$	$-\frac{1}{4}$	0	0	$\frac{3}{4}$	75
0	$-\frac{5}{8}$	$4\frac{5}{8}$	0	1	$\frac{1}{8}$	$12\frac{1}{2}$
0	$\frac{3}{8}$	$\frac{5}{8}$	1	0	$\frac{1}{8}$	$12\frac{1}{2}$

(f) Perform one further complete iteration.

(g) Explain how you can decide if your tableau is now final.

(h) Find the number of each type of table the restaurant should buy and their total cost. [E]

18 Kuddly Pals Co. Ltd make two types of soft toy: bears and cats. The quantity of material needed and the time taken to make each type of toy is given in the table below.

Toy	Material (m^2)	Time (minutes)
Bear	0.05	12
Cat	0.08	8

Each day the company can process up to 20 m^2 of material and there are 48 worker hours available to assemble the toys.

Let x be the number of bears made and y the number of cats made each day.

(a) Show that this situation can be modelled by the inequalities:

$$5x + 8y \leqslant 2000$$
$$3x + 2y \leqslant 720$$

in addition to $x \leqslant 0$, $y \leqslant 0$.

The profit made on each bear is £1.50 and on each cat £1.75. Kuddly Pals Co. Ltd wishes to maximise its daily profit.

(b) Set up an initial simplex tableau for this problem.

(c) Solve the problem using the simplex algorithm.

(d) A graphical representation of the feasible region is shown below.

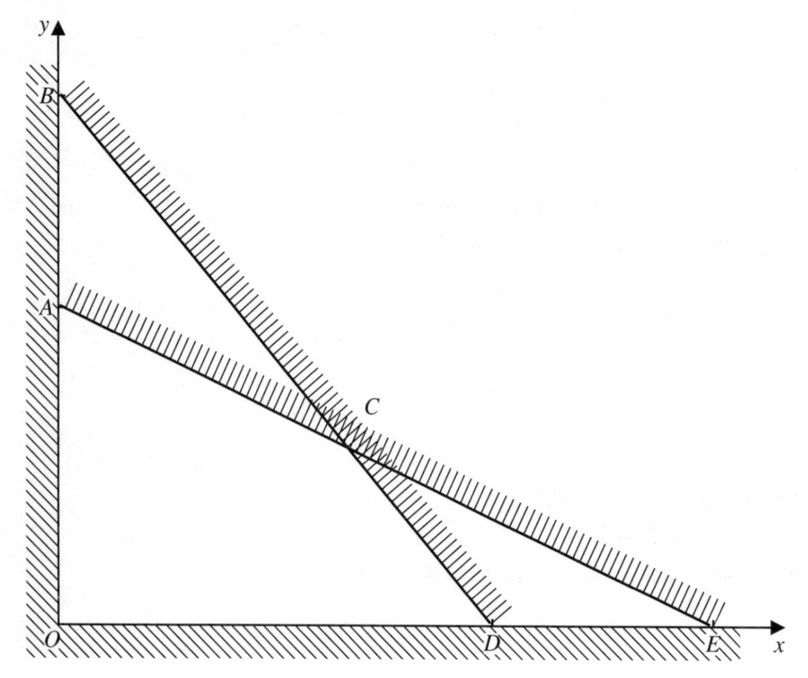

(e) Relate each stage of the simplex tableau to the corresponding point in the graph. [E]

19 A clocksmith makes three types of luxury wristwatch. The mechanism for each watch is assembled by hand by a skilled watchmaker and then the complete watch is formed, weatherproofed and packaged for sale by a fitter.

The table shows the times, in minutes, for each stage of the process.

Watch type	Watchmaker	Fitter
A	54	60
B	72	36
C	36	48

The watchmaker works for a maximum of 30 hours per week and the fitter for a maximum of 25 hours per week.

Let the number of type A, B and C watches made per week be x, y and z.

(a) Show that the above information leads to the two inequalities

$$3x + 4y + 2z \leqslant 100$$
$$5x + 3y + 4z \leqslant 125$$

The profits made on type A, B and C watches are £12, £24 and £20 respectively.

(b) Write down an expression for the profit, P, in pounds, in terms of x, y and z.

The watchmaker wishes to maximise his weekly profit. It is decided to use the simplex algorithm to solve this problem.

(c) Write down the initial tableau using r and s as the slack variables.

(d) Increasing y first, show that after two complete iterations of the simplex algorithm the tableau becomes

P	x	y	z	r	s	
1	14.8	0	0	3.6	3.2	760
0	0.2	1	0	0.4	−0.2	15
0	1.1	0	1	−0.3	0.4	20

(e) Give a reason why this tableau is optimal (final).

(f) Write down the numbers of each type of watch that should be made to maximise the profit. State the maximum profit. [E]

20 A craftworker makes three types of wooden animals for sale in wildlife parks. Each animal has to be carved and then sanded. Each lion takes 2 hours to carve and 25 minutes to sand. Each giraffe takes $2\frac{1}{2}$ hours to carve and 20 minutes to sand. Each elephant takes $1\frac{1}{2}$ hours to carve and 30 minutes to sand. Each day the craftworker wishes to spend at most 8 hours carving and at most 2 hours sanding.

Let x be the number of lions, y the number of giraffes and z the number of elephants he produces each day.

The craftworker makes a profit of £14 on each lion, £12 on each giraffe and £13 on each elephant. He wishes to maximise his profit, P.

(a) Model this as a linear programming problem, simplifying your expressions so that they have integer coefficients.

It is decided to use the simplex algorithm to solve this problem.

(b) Explaining the purpose of r and s, show that the initial tableau can be written as:

P	x	y	z	r	s	
1	−14	−12	−13	0	0	0
0	4	5	3	1	0	16
0	5	4	6	0	1	24

(c) Choosing to increase x first, work out the next complete tableau, where the x column includes two zeros.

(d) Explain what this first iteration means in practical terms. [E]

Matchings

7

7.1 Modelling using a bipartite graph

In this chapter you will consider a class of problems that can be modelled using a bipartite graph. You will recall from Chapter 2 that a **bipartite graph** G consists of two sets of vertices X and Y. The edges in G only join vertices in X to vertices in Y and not vertices within a set. An example of a bipartite graph is shown below.

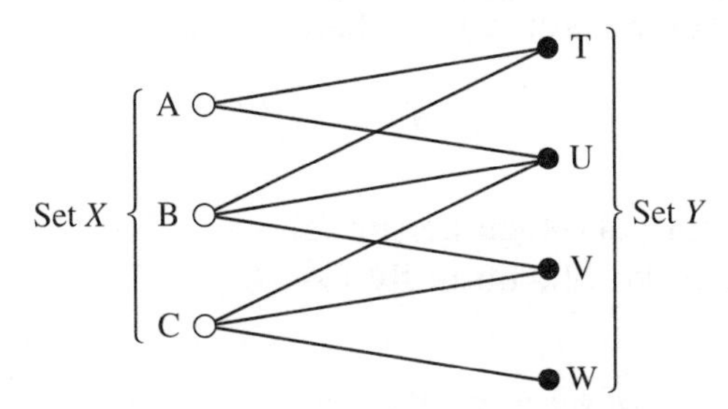

The two vertex sets are $X = \{A, B, C\}$ and $Y = \{T, U, V, W\}$. The edge set E only includes edges joining sets X and Y.

It is often useful to use a different symbol for the vertices in sets X and Y. Here we have used an open circle for vertices in set X and a solid circle for vertices in set Y.

An example of a situation that can be modelled in this way is that of allocating people to jobs for which they are qualified. The set X then represents people and the set Y represents jobs. Each person is qualified to do some of the jobs. The edges represent 'being qualified'.

Example 1
A school has five teachers available to teach mathematics (M), statistics (S), physics (P) and economics (E). The subjects each teacher is qualified to teach are shown in the table at the top of page 196.

Miss Ahmed (A)	Physics
Mr Brown (B)	Mathematics, statistics and physics
Mrs Croft (C)	Statistics and economics
Ms Dugal (D)	Mathematics, statistics and economics
Dr French (F)	Mathematics and physics

Show this information on a bipartite graph.

This information can be modelled using the bipartite graph shown opposite, where set X represents teachers, set Y represents subjects and an edge represents 'qualified to teach.'

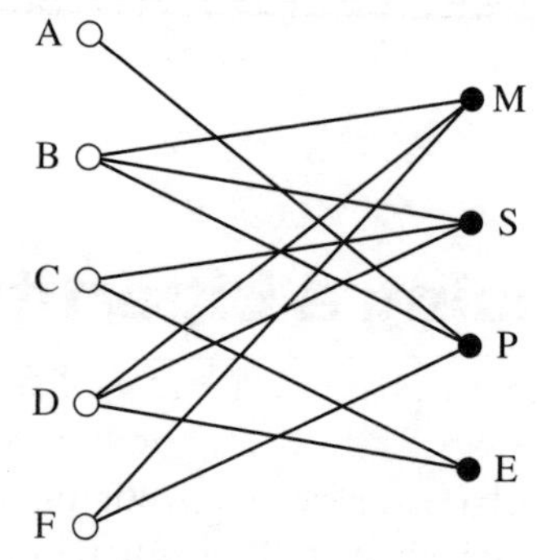

If one teacher is to be allocated to each subject it is immediately clear that, as there are five vertices in set X and only four vertices in set Y, one teacher will not be used.

The examples that follow illustrate the various ways in which data may be presented.

Example 2

A school requires a team of four to swim in a gala. One person is required for each of breaststroke (1), backstroke (2), butterfly (3) and freestyle (4).

Five girls were asked to indicate on a form which of these events they would be willing to take part in. The results are summarised below.

	Breaststroke (1)	Backstroke (2)	Butterfly (3)	Freestyle (4)
Mina		×		×
Shamina		×	×	
Rita	×	×		×
Elizabeth	×		×	
Nicole			×	

× indicates 'willing to swim this event'.

(a) Draw a bipartite graph to show 'willing to swim each event'.
(b) Draw a bipartite graph to show 'unwilling to swim each event'.

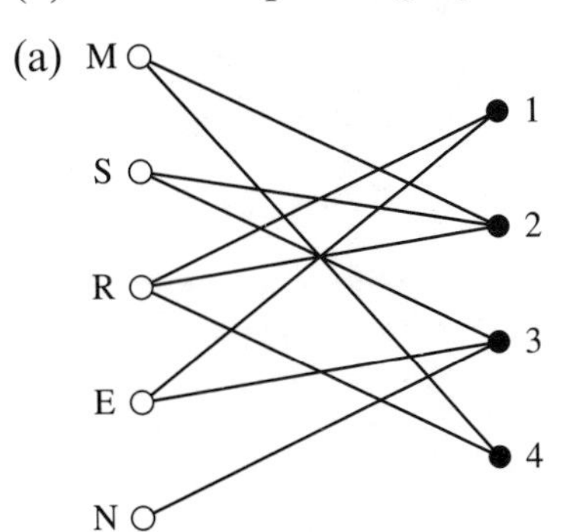

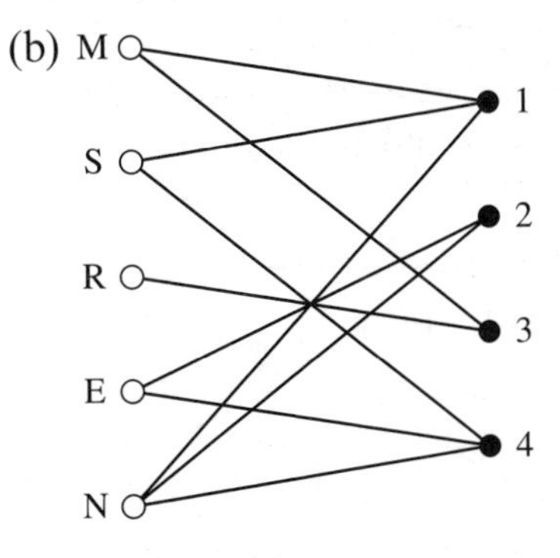

Example 3
Avril, Calleen, Diane and Eva live in the same row of houses. Their houses are to be painted. The choice of colour for the front doors is red, green, yellow or blue. There must be one door of each colour. The girls' colour preferences were given in the following way:

Avril	not red or blue
Calleen	green or blue
Diane	any colour except green
Eva	not green or yellow

Express these preferences as a bipartite graph.

It is a good idea to rewrite the preferences so that all the information is positive:

Avril	green or yellow
Calleen	green or blue
Diane	red, yellow or blue
Eva	red or blue

The bipartite graph showing this information is then:

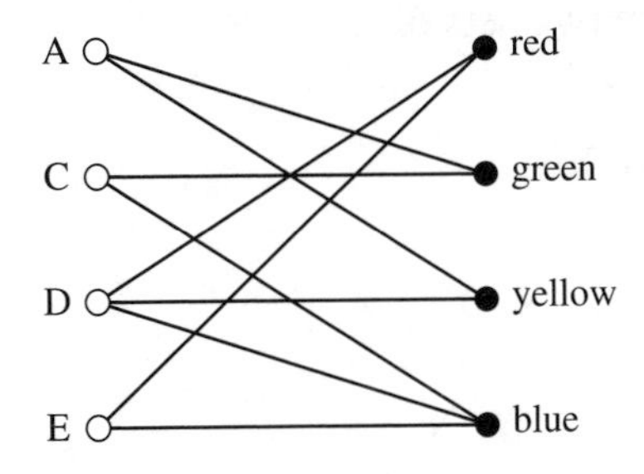

Example 4
Five gardeners, Will (x_1), Nadia (x_2), Fred (x_3), Daniel (x_4) and Clare (x_5), are questioned as to their specialities amongst dahlias (y_1), sweet peas (y_2), gladioli (y_3), roses (y_4) and lilies (y_5).

A garden centre wants each gardener to supply one of their specialities. The answers obtained are summarised below.

x_1	x_2	x_3	x_4	x_5
y_1 y_3	y_1 y_3	y_1 y_3 y_5	y_2 y_4 y_5	y_3 y_5

Show this information in a bipartite graph.

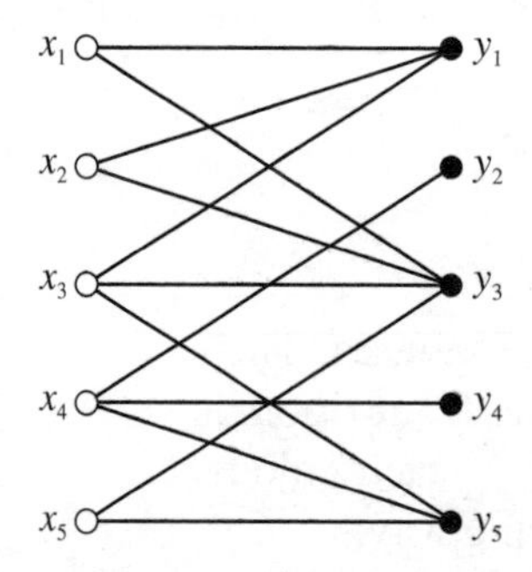

Exercise 7A

1 A graph G is represented by the following adjacency matrix:

	A	B	C	D	E	F
A	0	0	0	1	0	1
B	0	0	1	0	1	0
C	0	1	0	1	0	1
D	1	0	1	0	1	0
E	0	1	0	1	0	0
F	1	0	1	0	0	0

Show that this is a bipartite graph. Identify the two sets of vertices X and Y.

2 Which of the following describe a bipartite graph?

(a)

	A	B	C	D
A	0	1	1	1
B	1	0	1	1
C	1	1	0	0
D	1	1	0	0

(b) Vertex set {A, B, C, D, E}
Edge set {(A, B), (C, D), (D, B), (A, C), (E, A)}

3 Five people in an office were asked which of five available jobs they would like to do. Their replies are summarised in the table.

Person	Jobs
1	A, C
2	B, D, E
3	C, D
4	A, C
5	A, C, E

Show this information in a bipartite graph.

4 Mr Dixon, a gardener, has four plots of land available to grow potatoes (P), carrots (C), beans (B) and kale (K). The soil in each of these plots is of a different type and so the plots can only be used for certain vegetables. Which plot can be used for which vegetables is shown in the table.

	Potatoes (P)	Carrots (C)	Beans (B)	Kale (K)
Plot 1	×	×		
Plot 2			×	×
Plot 3	×		×	
Plot 4		×	×	×

Represent this information as a bipartite graph.

5 Five mathematics teachers (1, 2, 3, 4 and 5) are available to teach the sixth form. The areas to be taught are pure maths (P), mechanics (M), statistics (S) and decision maths (D). The teachers are asked to state their subject preferences. The table summarises the information obtained:

1	2	3	4	5
P S	M D	M	P S	S

Represent this information as a bipartite graph.

6 Four classes, C1, C2, C3 and C4, are to be timetabled into four rooms, R1, R2, R3 and R4. Room facilities, such as size and computing facilities, produce the following restrictions:

C1 must be in R4
C2 can be in any room
C3 cannot be in R1 or R4
C4 cannot be in R3 or R4

Summarise this information in a table and then model it using a bipartite graph.

7.2 Matchings

Having modelled a situation using a bipartite graph, we now look at the possibility of pairing some or all of the vertices of set X, in a one-to-one way, with vertices of the second set Y. This is called a **matching**. In Example 1 the pairing of a teacher with a subject he or she is qualified to teach is a matching.

- **A matching in a bipartite graph, G, is a subset M of the edges E of the graph such that no two edges in M have a common vertex.**

A matching is usually shown by marking the edges involved in a distinctive way. In this book thickened lines will be used, but in your work you can use coloured pens or pencils.

Example 5

Show four possible matchings in the bipartite graph obtained in Example 1.

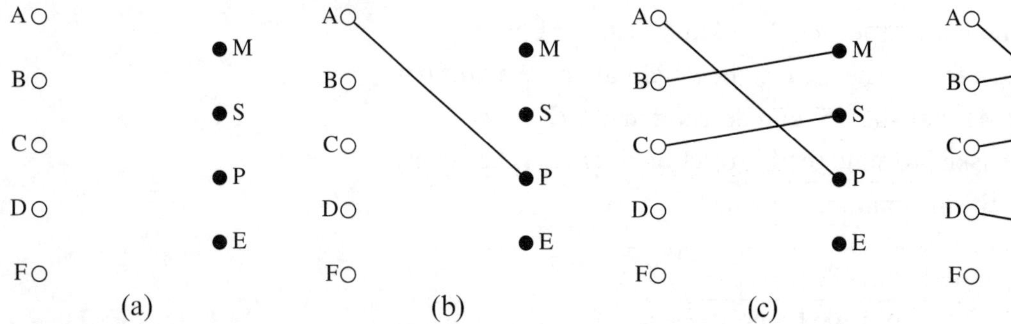

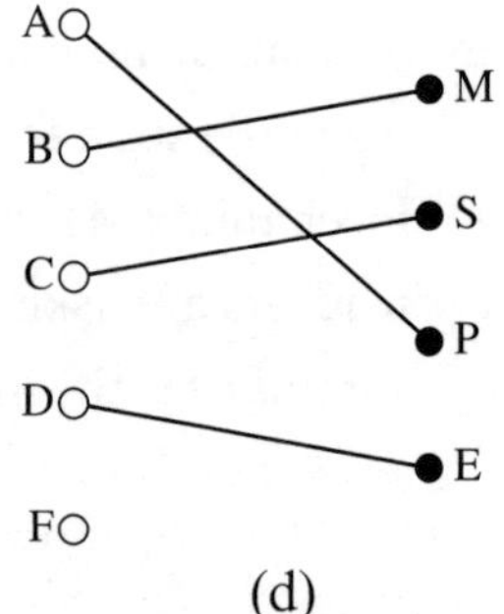

Answer (a) may seem odd as it has no edges. However, it is still a matching according to the above definition. It is often called the **trivial matching**. All the other matchings shown are non-trivial.

On the other hand, answer (d) is a matching with four edges. This is the largest number of edges that a matching can have in this bipartite graph as there are only four vertices in the set Y (i.e. four subjects to be taught).

- **A maximal matching M is a matching in which the number of edges is as large as possible.**

Suppose the first vertex set X in a bipartite graph has m vertices and the second vertex set Y has n vertices. The number of edges in a maximal matching cannot exceed the *smaller* of m and n.

For instance, in Example 1 $m = 5$, that is there are five teachers available, and $n = 4$, that is there are four subjects to be taught. The number of edges in a maximal matching therefore cannot exceed four. In fact answer (d) in Example 5 shows that a maximal matching with four edges is possible.

However, it should be noted that there may not be a maximal matching with the number of edges equal to the smaller of m and n. Example 4 has five vertices in set X and five vertices in set Y, but, as we shall see later, a maximal matching in this case contains only four edges.

- **In a bipartite graph with n vertices in each set a complete matching M is a matching in which the number of edges is also n.**

This means that in such a bipartite graph a complete matching pairs every vertex in X with a vertex in Y.

Example 6

Indicate a complete matching in the bipartite graph obtained in Example 3 and explain how you found it.

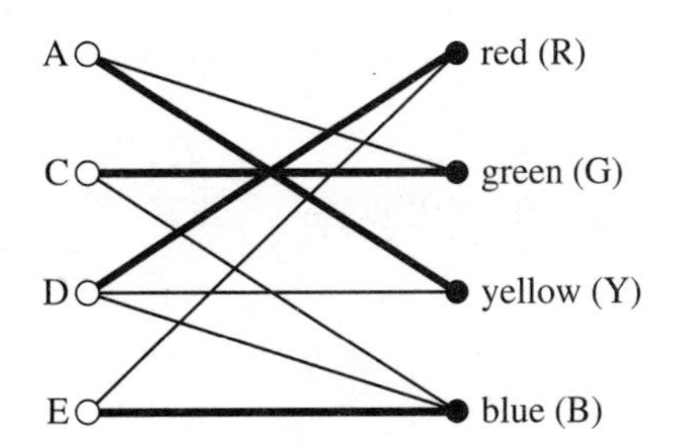

Begin by pairing E with B.
Then C *must* be paired with G (B already paired).
A *must* be paired with Y (G already paired).
Finally D *must* be paired with R (Y and B already paired).

This is not the only complete matching for this bipartite graph (see Exercise 7B, question 5).

Exercise 7B

1 Find a maximal matching in the following bipartite graph.

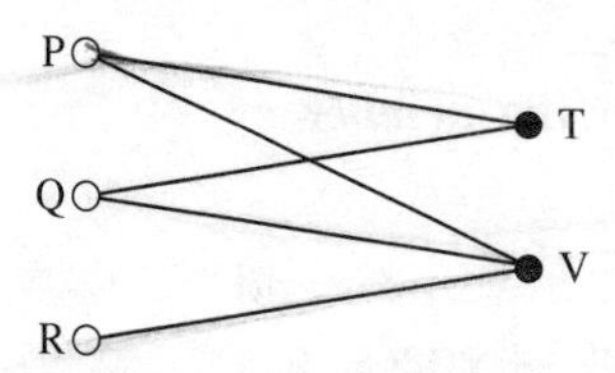

How many different maximal matchings are there?

2 The bipartite graph obtained in Example 2 is shown below.

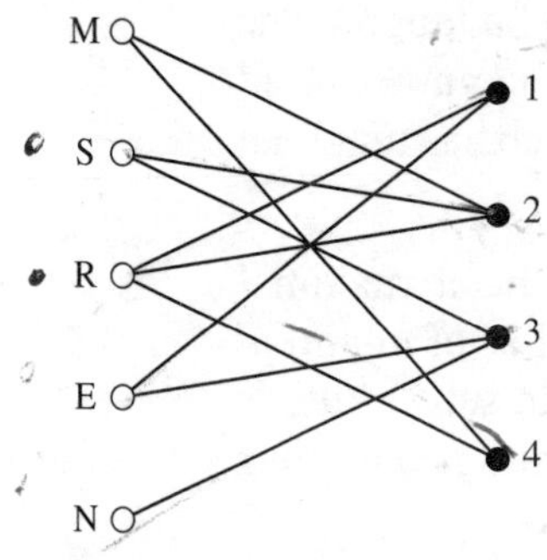

Obtain a maximal matching involving

(a) S, R, E and N

(b) S, M, E and N.

Explain your reasoning.

3 Find a complete matching for the following bipartite graph and explain why there is only one complete matching.

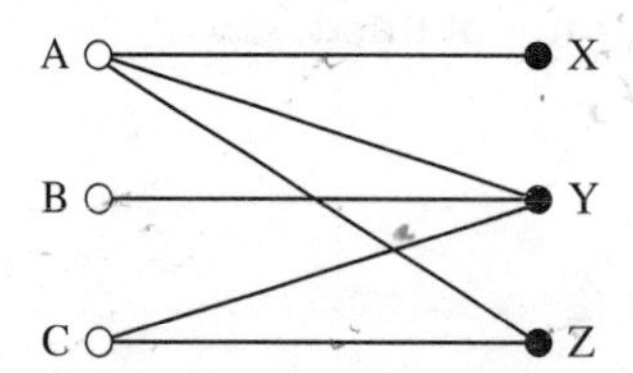

4 Show that there is only one complete matching for the following bipartite graph.

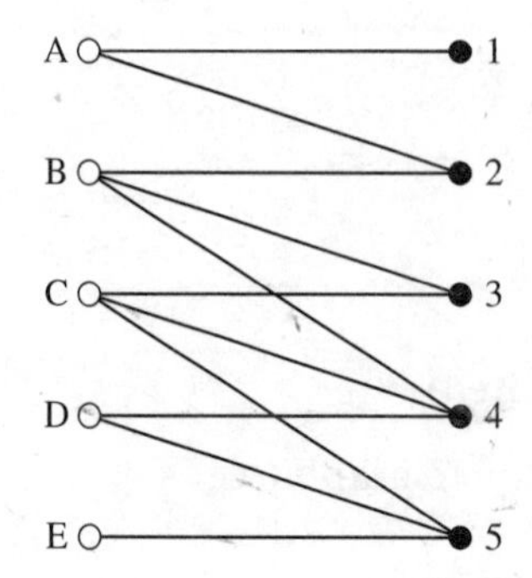

5 In Example 6 a complete matching was obtained for the following bipartite graph.

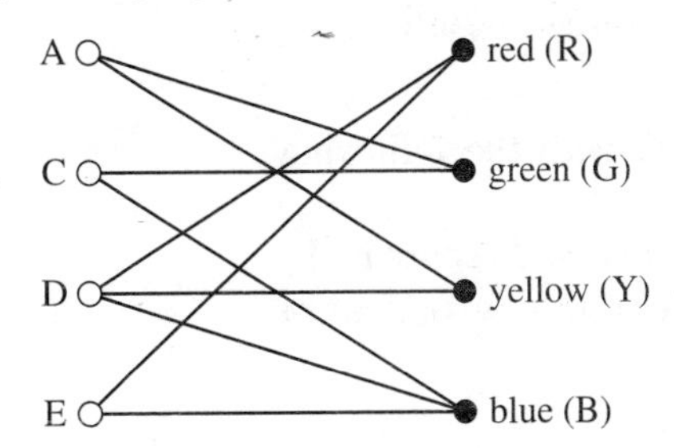

Find two other complete matchings.

6

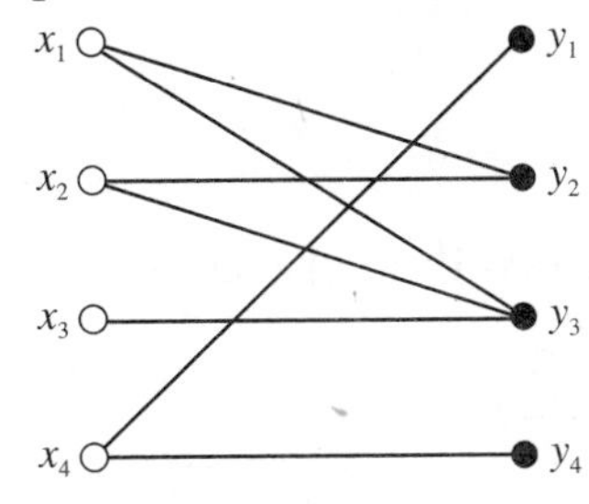

Consider the above bipartite graph.

(a) Explain why a complete matching is not possible.

(b) Obtain a maximal matching containing three edges.

(c) Show that it is not possible to obtain a maximal matching in which x_1, x_2 and x_3 are all paired.

7.3 Improving a matching to obtain a maximal matching

In situations that can be modelled by bipartite graphs the usual problem is to pair as many vertices as possible in set X with vertices in set Y. If, for example, set X represents people and set Y represents jobs, then we want to determine how to assign people to jobs so that the maximum number of people get jobs for which they are qualified.

In the language used in section 7.2, we are looking for a maximum matching. It is usually quite easy to obtain a matching with a given number of edges. However, we need a systematic way of

(i) improving a given matching, if this is possible
(ii) deciding if a given matching is maximal.

Before discussing an algorithm for achieving these aims it is necessary to introduce a special kind of path in a bipartite graph which is a fundamental part of this algorithm. This path is called an **alternating path**.

Suppose we have a bipartite graph G and an initial matching M consisting of certain edges, then:

- **An alternating path for M in G is a path in G with the following properties:**
  (i) it joins an *unmatched vertex* in X to an *unmatched vertex* in Y
  (ii) it is such that the edges in the path are alternately *in* and *not in* the matching M.

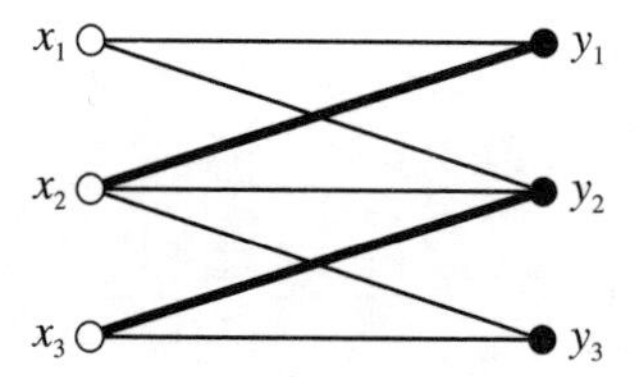

For the graph G above and the matching M shown with thickened lines, an alternating path is:

$$x_1 \text{——} y_1 \text{▬▬} x_2 \text{——} y_2 \text{▬▬} x_3 \text{——} y_3$$

since

x_1 and y_3 are not matched in M
edges (x_1, y_1), (x_2, y_2) and (x_3, y_3) are *not in* M
edges (y_1, x_2) and (y_2, x_3) are *in* M

Example 7

Obtain two alternating paths for M in G when G is the bipartite graph in Example 1 and M is the matching shown in Example 5, answer (c).

The diagram below shows the graph G with the matching M shown by thickened lines.

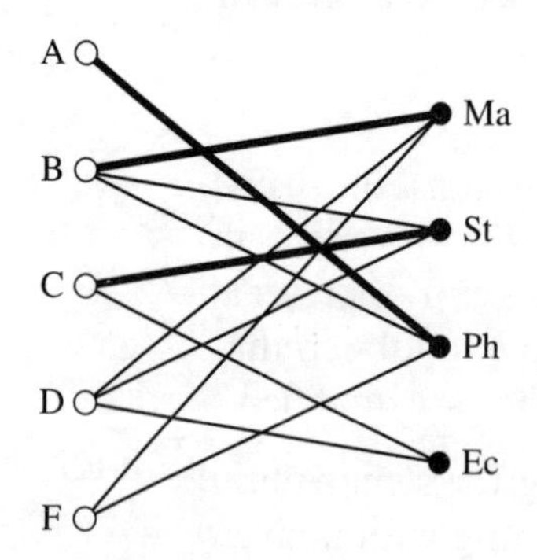

(a) The vertex F is unmatched in M.
Take first edge as F —— Ma.
Ma is matched with B (second edge), Ma ▬▬ B.
Take third edge as B —— St.
St is matched with C (fourth edge), St ▬▬ C.
Take fifth edge as C —— Ec.

As Ec is unmatched we now have an alternating path:

F —— Ma ▬▬ B —— St ▬▬ C —— Ec

When we reach an unmatched vertex in Y, such as Ec, we are said to have achieved **breakthrough**.

(b) A rather easier alternating path for M in G is

D —— Ec (breakthrough)

This path has only one edge but satisfies all the conditions for an alternating path.

Finding alternating paths

In Example 3 we obtained the following bipartite graph:

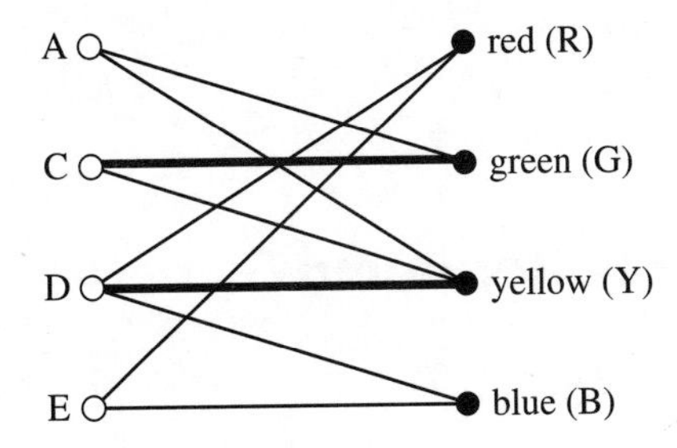

Take as an initial matching C with G and D with Y, as shown on the figure by the thickened lines.

To find an alternating path we start at vertex A, which is unmatched in set X. There is no reason why you should not start with *any unmatched vertex* in set Y.

The edges that have A as a vertex are AG and AY.

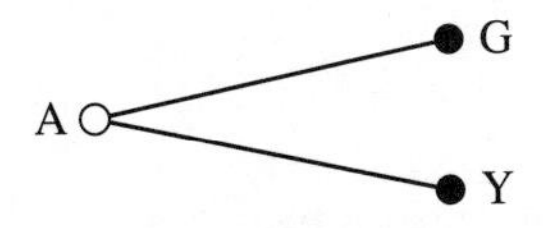

For each vertex reached so far there is an edge in the current matching, GC and YD respectively.

You can now draw in these edges, using thick lines.

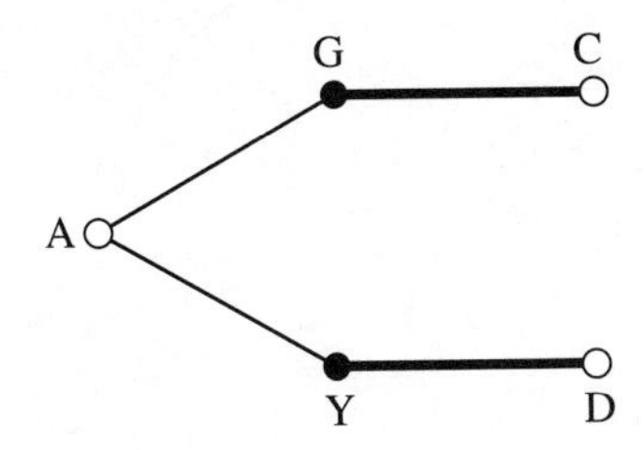

We continue by adding unused edges at C and D as long as they do not lead to vertices already included in the tree. We obtain:

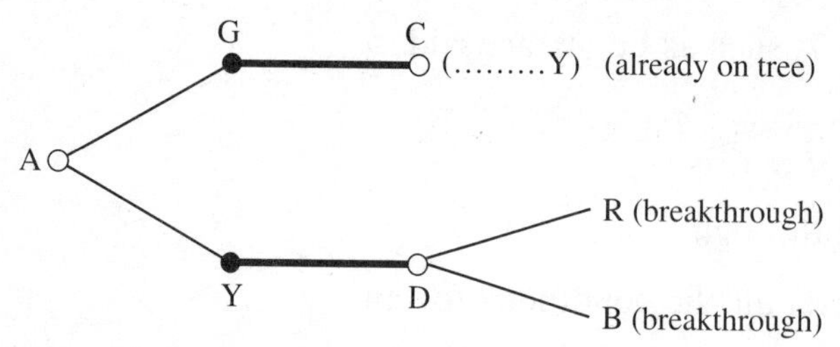

Neither R nor B is matched in M and so we have breakthrough in both cases and we may choose either.

(a) If we choose R then we have the alternating path

A —— Y ━━ D —— R

(b) If we choose B then we have the alternating path

A —— Y ━━ D —— B

Changing the status of edges on an alternating path

It is clear from the definition given above that:

- **on an alternating path the number of edges not in M is one more than the number of edges in M.**

Therefore, if we change (*all edges on the alternating path not in M*) to (*edges in a new matching M'*) and (*all edges on the alternating path in M*) to (*edges not in M'*), i.e.

○——● (not in M) → ○━━● (in M')

and

●━━○ (in M) → ●——○ (not in M')

then we can form a new matching M', consisting of the changed edges together with the edges in the existing matching that were not used in the alternating path. This new matching will have one more edge than the old matching M. The process we have just described is called **changing the status of the edges**.

Example 8

Change the status of the edges on the alternating paths found in Example 7 and state the new matching M' obtained in each case.

(a) The alternating path obtained was

F —— Ma ━━ B —— St ━━ C —— Ec

Changing the status of the edges we obtain

F ━━ Ma —— B ━━ St —— C ━━ Ec

The new matching M' is then

F with Ma, B with St, C with Ec, A with Ph (unchanged)

(b) The alternating path obtained was

D —— Ec

Changing the status of the edges we obtain

D ━━ Ec

The new matching M'' is

D with Ec, A with Ph, B with Ma, C with St

The matching improvement algorithm (alternating path algorithm)

The **matching improvement algorithm** improves an existing matching, if this is possible, by first establishing an alternating path between vertices not in the current matching. The status of the edges on this path are then changed to produce the improved matching. If the current matching is maximal then no alternating path will be found.

Example 8 illustrates the use of this algorithm and in both (a) and (b) produces a matching with four edges. Both of the matchings obtained, M' and M'', are therefore **maximal**.

There is no unmatched vertex in set Y and so no further alternating path exists.

The matching improvement algorithm is based on the theorem:

> 'If a matching M in a bipartite graph G is not a maximal matching, then G contains an alternating path for M.'

A formal statement of the matching improvement algorithm is as follows.

Step 1 Start with any non-trivial matching M in G (one edge will do).

Step 2 Search for an alternating path for M in G.

Step 3 If an alternating path is found, construct a better matching M' by changing the status of the edges in the alternating path and return to step 2 with M' replacing M.

Step 4 Stop when no alternating path can be found. The matching obtained is then maximal.

Example 9

Apply the maximum matching algorithm to find a maximal matching for the bipartite graph obtained in Example 4.

Take as an initial matching

$$(x_1, y_1), (x_3, y_3), (x_4, y_4)$$

This is shown on the bipartite graph:

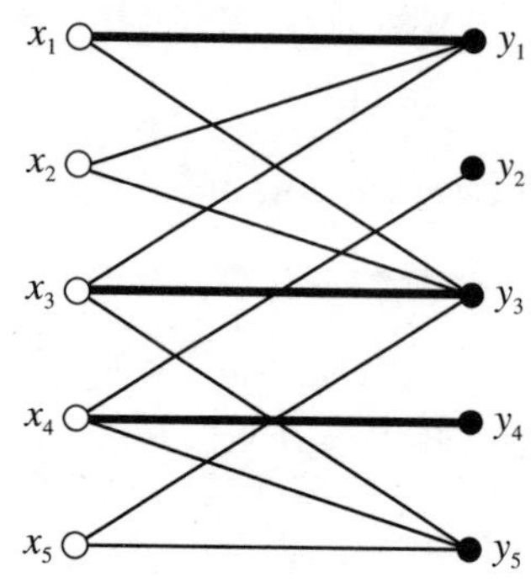

As x_2 is unmatched, we can search for an alternating path starting at x_2. Proceeding as described above we obtain:

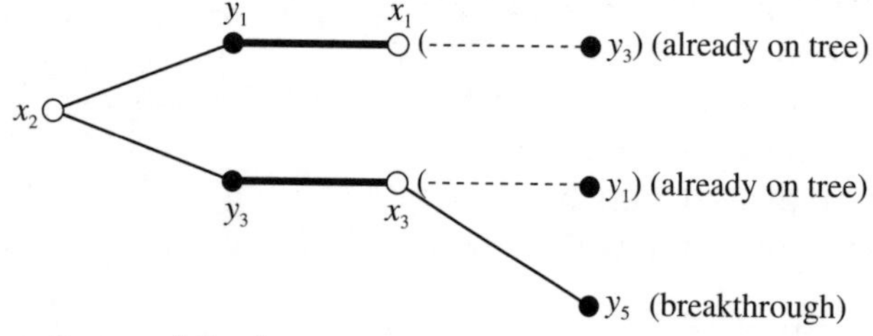

An alternating path is then

x_2 y_3 x_3 y_5

Changing the status of the edges gives

x_2 y_3 x_3 y_5

M', the improved matching, is therefore obtained by:

removing (x_3, y_3)
adding (x_2, y_3) and (x_3, y_5)

The matching is then (x_1, y_1), (x_4, y_4), (x_2, y_3) and (x_3, y_5). This is shown below.

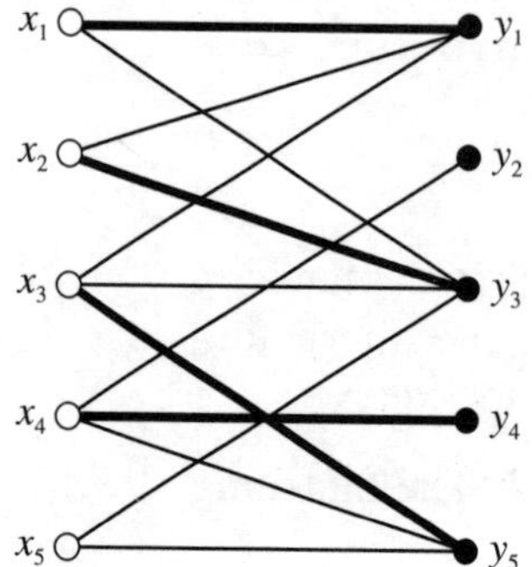

As x_5 is still unmatched, we search for an alternating path starting at x_5. Proceeding as described above we obtain:

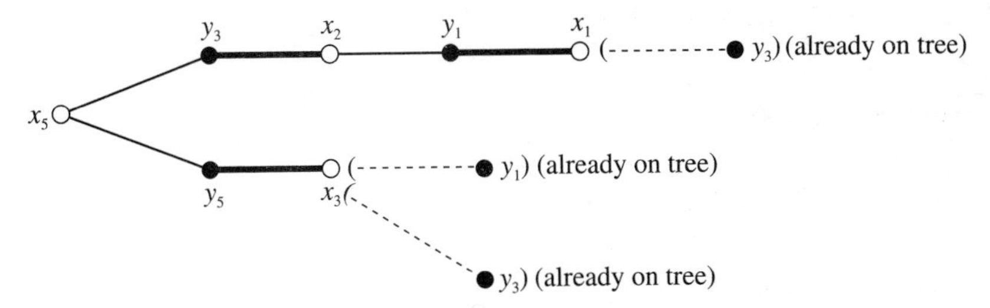

It is therefore not possible to obtain an alternating path. A maximal matching contains just four edges. The matching M' shown above is one such maximal matching.

We can actually see from the original table that a complete matching will not be possible since both y_2 and y_4 are only listed below x_4.

Exercise 7C

1 Consider the following bipartite graph:

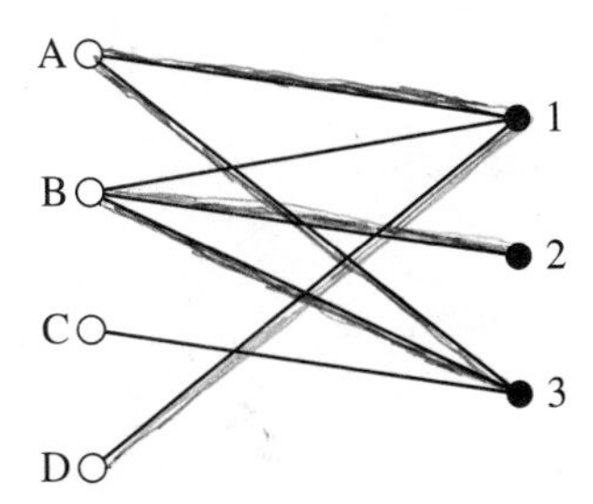

Taking as the initial matching (A paired with 1) and (B paired with 3), obtain two maximal matchings by using the matching improvement algorithm.

2 Consider the following bipartite graph:

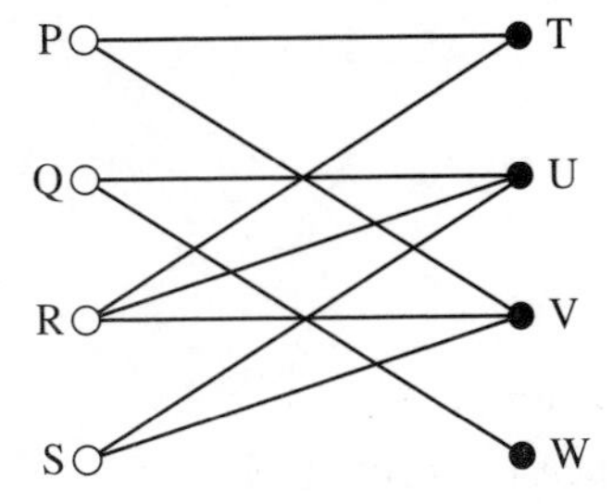

Taking as your initial matching edges PV and QU obtain a complete matching. State clearly the alternating paths used in applying the matching improvement algorithm.

3

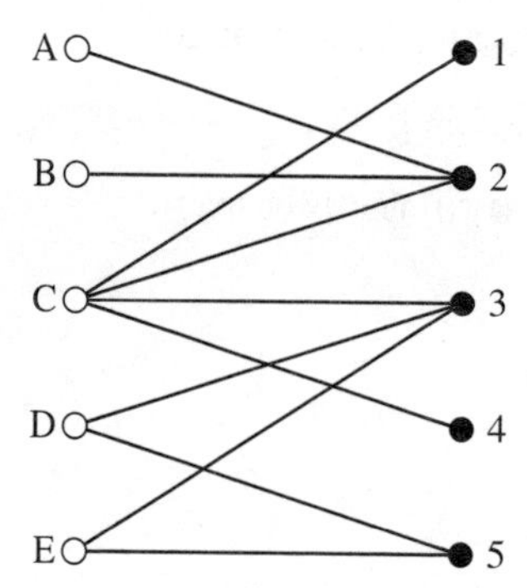

Show that there is no complete matching for the above graph.

4 In question 3 of Exercise 7A the following table of information was given, summarising people's preferences for jobs.

Person	Jobs
1	A, C
2	B, D, E
3	C, D
4	A, C
5	A, C, E

(a) Draw a bipartite graph summarising this information.

(b) Write down a matching containing four edges.

(c) Apply the matching improvement algorithm to obtain a complete matching.

5

	Potatoes (P)	Carrots (C)	Beans (B)	Kale (K)
Plot 1	×	×		
Plot 2			×	×
Plot 3	×		×	
Plot 4		×	×	×

The above table of information was given in question 4 of Exercise 7A. Represent this information as a bipartite graph and obtain a maximal matching using the matching improvement algorithm. State clearly your initial matching and any alternating paths used.

6 The problem of allocating mathematics teachers to areas is considered in question 5 of Exercise 7A. The teachers' preferences are summarised in the table below:

1	2	3	4	5
P S	M D	M	P S	S

As an initial matching, teachers 1, 2 and 5 choose the first area on their lists.

(a) Apply the matching improvement algorithm to obtain a maximal matching.

(b) Find another maximal matching.

7 In question 6 of Exercise 7A you obtained a bipartite graph modelling the information given there on classes and rooms. Show that there are three different complete matchings possible.

8 A store manager wishes to appoint supervisors to four of his departments: electrical, furniture, stationery and clothing. He has four candidates for these positions. The table shows which departments each candidate is qualified to supervise.

Amy	Stationery, clothing
Bhavana	Electrical, furniture, stationery
Tina	Furniture, stationery, clothing
Dylan	Electrical, stationery

(a) Model this situation with a bipartite graph.

The manager allocates Amy, Tina and Dylan to the first department on their individual lists.

(b) Starting from this matching show clearly how a matching in which each candidate is assigned to a department can be obtained.

SUMMARY OF KEY POINTS

1. A **bipartite graph** consists of two sets of vertices X and Y. The edges only join vertices in X to vertices in Y, not vertices within a set.
2. A **matching** in a bipartite graph is a subset M of the edges of the graph G such that no two edges in M have a common vertex.
3. A **maximal matching** M is a matching in which the number of edges is as large as possible.
4. In a bipartite graph with n vertices in each set a **complete matching** M is a matching in which the number of edges is also n.
5. An **alternating path** for a matching M in G is a path in G with the following properties:
   (i) it joins an unmatched vertex in X to an unmatched vertex in Y
   (ii) it is such that the edges in the path are alternately **in** and **not in** the matching M.
6. **The matching improvement algorithm**
   Step 1 Start with any non-trivial matching M in G.
   Step 2 Search for an alternating path for M in G.
   Step 3 If an alternating path is found, construct a better matching M' by changing the status of the edges in the alternating path and return to step 2 with M' replacing M.
   Step 4 Stop when no alternating path can be found. The matching obtained is maximal.

Flows in networks 8

8.1 Sources, sinks and flows

In Chapter 2 we defined a **digraph** as a graph with directed arcs. In this chapter we consider situations that can be modelled by a weighted digraph in which the weights are the **capacities** of these arcs. The arcs may then be thought of as pipelines along which some commodity – water, oil, cars – can flow. The capacity of an arc is then the practical limit to the amount of the commodity that can flow along that arc. We should really call such a model a 'capacitated network' but for simplicity in this chapter we will call it a network.

An obvious example of a situation that can be modelled in this way is a system of pipelines carrying water or oil, however it could also be a system of one-way streets along which cars can drive or an electrical distribution system. Examples of systems that can be modelled in this way will be found in the exercises.

- **A vertex S is called a *source* if all arcs containing S are directed *away from S*:**

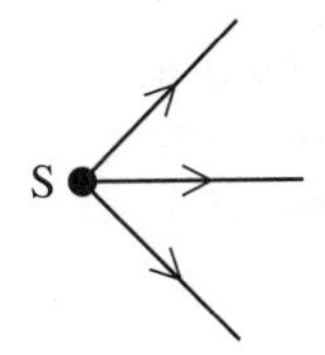

- **A vertex T is called a *sink* if all arcs containing T are *directed towards T*.**

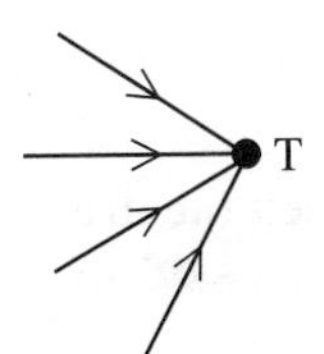

We will concentrate, at least initially, on what is sometimes called the basic flow problem. In this we assume that there are **no capacity restrictions at the vertices** and there is **only one source and one sink**. The general case of several sources and sinks is easily reduced to this special case and will be considered in the last section of this chapter.

Here is a typical network representing a system of one-way streets:

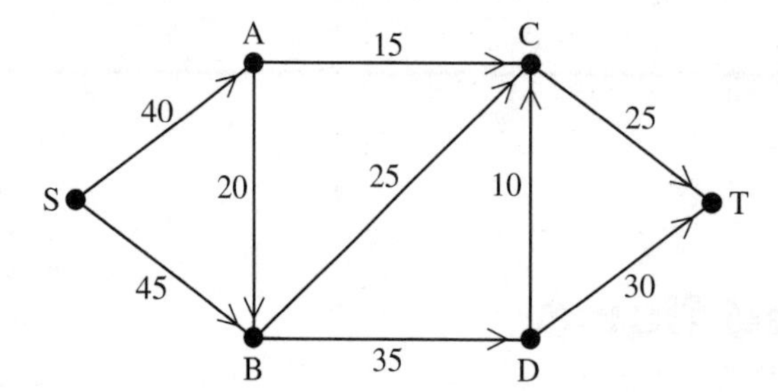

The number next to each arc is the maximum flow of traffic along that street, in vehicles per hour, which is the capacity. S is the source and T is the sink.

The question we wish to answer is 'what is the maximum amount of the commodity that can pass through the network from source to sink in a given time?'

Flows

- **A *flow* in a network N with a single source S and a single sink T is an assignment to each arc e of N of a non-negative number (which could be zero) called the *flow along the arc e*.**

A flow must satisfy two conditions:

1. the **feasibility condition**, which simply says that the flow along each arc **cannot exceed the capacity of the arc**
2. the **conservation condition**, which expresses the fact that nothing is allowed to build up at an **intermediate vertex**, that is a vertex other than S or T. Therefore for every vertex V, other than S or T:

   (*sum of flows along arcs into V*) = (*sum of flows along arcs out of V*)

   An immediate consequence of condition 2 is that:

   (*sum of flows out of S*) = (*sum of flows into T*)

   This sum is called the **value of the flow**.

- **If the flow along arc e is equal to the capacity of arc e then the arc is said to be *saturated*. If an arc is not saturated then it is said to be *unsaturated*.**

A very special flow is the **zero flow**, in which the flow in every arc is zero. Any other flow is a non-zero flow.

> **WARNING**
>
> Mistakes are often made by confusing **capacity** and **flow**. It is suggested that you put your flows in circles, e.g. (5). One way to remember this is to think of the number in the circle as being inside a pipe. In what follows make certain that you distinguish between statements relating to capacities and statements relating to flows.

Example 1

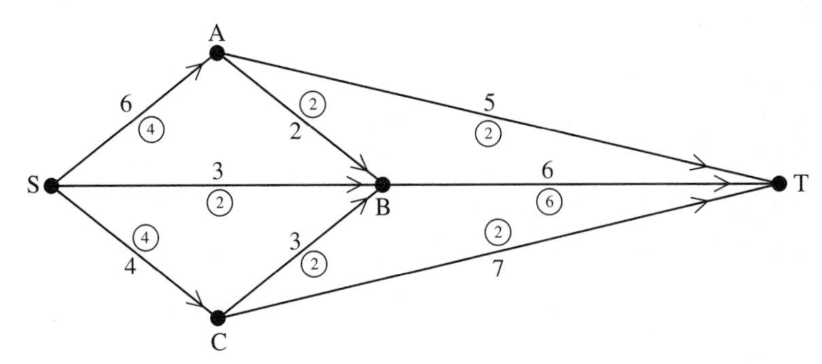

In the above network the unringed numbers represent the capacities of the arcs. The numbers in rings are a set of flows. Show that this set of flows satisfies the feasibility and conservation conditions.

Consider first the arcs. We must check that for each arc the flow is less than or equal to the capacity:

arc SA: $4 \leqslant 6$
arc SB: $2 \leqslant 3$
arc SC: $4 \leqslant 4$
arc AB: $2 \leqslant 2$
arc CB: $2 \leqslant 3$
arc AT: $2 \leqslant 5$
arc BT: $6 \leqslant 6$
arc CT: $2 \leqslant 7$

The feasibility condition is therefore satisfied.

Consider now the intermediate vertices. We must check that at each vertex (flow in) = (flow out):

vertex A: in = (4), out = (2) + (2) = (4)

vertex B: in = (2) + (2) + (2) = (6), out = (6)

vertex C: in = (4), out = (2) + (2) = (4)

The conservation condition is therefore satisfied at intermediate vertices.

Finally, consider the source S and the sink T. We must check that (flow out of S) = (flow into T):

source S: flow out of S $= \textcircled{4} + \textcircled{2} + \textcircled{4} = \textcircled{10}$

sink T: flow into T $= \textcircled{2} + \textcircled{6} + \textcircled{2} = \textcircled{10}$

This condition is therefore satisfied and the value of the flow is 10 units.

Arcs SC, AB and BT are saturated. All other arcs are unsaturated.

Example 2

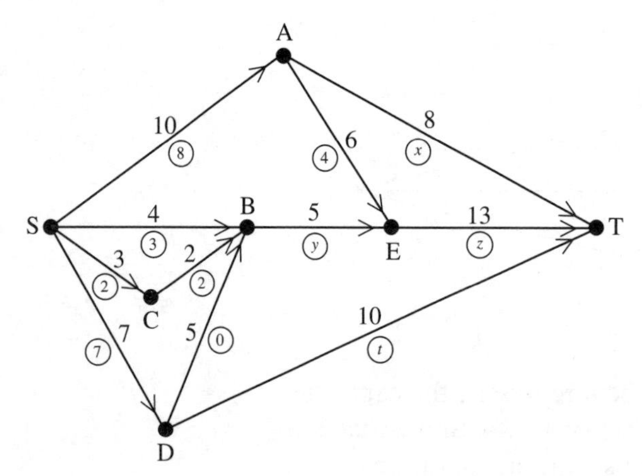

In the above network the arcs have the capacities shown by the unringed numbers. Given that the flow shown by the circled numbers satisfies the feasibility and conservation conditions, find the values of x, y, z and t.

Consider first vertex A. As only one unknown x occurs there we have by conservation:

$$\text{in } \textcircled{8}, \text{ out } \textcircled{4} + \textcircled{x}$$

so

$$x = 8 - 4 = 4 (\leqslant 8)$$
$$x = 4$$

Similarly, at vertex B:

$$\text{in } \textcircled{3} + \textcircled{2} + \textcircled{0}, \text{ out } \textcircled{y}$$

so

$$y = 3 + 2 = 5 (\leqslant 5)$$
$$y = 5$$

At vertex D:

$$\text{in } \textcircled{7} + \textcircled{0}, \text{ out } \textcircled{t}$$

so $\quad t = 7 (\leqslant 10)$

$$t = 7$$

At vertex E:

in $\quad \textcircled{y} + \textcircled{4}$, out $\textcircled{z}$

so $\quad y + 4 = z$

Using the value of y found above:

$$z = 5 + 4 = 9 (\leqslant 13)$$

As a check:

sum of flows out of S $= \textcircled{8} + \textcircled{3} + \textcircled{2} + \textcircled{7} = \textcircled{20}$

sum of flows into T $= \textcircled{x} + \textcircled{z} + \textcircled{t}$

$= \textcircled{4} + \textcircled{9} + \textcircled{7} = \textcircled{20}$

The value of this flow is therefore 20.

The method of constructing flows is illustrated in the following example.

Example 3

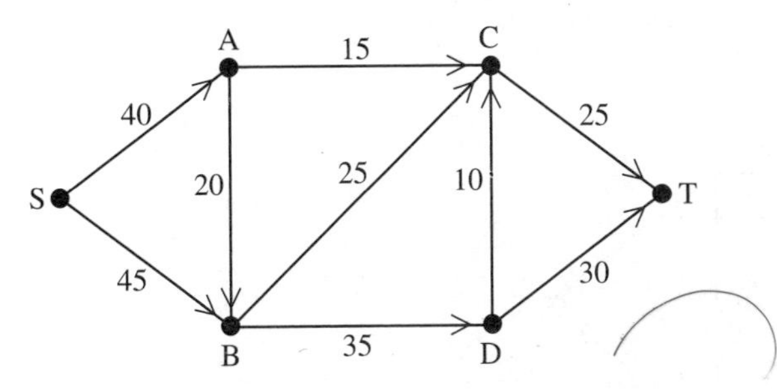

The above is the network of one-way streets introduced at the beginning of this chapter. Let us find some elementary flows for this network.

(i) Consider the path SBDT. There are three arcs on this path: SB of capacity 45, BD of capacity 35 and DT of capacity 30. The smallest of the capacities is minimum(45, 35, 30) = 30. A flow of 30 units is therefore possible along this path:

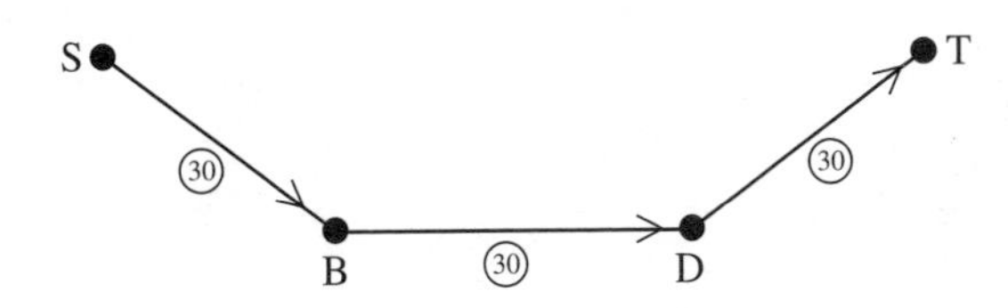

Arc DT is saturated. The excess capacities of the other arcs are:

SB: $45 - 30 = 15$; BD: $35 - 30 = 5$

We can now consider a modified network formed by subtracting the flow along each arc from the capacity of that arc:

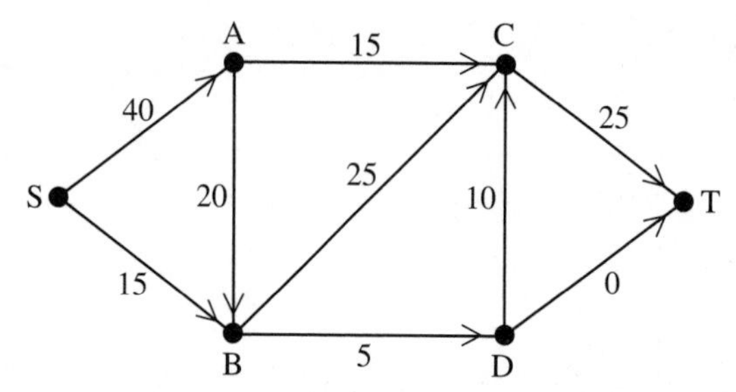

(ii) Consider now the path SACT. Proceeding as above, the smallest capacity on this path is minimum(40, 15, 25) = 15. A flow of 15 units is therefore possible along this path:

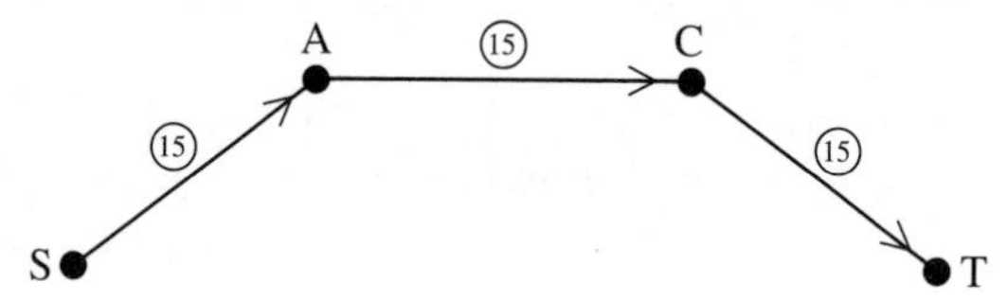

Arc AC is saturated and the excess capacities of the other arcs are:

SA: 40 – 15 = 25; CT: 25 – 15 = 10

We then have the modified network:

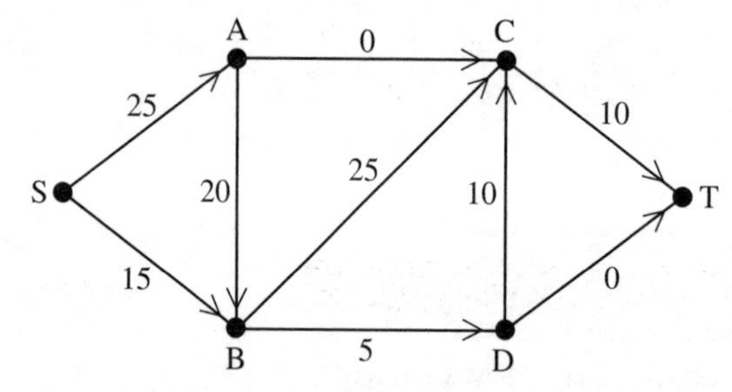

(iii) We now look for a path from S to T on which the excess capacity is as large as possible. In this case the path is SBCT. On this path the smallest capacity is minimum(15, 25, 10) = 10. A flow of 10 units is therefore possible along this path:

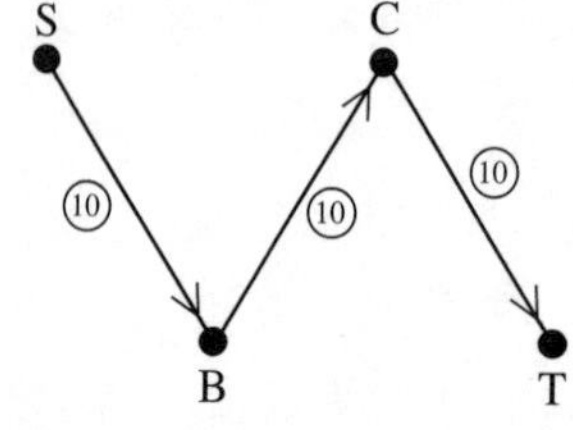

The arc CT is saturated and the excess capacities of the other arcs are:

SB: 15 – 10 = 5; BC: 25 – 10 = 15

The modified network is then:

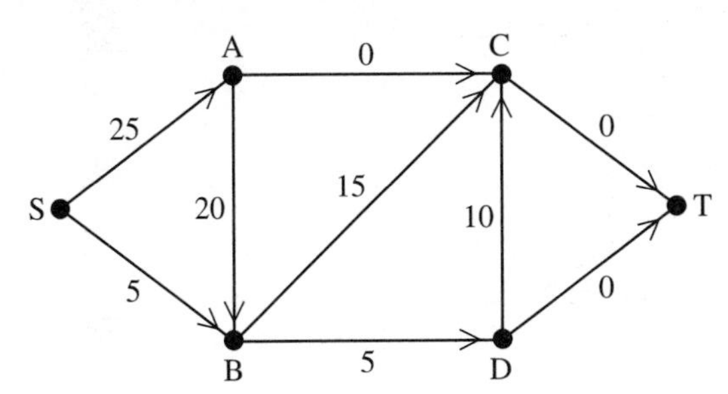

As both arcs CT and DT are saturated, there are clearly no further flows possible. If we combine the three flows obtained in (i), (ii) and (iii) we obtain the following flow pattern:

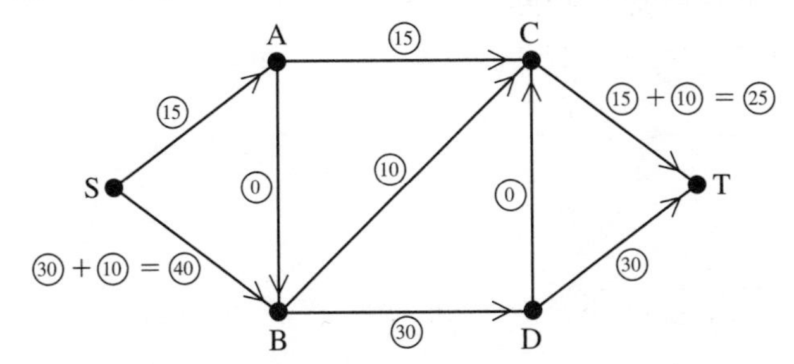

The value of this flow is equal to the flow out of S = 40 + 15 = 55 or the flow into T = 30 + 25 = 55.

There may be other flow patterns that have the same value of 55 units. The pattern obtained depends on how the flow is constructed.

Exercise 8A

1

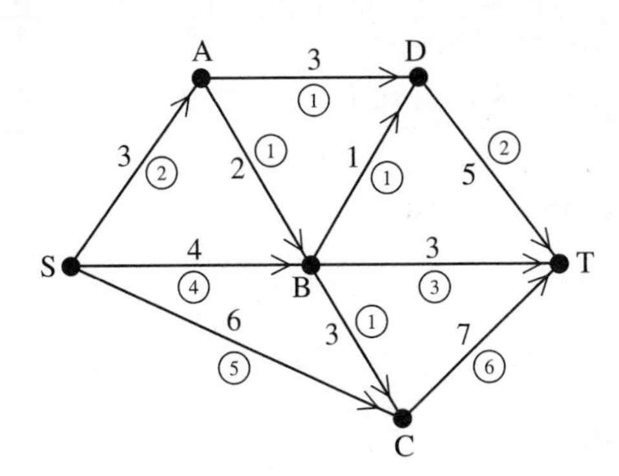

In the above network the unringed numbers give the arc capacities and the ringed numbers are a set of flows. Show that the feasibility condition and the conservation condition are satisfied by this set of flows.

2

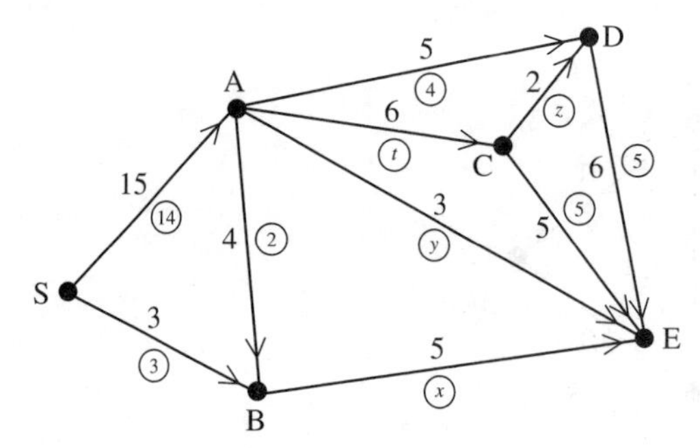

In the above network the unringed numbers are the arc capacities and the ringed numbers are a set of flows. Given that this set of flows satisfies the feasibility and conservation conditions find x, y, z and t.

3

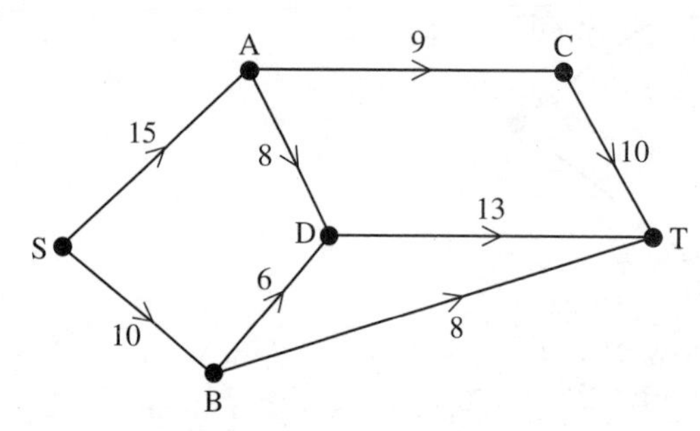

The figure shows a capacitated, directed network. The number on each arc indicates the capacity of that arc.

(a) Find the maximal flow along

(i) SACT (ii) SBT

Consider the modified network when the flows in (a) have been taken into account.

(b) Find another flow path that begins with arc SA.

(c) Find a further flow such that both SA and SB are saturated.

(d) Draw the final flow pattern.

4 Natural gas is produced at S and transported by a network of underwater pipelines to a refinery at T. The maximum capacity of each pipe, in appropriate units, is given on the arcs.

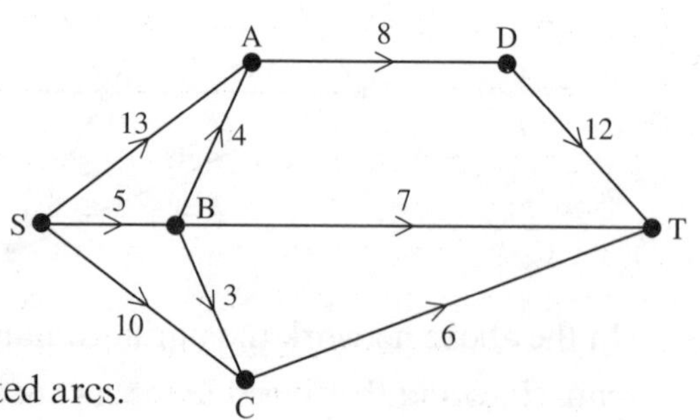

(a) Find the maximum flow along the paths

(i) SADT (ii) SCT (iii) SBT

(b) Write down the flow on each path and the saturated arcs.

(c) Explain why no further increase in the flow is possible. State the total flow.

5

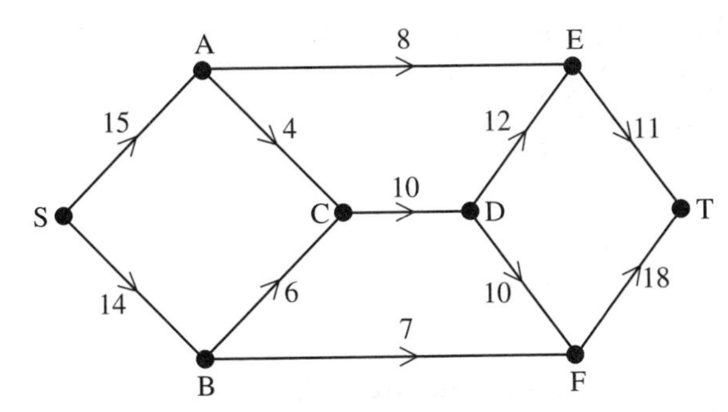

The above network represents a one-way road system through a town. The number on each arc represents the maximum number of vehicles that can pass along that road every minute (or the capacity of the road). It is possible for 29 vehicles to leave S every minute and for 29 vehicles to enter T every minute. Explain why it is not possible to achieve a flow of 29 through the network from S to T.

8.2 The labelling procedure

The basic approach employed when looking for a maximum flow is to first find a flow by inspection and then to increase it step by step until it can be increased no further. Finding an initial flow is not difficult. For example, we can always start with the zero flow, in which the flow in each arc is zero. However, it saves time if we start with a non-zero flow.

As we saw in section 8.1, it is tedious to have to keep drawing modified networks after each additional flow has been found. A **labelling procedure** has therefore been developed which allows all calculations to be done on a single diagram.

At each step every arc of the network is labelled with two numbers, which are each associated with an arrow alongside the arc.

The numbers are:

(i) the remaining **excess capacity** of the arc – the amount by which flow along the arc may be **increased**
(ii) the **flow** along the arc, which is also called the **back capacity** – the amount by which flow along the arc may be **reduced**. (This will be explained further when we look at flow-augmenting paths and the concept of back flow.)

Let us look again at Example 3(i) on page 217. The arc BD has a capacity of 35 units from B to D and a flow of 30 units. The remaining excess capacity is then $(35 - 30) = 5$ units. All this information may be recorded using the labelling procedure in the following way:

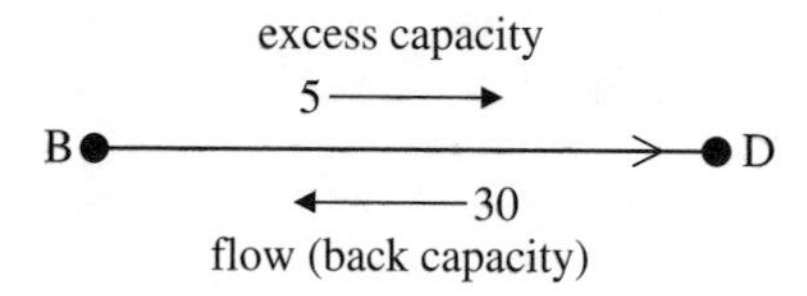

The excess capacity arrow is always in the direction of the arc and the flow arrow is always in the opposite direction, as shown above.

Notice that the sum of the two numbers is equal to the original capacity of the arc. This provides us with a check that should be used each time an arc is labelled.

To illustrate the labelling procedure let us look again at Example 3 on page 217. A zero flow using the labelling procedure for this network gives:

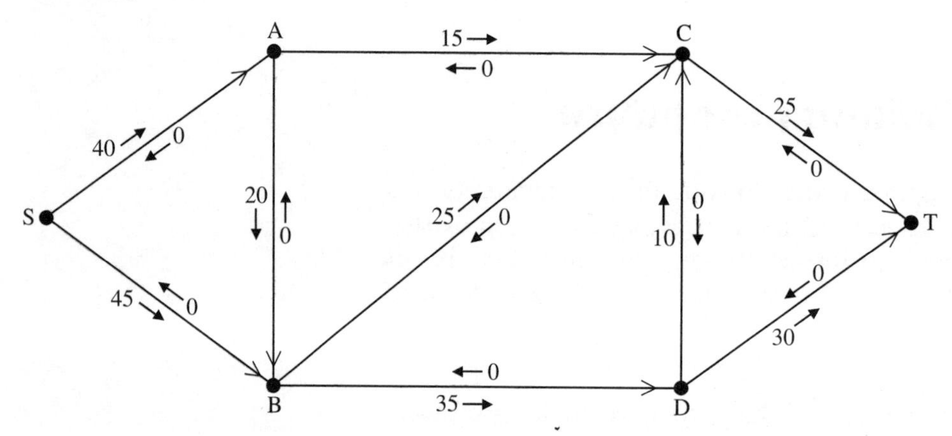

(i) The flow of 30 units along SBDT may be shown using the labelling procedure as:

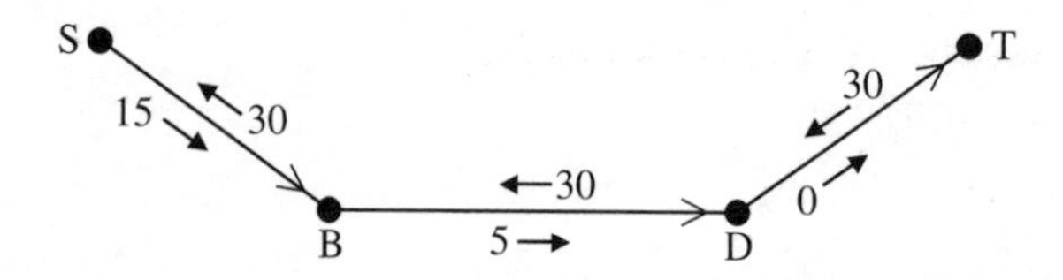

(ii) The flow of 15 units along SACT may be shown using the labelling procedure as:

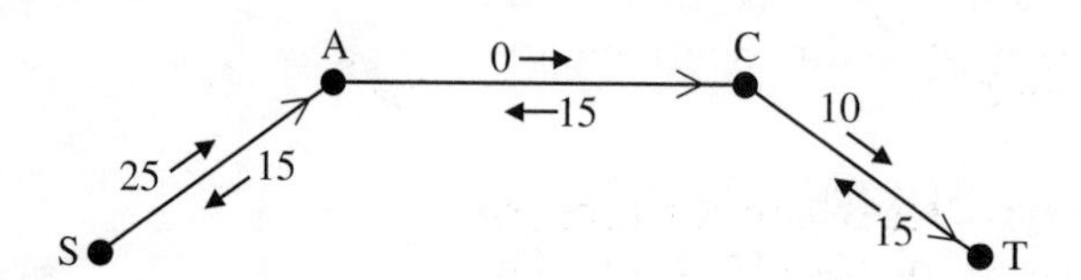

(iii) The flow of 10 units along SBCT may be shown using the labelling procedure as:

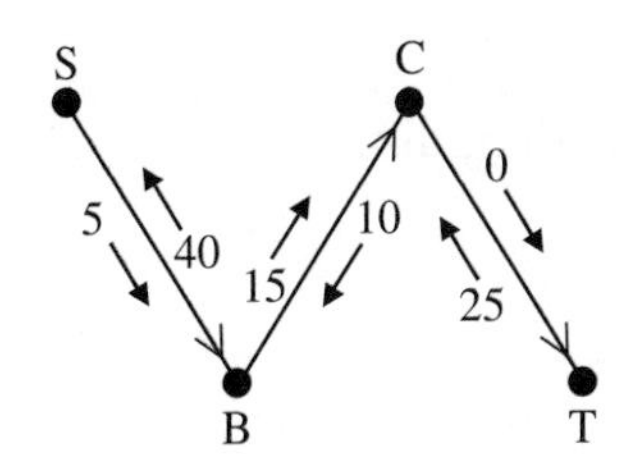

Notice that we take into account the previous flows when recording the capacities on SB and CT.

All of the above calculations can be carried out on a single (large) diagram by updating the numbers on the arcs. When updating a number simply put a single line through it. We then have:

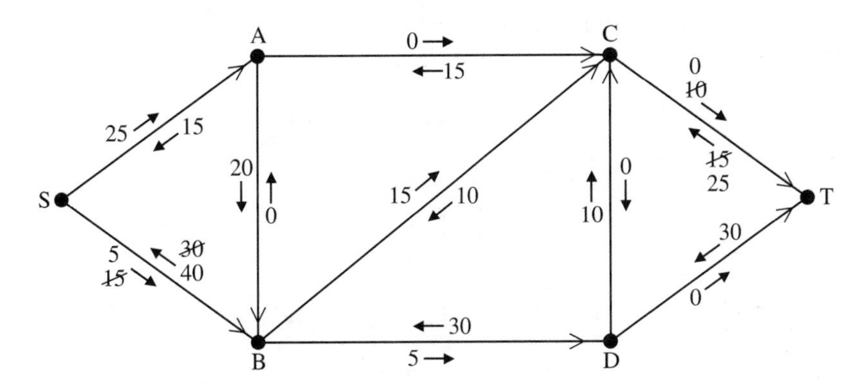

It is recommended that you list the flows that you use together with the saturated arcs created at each stage. In this case we have:

(i) SBDT: flow 30 units, DT saturated
(ii) SACT: flow 15 units, AC saturated
(iii) SBCT: flow 10 units, CT saturated

The total flow is $30 + 15 + 10 = 55$.

As before the final flow pattern can be obtained simply by using the flows found. We thus obtain:

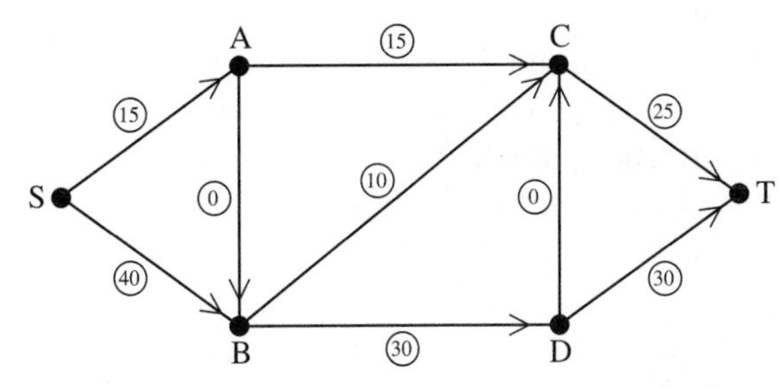

It is also easy to identify the saturated arcs. These are the arcs for which the excess capacities are zero, that is AC, CT and DT.

Exercise 8B

1 Consider the network in question 3 of Exercise 8A.

(a) Using the labelling procedure show on a single diagram:

(i) the flows along SACT and SBT

(ii) the flow you found in part (b) of question 3

(iii) the further flow you found.

(b) Draw the final flow pattern and state the saturated arcs.

2 Consider the network in question 4 of Exercise 8A. Using the labelling procedure:

(a) show the maximum flows along SADT, SCT and SBT

(b) draw the final flow pattern and state the saturated arcs.

3 In this network the following flows were identified:

SACET flow 10 units

SBDFT flow 22 units

SABCFET flow 10 units

SBDFET flow 4 units

SBCET flow 2 units

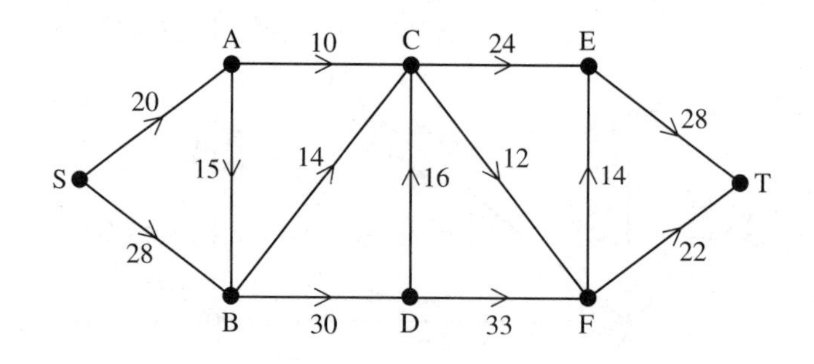

(a) Use the labelling procedure to record these on a single diagram.

(b) Hence obtain a maximum flow. Give the value of this maximum flow and your flow pattern. Identify the saturated arcs.

8.3 Flow-augmenting paths

In Example 3 on page 217 we obtained the maximum flow by finding three paths from S to T consisting entirely of unsaturated arcs. These paths are usually called **flow-augmenting paths**. Finding flow-augmenting paths is an essential part of the process of finding maximal flows. However, the situation is not always quite as simple as that encountered in this example. Let us look at the problem of finding a maximal flow through the network given in Example 1. We will use the flow shown there as our initial flow.

Example 4

The information given in Example 1 may be summarised using the labelling procedure.

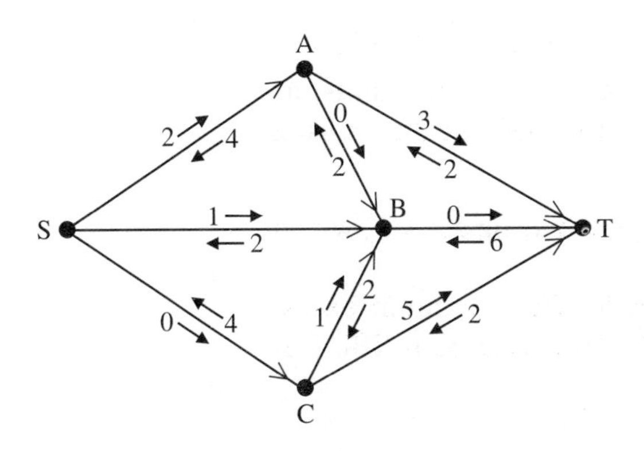

arc SA: capacity 6, flow 4, excess capacity 2
arc SB: capacity 3, flow 2, excess capacity 1
arc SC: capacity 4, flow 4, excess capacity 0
arc AB: capacity 2, flow 2, excess capacity 0
arc CB: capacity 3, flow 2, excess capacity 1
arc AT: capacity 5, flow 2, excess capacity 3
arc BT: capacity 6, flow 6, excess capacity 0
arc CT: capacity 7, flow 2, excess capacity 5

Having drawn the diagram, you should check that the sum of the numbers on the arrows on any given arc is equal to the initial capacity of that arc.

We now look for flow-augmenting paths. A path from S to T consisting entirely of unsaturated arcs is SAT:

arc SA has excess capacity of 2 units
arc AT has excess capacity of 3 units

SAT is therefore a flow-augmenting path with value equal to minimum $(2, 3) = 2$ units. The modified path SAT is then:

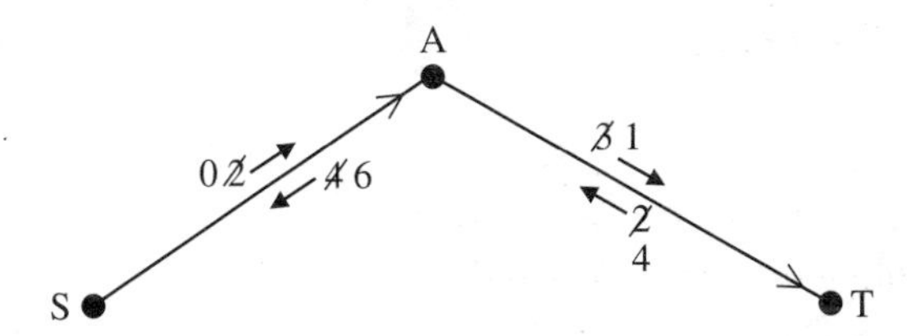

This can be included on the original figure if you wish, but is shown here separately for clarity.

The saturated arcs at this stage are SA, SC, AB and BT, so there can be no further flow-augmenting paths using arcs SA or SC as their first arc or arc BT as their last arc.

You may be tempted to think there are no further flow-augmenting paths, *but* there is one more:

arc SB has excess capacity of 1 unit
arc BT has zero excess capacity

But, we can *decrease* the existing flow on CB by up to 2 units, as indicated by the **back capacity** of 2 units. This is often called a **back flow**.

A flow-augmenting path can then be completed using arc CT, which has a non-zero excess capacity.
SBCT is therefore a flow-augmenting path. The minimum excess capacity is that of SB, namely 1 unit. The modified path SBCT is then:

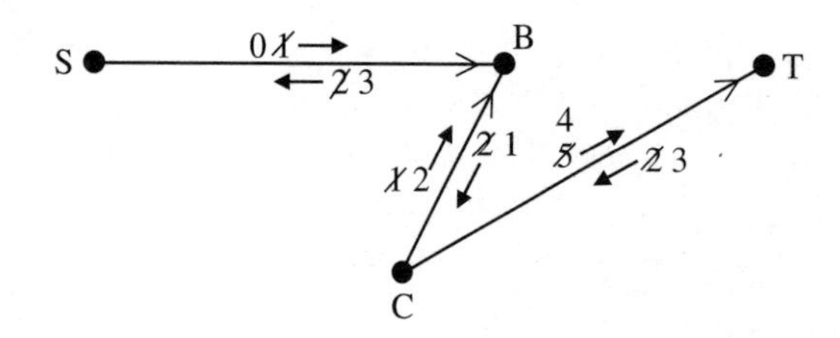

All the above can be shown on a single diagram:

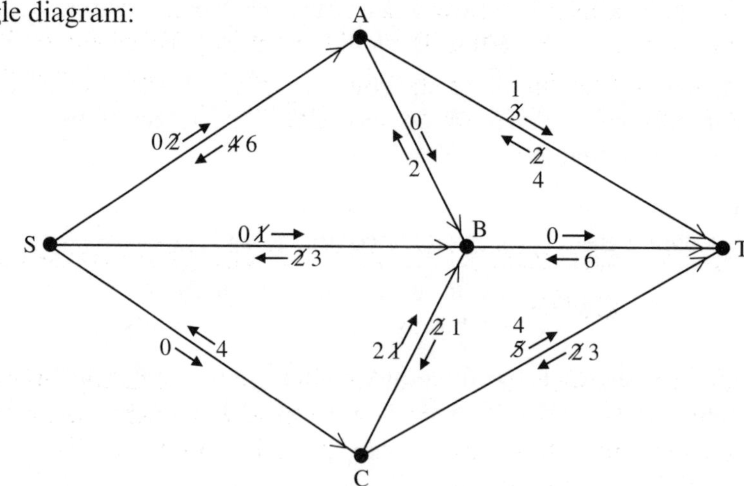

The flow pattern is then:

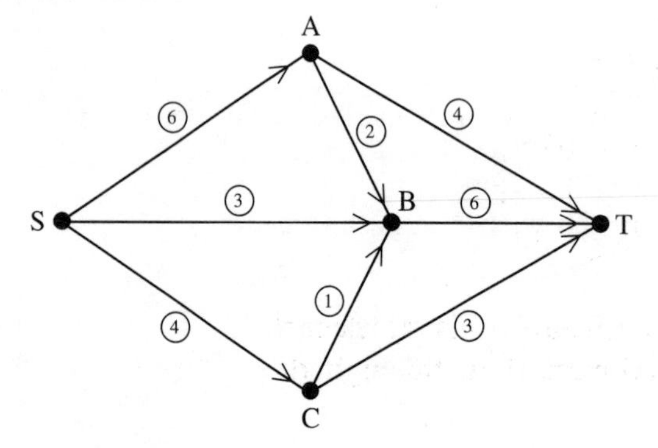

As arcs SA, SB and SC are all saturated, there can be no further flow-augmenting paths. The maximum flow is then ⑥ + ③ + ④ = 13 units.

Note: Choosing flow-augmenting paths in a different order may lead to a different flow pattern, but the value of the maximal flow will always be the same.

In the light of our experience in previous examples we will now make a formal definition of a flow-augmenting path.

- **A flow-augmenting path in a capacitated network with a single source S and a single sink T is a path from S to T consisting of:**
  (i) forward arcs – unsaturated arcs directed along the path
  (ii) backward arcs – arcs directed against the direction of the path and carrying a non-zero flow.

In Examples 3 and 4 it was quite easy to decide when a maximal flow had been obtained. In Example 3 it was when all arcs entering T were saturated. In Example 4 it was when all arcs leaving S were saturated. In general you may not have either of these situations.

We have a maximal flow when it is not possible to find any further flow-augmenting paths. It would obviously be useful to have an alternative method for determining whether or not a given flow is a maximal flow. In section 8.4 we consider such a method.

8.4 Maximum flows and minimum cuts

Let us consider for a moment a network consisting of pipelines which carry water. If there is a bottleneck in such a network, then the amount of flow through the network is limited by the amount of flow through that bottleneck.

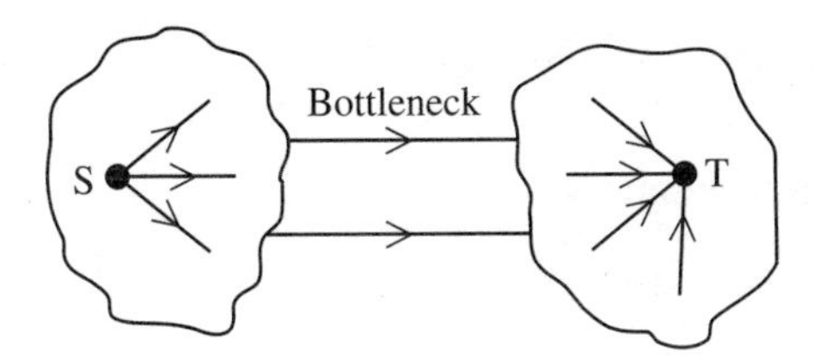

It is clear that information about flow through this bottleneck will give us information about the maximum flow through the network.

Example 5

Consider the following four networks, where the numbers give the capacities of the arcs, and find the bottleneck in each case.

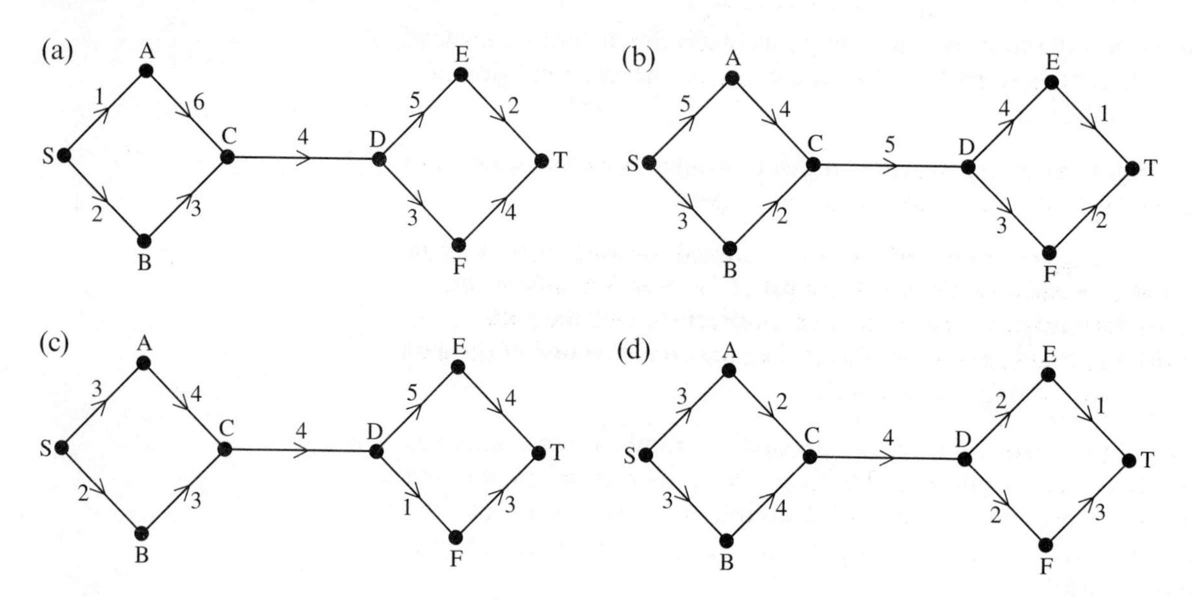

In (a) the bottleneck is the arcs leaving S, which have a combined capacity of 3. (cf Example 4)

In (b) the bottleneck is the arcs entering T, which have a combined capacity of 3. (cf Example 3)

In (c) the bottleneck is the arc CD, which has a capacity of 4.

In (d) the bottleneck is the arcs DF and ET, which have a combined capacity of 3.

It is clear therefore that the bottleneck may consist of more than one arc. We make this idea more precise by introducing the idea of a **cut**.

- **A *cut* in a network, with source S and sink T, is a set of arcs A whose removal separates the network into two parts *X* and *Y*, where S is contained in *X* and T is contained in *Y*.**
- **The *capacity of a cut* is the sum of the capacities of those arcs in the cut that are *directed from X to Y*.**
- **A *minimum cut* is a cut of the smallest possible capacity.**

An alternative definition of a cut is that it is a set of arcs A such that every path from S to T includes an arc in A.

A cut may be written down in several different ways. For example, we can write down the arcs in A or we can write down the vertices in *X* and the vertices in *Y*. This is illustrated in Example 6.

Example 6

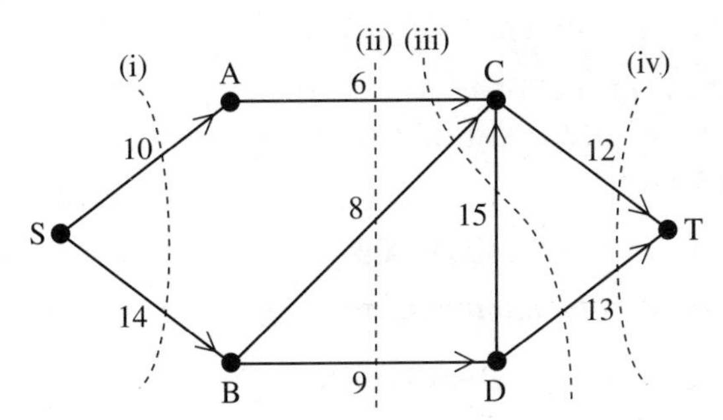

Four cuts are shown by the dotted lines in the above capacitated network. Write down the source set, the sink set and the capacity in each case.

(i) The cut consists of arcs SA and SB:

the source set X = {S}
the sink set Y = {A, B, C, D, T}
the capacity of the cut = 10 + 14 = 24

(ii) The cut consists of arcs AC, BC and BD:

the source set X = {S, A, B}
the sink set Y = {C, D, T}
the capacity of the cut = 6 + 8 + 9 = 23

(iii) The cut consists of arcs AC, BC, DC and DT:

the source set X = {S, A, B, D}
the sink set Y = {C, T}
the capacity of the cut = 6 + 8 + 15 + 13 = 42

(iv) The cut consists of arcs CT and DT:

the source set X = {S, A, B, C, D}
the sink set Y = {T}
the capacity of the cut = 12 + 13 = 25

A close inspection of the network shows that (ii) is the minimum cut with capacity 23 units.

Suppose now that *instead* of having a capacity of 15 from D to C the network were to have a capacity of 15 from C to D. Then:

(i) is unchanged,
(ii) is unchanged,
(iv) is unchanged,

but in (iii) the arcs going *from X to Y* are AC, BC and DT. So the capacity of this cut is now 6 + 8 + 13 = 27.

It is important to note that in calculating the capacity of a cut we only sum the capacities of those arcs *which are directed from X to Y*.

The Maximum Flow–Minimum Cut Theorem

Having introduced the concept of a cut we can now state the fundamental theorem that provides us with a means of testing if a flow is maximal. It was first proved in 1955 by Ford and Fulkerson, and is usually called the **Max Flow–Min Cut Theorem**.

- **In any capacitated network with a single source S and a single sink T**
  (*the value of a maximal flow*) = (*the capacity of a minimum cut*)

This theorem has an immediate consequence which enables you to tell if a flow, obtained using the labelling procedure, is maximal.

- **A flow through a network is maximal if and only if a cut can be found with capacity equal to the value of the flow.**

When looking for a cut with capacity equal to the value of your flow, it is a good idea to consider the saturated arcs.

The **maximum flow algorithm** may now be stated.

Step 1 Obtain an initial flow by inspection.

Step 2 Find flow-augmenting paths using the labelling procedure until no further flow-augmenting paths can be found.

Step 3 Check that the flow obtained is maximal by using the max flow–min cut theorem and finding a cut whose capacity is equal to the value of the flow.

Example 7

Use the maximum flow algorithm to obtain the maximum flow through the capacitated network shown below.

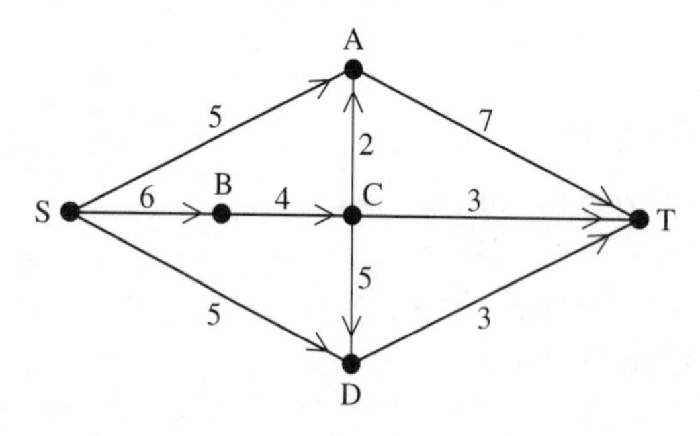

There is an obvious flow along SAT, value minimum$(5, 7) = 5$.

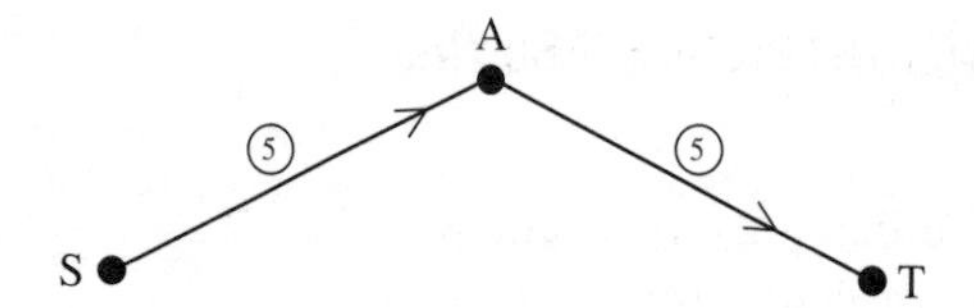

Another obvious flow is along SDT, value minimum(5, 3) = 3.

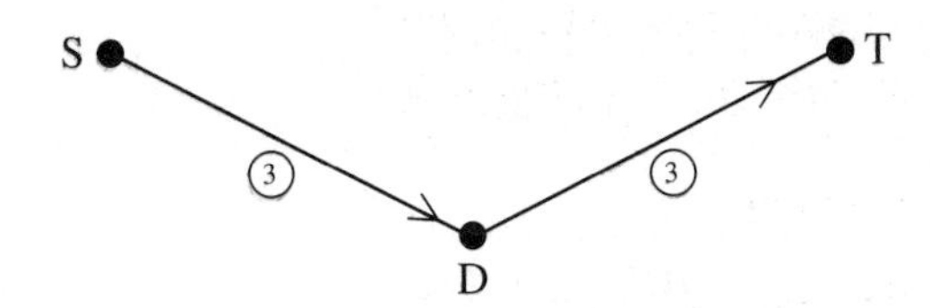

A further flow is along SBCT, value minimum(6, 4, 3) = 3.

Using the labelling procedure we then have at this stage:

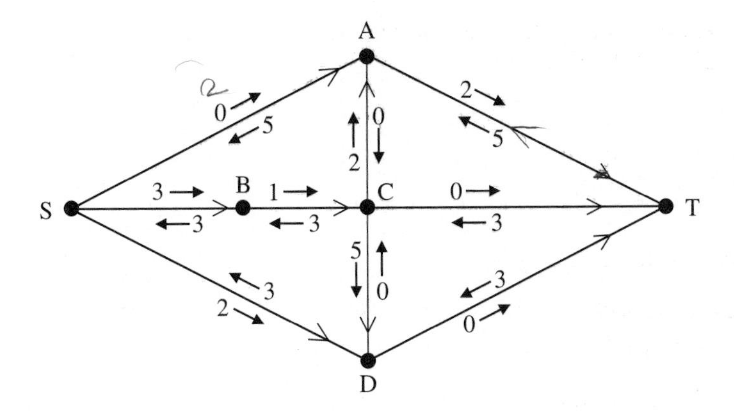

The saturated arcs at this stage are SA, DT and CT.

From the above diagram we can see that flow is possible along the path SBCAT, value minimum(3, 1, 2, 2) = 1. The updated diagram is then:

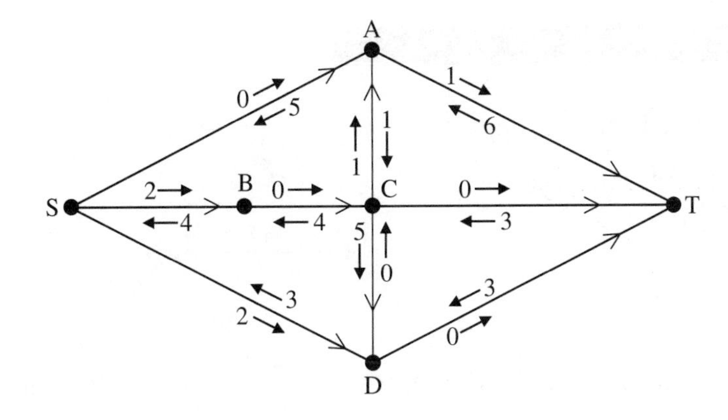

There is now an additional saturated arc BC. The total flow so far is (5 + 3 + 3 + 1) = 12.

There are two ways of seeing that this is the maximum flow:

(i) CT and DT are saturated so no further flow can reach T from C and D. In addition, since SA and BC are saturated no further flow can reach A, and hence T, along AT.
(ii) By looking at the saturated arcs you can obtain a minimum cut of capacity 12. This cut consists of arcs SA, BC and DT (capacity $5 + 4 + 3 = 12$).

The obvious cut of SA, SB and SD has capacity 16 and the other obvious cut AT, CT and DT has capacity 13.

In summary, the maximal flow has value 12 and a possible flow pattern with this value is:

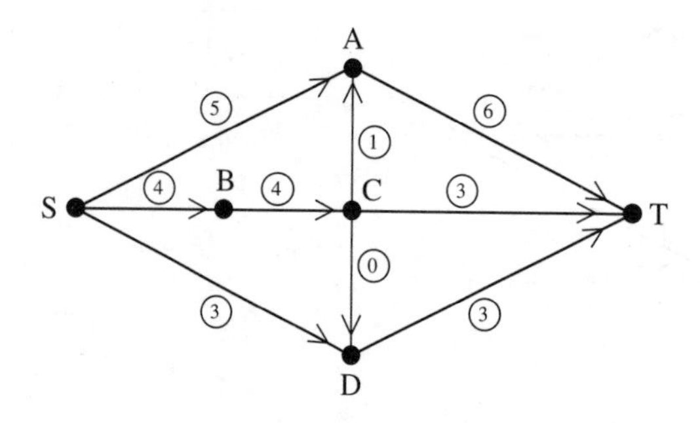

There are other flow patterns with value 12.

Exercise 8C

1 Four cuts are shown on the capacitated network opposite. For each of the cuts shown write down:

(a) the corresponding sets X and Y
(b) the arcs in the cut
(c) the capacity of the cut.

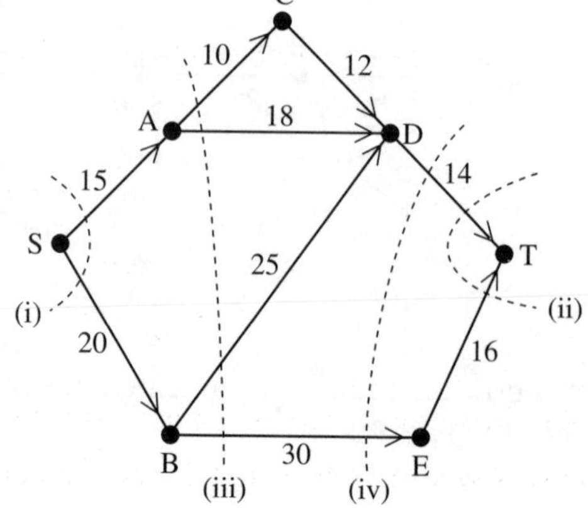

2

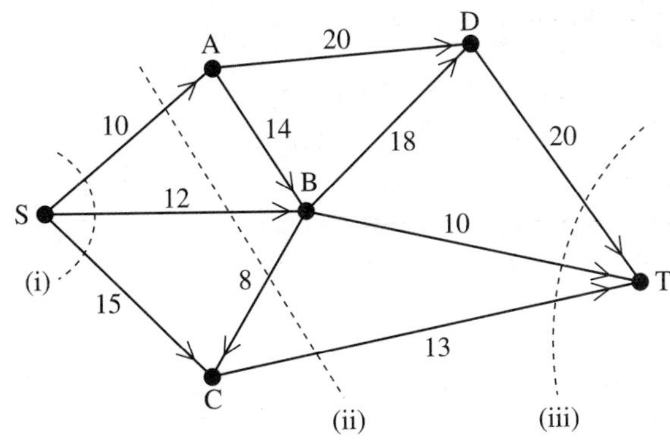

Three cuts are shown on the above capacitated network. For each of the cuts shown write down:

(a) the corresponding sets X and Y

(b) the arcs in the cut

(c) the capacity of the cut.

3

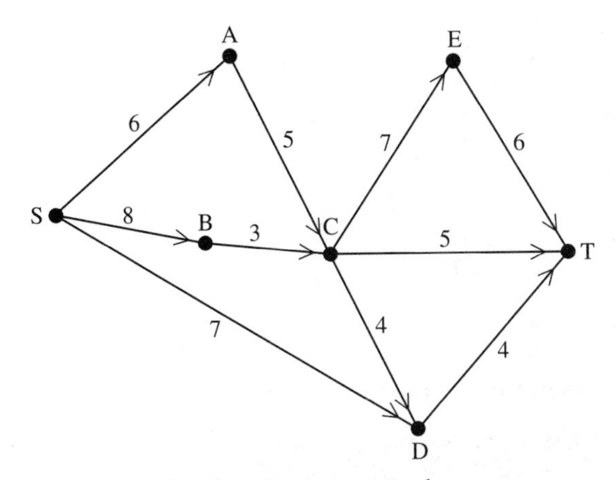

Find a minimum cut in the above network.

4

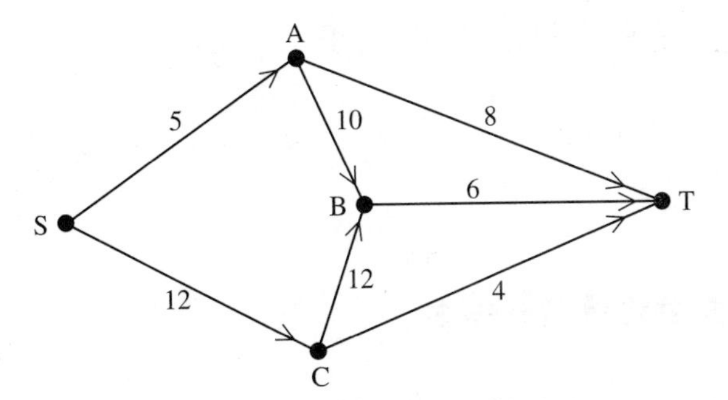

For the above network list all possible cuts, and write down the corresponding sets X and Y and the capacity of the cut. Hence determine the minimum cut.

5

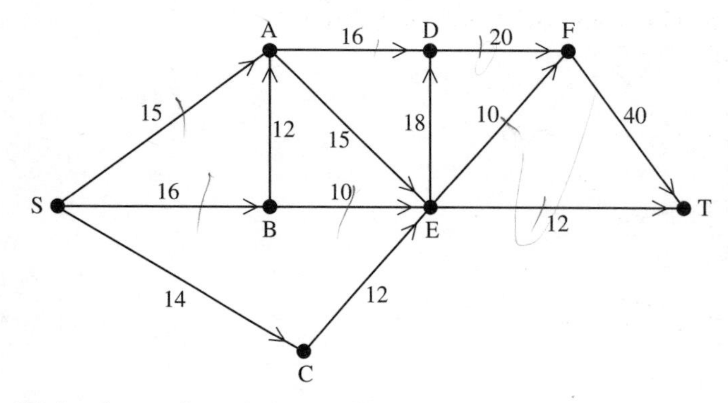

(a) Write down the maximum flows along

(i) SBET (ii) SADFT

(b) Using the flows found in (a) as your initial flow, find flow-augmenting paths and hence obtain the maximum flow.

(c) Show your final flow pattern on a diagram.

(d) Prove that your flow is maximal by finding a minimum cut.

6

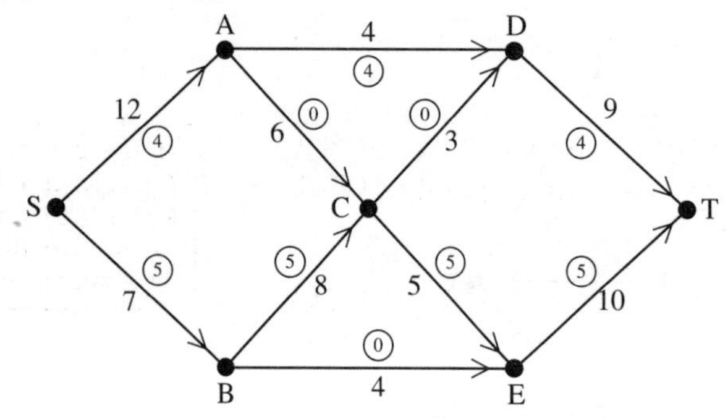

In the above network unringed numbers are the capacities of the arcs and ringed numbers are the flows along the arcs.

(a) Use the labelling procedure to show all this information on a diagram.

(b) Find flow-augmenting paths and hence obtain a maximum flow. List the paths you used.

(c) Show your final flow pattern on a diagram.

(d) Prove that your flow is maximal.

8.5 Multiple sources and sinks

So far in this chapter we have been concerned with networks with a *single* source S and a *single* sink T. In this section we will show how to extend that work to situations which often occur in practice where there are several sources $S_1, S_2, \ldots$ and several sinks $T_1, T_2, \ldots$.

Consider the network shown below, which has two sources, S_1 and S_2, and three sinks, T_1, T_2 and T_3.

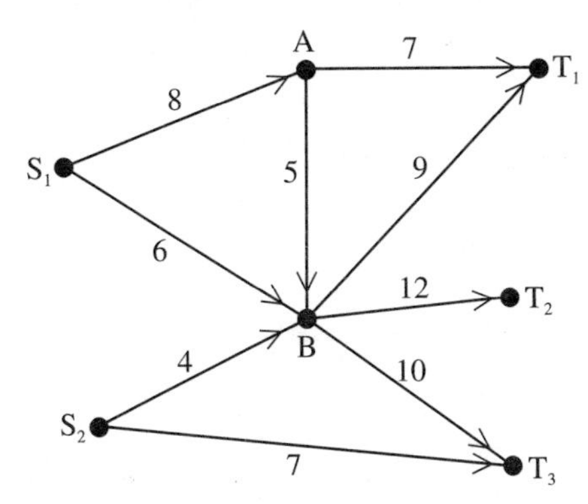

We now introduce a supersource S and a supersink T. Arcs SS_1 and SS_2 are added from the supersource S to the sources S_1 and S_2. Arcs T_1T, T_2T and T_3T are added from the sinks T_1, T_2 and T_3 to the supersink T.

Since no more than $(8 + 6) = 14$ units can leave S_1, the capacity of SS_1 is set at 14. Similarly, since no more than $(7 + 4) = 11$ units can leave S_2, the capacity of SS_2 is set at 11.

Since T_1 cannot have more than $(7 + 9) = 16$ units entering it, the capacity of T_1T is set at 16 and in a similar way the capacity of T_2T is set at 12 and the capacity of T_3T is set at $(10 + 7) = 17$.

In fact any capacities can be assigned to the arcs SS_i and T_jT as long as they are not less than the capacities given here.

You now have the following network to which you can apply the maximum flow algorithm:

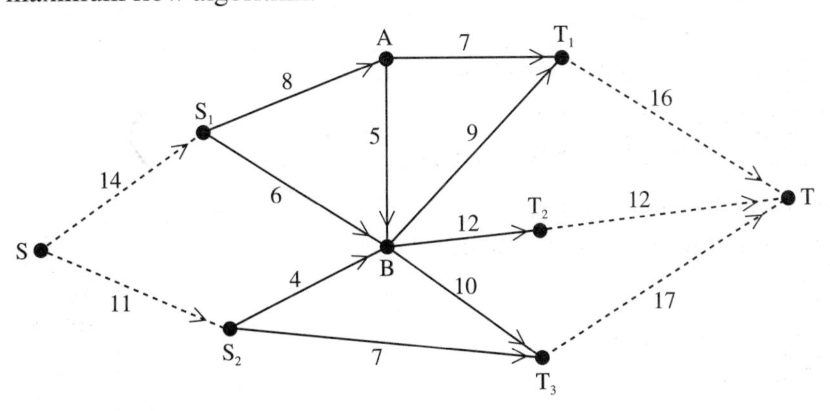

The following flow-augmenting paths can be identified by inspection:

SS_1AT_1T, 7 units	(AT_1 saturated)
SS_2T_3T, 7 units	(S_2T_3 saturated)
SS_1BT_2T, 6 units	(S_1B saturated)
SS_2BT_2T, 4 units	(SS_2 and S_2B saturated)
SS_1ABT_2T, 1 unit	(SS_1 saturated)

Since both SS_1 and SS_2 are saturated, this is a maximum flow. The corresponding flow pattern is shown below. The value of the flow is 25 units.

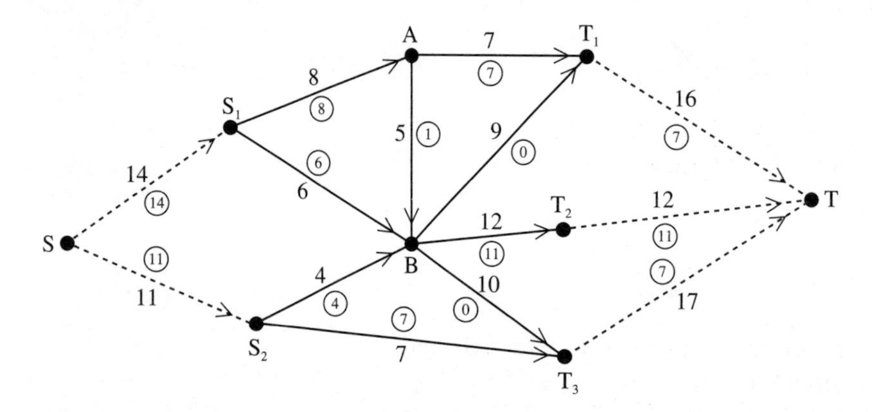

The maximum flow pattern for the original network is found simply by removing the supersource, the supersink and the added arcs:

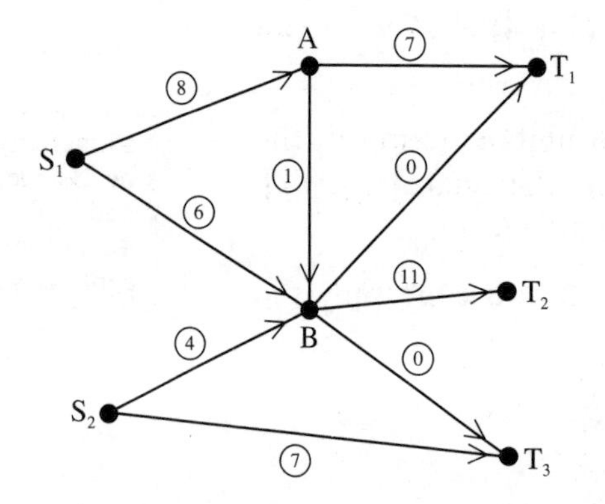

The minimum cut here is S_1A, S_1B, S_2B and S_2T_3 of capacity 25.

From the above diagram we can see that in this solution

(i) (8) + (6) = 14 units leave S_1
(ii) (4) + (7) = 11 units leave S_2
(iii) (7) + (0) = 7 units arrive at T_1
(iv) 11 units arrive at T_2
(v) (7) + (0) = 7 units arrive at T_3

The number of units of flow passing through B is
(1) + (6) + (4) = 11 units.

Exercise 8D

1

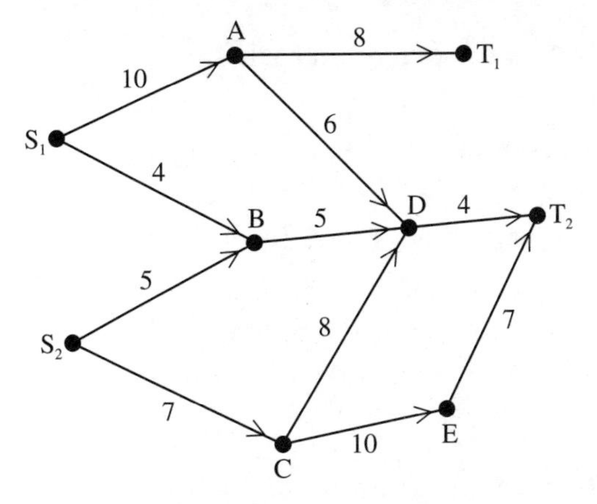

Natural gas is produced at two sources, S_1 and S_2, and is transported by a network of underground pipelines to two refineries, T_1 and T_2. The capacity of the pipelines in appropriate units is shown by the numbers on the arcs. Determine the maximum flow through the network. State your flow-augmenting paths and draw a diagram to show your maximum flow pattern. Which arcs are saturated in your solution?

2

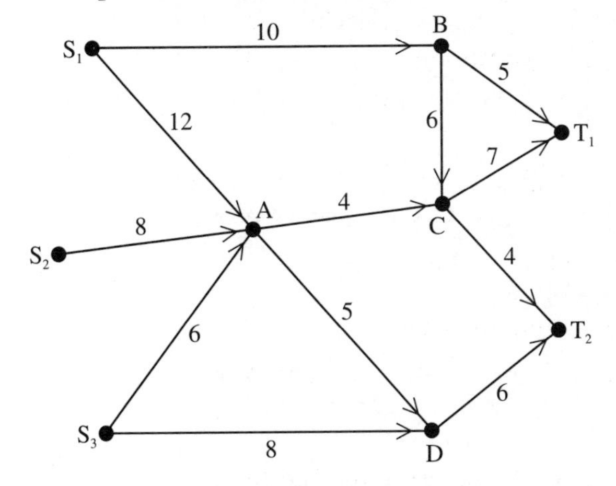

Determine the maximum flow through the above network. State your flow-augmenting paths and draw a diagram of your maximum flow pattern. Explain why your flow is maximal. For your flow find:

(a) the total flow out of S_1, S_2 and S_3

(b) the total flow into T_1 and T_2

(c) the flow through vertex C.

SUMMARY OF KEY POINTS

1 A **network** is a weighted digraph in which the weight on each arc represents the capacity of that arc.

2 A vertex S is called a **source** if all arcs containing S are directed away from S.

3 A vertex T is called a **sink** if all arcs containing T are directed towards T.

4 A **flow** in a network N with a single source S and a single sink T is an assignment to each arc e of N of a non-negative number called the flow along the arc e.
A flow must satisfy two conditions:
- The **feasibility condition**, which states that the flow along each arc cannot exceed the capacity of that arc.
- The **conservation condition**, which states that for every intermediate vertex, V (not S or T) the sum of the flows into V is equal to the sum of the flows out of V.

5 If the flow along an arc is equal to the capacity of that arc the arc is said to be **saturated**. If an arc is not saturated it is said to be **unsaturated**.

6 A **flow-augmenting path** in a capacitated network with a single source S and a single sink T is a path from S to T consisting of:
- **forward arcs** – unsaturated arcs directed along the path.
- **backward arcs** – arcs directed against the direction of the path and carrying a non-zero flow.

7 In any capacitated network with a single source S and a single sink T, **the value of a maximal flow is equal to the capacity of a minimum cut**.

8 The maximum flow algorithm is as follows:
Step 1 Obtain an initial flow by inspection.
Step 2 Find flow-augmenting paths using the labelling procedure until no further flow-augmenting paths can be found.
Step 3 Check that the flow obtained is maximal by finding a cut whose capacity is equal to the value of the flow.

9 A network with multiple sources $S_1, S_2, \ldots, S_m$ and multiple sinks $T_1, T_2, \ldots, T_n$ can be reduced to a network with only one source and one sink by introducing a **supersource** S and a **supersink** T. Arcs $SS_1, SS_2, \ldots, SS_m$ are added from the supersource S to the sources $S_1, S_2, \ldots, S_m$. Arcs $T_1T, T_2T, \ldots, T_nT$ are added from sinks $T_1, T_2, \ldots, T_n$ to the supersink. The capacities of the new arcs can be considered to be infinite.

Review exercise 3

1 The tour director of a museum needs to allocate five of his guides to parties of tourists from France, Germany, Italy, Japan and Spain. The table shows the languages spoken by the five guides.

Ruth	French	Spanish	
Steve	German	Japanese	
Tony	French	German	
Ursula	Spanish	Italian	
Victoria	Italian	Spanish	Japanese

(a) Draw a bipartite graph to model this situation.

The director allocates Ruth, Steve, Ursula and Victoria to the parties who speak the first language in their individual lists.

(b) Starting from this matching use the maximum matching algorithm to find a complete matching. Indicate clearly how the algorithm has been applied in this case. [E]

2 In order to help new A-level students to select their courses a college organises an open evening. Some students already studying A-level courses have agreed to talk about one of their A-level courses. Six of these students, Ann, Barry, David, Gemma, Jasmine and Nickos, are between them following six A-level courses in Chemistry, English, French, History, Mathematics and Physics.

The table below shows the courses being followed by each student:

Ann	English	French	History
Barry	History	English	
David	French	Chemistry	Mathematics
Gemma	Physics	Mathematics	History
Jasmine	Mathematics	Physics	Chemistry
Nickos	English	Mathematics	French

(a) Draw a bipartite graph to model this situation.

Initially Ann, Barry, David, Gemma and Jasmine are allocated to the first subject in their lists.

(b) Starting from this matching use the maximum matching algorithm to find a complete matching. Indicate clearly how the algorithm has been applied in this case.

(c) Explaining your reasoning carefully, determine whether or not your answer to (b) is unique.

3 Five coach drivers, Mihi, Pat, Robert, Sarah and Tony, have to be assigned to drive five coaches for the following school trips:

Adupgud Senior School is going to the Lake District
Brayknee Junior School is going to the seaside
Korry Stur Junior School is going to a concert
Learnalott Senior School is going to the museum (two coaches needed)

Mihi and Sarah would like to drive senior school children. Robert and Pat would like to go on the seaside trip. Pat and Tony would like to attend the concert. Robert and Pat would like to visit the museum. Pat and Tony would like to visit the Lake District.

The driver manager wishes to assign each driver to a trip they would like to do.

(a) Draw a bipartite graph to show the trips that the drivers would like to take.

Initially Mihi and Pat are assigned to Learnalott Senior School, Robert is assigned to Brayknee Junior School and Tony is assigned to Adupgud Senior School.

(b) Starting from this matching use the maximum matching algorithm to find a complete matching. You must indicate clearly how the algorithm has been applied in this case. State your alternating path and the final matching. [E]

4

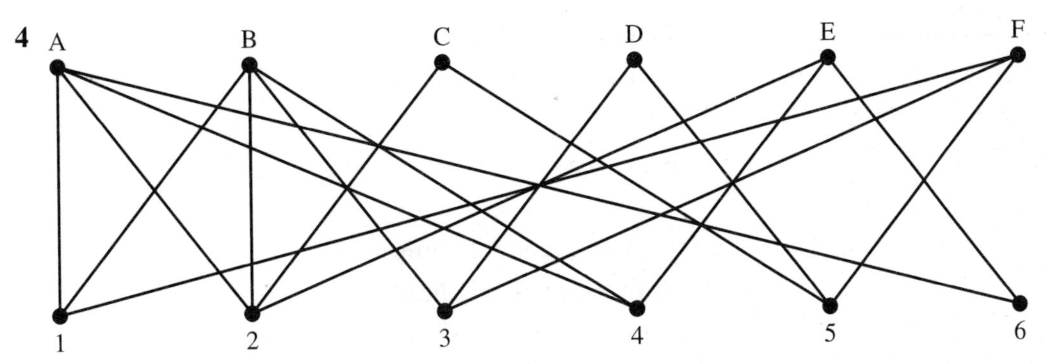

The bipartite graph above shows a mapping between six volunteers, A, B, C, D, E and F, and six tasks, 1, 2, 3, 4, 5 and 6. The lines indicate which tasks each volunteer is qualified to do.

The initial matching is A – 2, B – 1, C – 5, D – 3 and E – 4.

(a) Starting from this matching, use the maximum matching algorithm to find a complete matching. You must indicate clearly how the algorithm has been applied in this case. State your alternating path and your final matching.

Volunteer E now insists on doing task 2.

(b) State the changes that need to be made to the initial model to accommodate this. [E]

5 A college dramatic society has six helpers: Andrew (A), Deepa (D), Henry (H), Karl (K), Nicola (N) and Yi-Ying (Y). They are to be matched to six tasks: Props (P), Lighting (L), Make-up (M), Sound (S), Tickets (T) and Wardrobe (W).

This table indicates which tasks each person is able to do.

Name	Tasks
Andrew	Wardrobe, Props, Tickets
Deepa	Tickets, Make-up
Henry	Lighting, Make-up
Karl	Sound, Wardrobe, Lighting
Nicola	Sound
Yi-Ying	Lighting, Tickets

(a) Draw a bipartite graph to model this situation.
Initially Andrew, Deepa, Henry and Karl are matched to the first task in their individual lists.
(b) Indicate this initial matching in a distinctive way on the bipartite graph drawn in (a).
(c) Starting from this matching, use the maximum matching algorithm to find a complete matching. Indicate clearly how the algorithm has been applied, listing any alternating paths used. [E]

6 At a school fete six teachers, A, B, C, D, E and F, are to run six stalls, R, S, T, U, V and W.

A would prefer to run T but is willing to run R
B would prefer to run S but is willing to run R or W
C would prefer to run U but is willing to run S
D would prefer to run V but is willing to run R
E is willing to run T or V
F is willing to run V

(a) Draw a bipartite graph to model this situation.
Initially, A, B, C and D are matched to their preferred choices.
(b) Indicate this initial matching in a distinctive way on the bipartite graph drawn in (a).
(c) Use the maximum matching algorithm to find a maximum matching, listing clearly your alternating path.
(d) Explain why it is not possible to find a complete matching. You should make specific reference to individual stalls and teachers.

Teacher A now offers to run stall S.

(e) Draw a new bipartite graph. Hence, using the previous initial matching and the maximum matching algorithm, determine if it is now possible to obtain a complete matching. If it is possible, give the matching, stating clearly your alternating path; if it is still not possible explain why. [E]

7

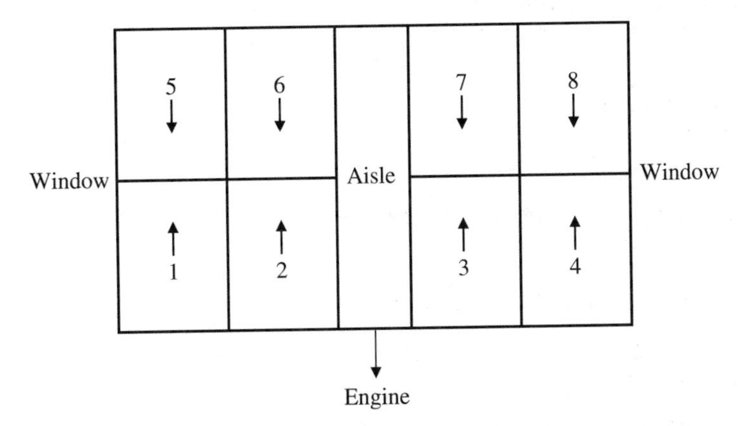

The arrows indicate the direction each seat is facing

The diagram above represents eight seats in a railway carriage, numbered 1, 2, 3, 4, 5, 6, 7 and 8. These are the last eight seats available on a special sightseeing trip. The booking clerk has to arrange the seating for the final customers. Six customers make the following requests:

Ms A wants an aisle seat facing the engine (6 or 7)
Mr B wants a window seat (1, 4, 5 or 8)
Rev C wants a seat with his back to the engine (1, 2, 3 or 4)
Mrs D wants an aisle seat (2, 3, 6 or 7)
Miss E wants a seat facing the engine (5, 6, 7 or 8)
Dr F wants a window seat with her back to the engine (1 or 4)

Initially the clerk assigns the seats as follows:

A to 6, B to 5, C to 4, D to 2, E to 7 and F to 1

The day before departure Mr and Mrs G join the trip. They ask to sit next to each other (1 and 2, 3 and 4, 5 and 6 or 7 and 8). The clerk reassigns the seats, using as far as possible the original seat assignments as the initial matching.

(a) Choose two seats for Mr and Mrs G and, using a bipartite graph, model the possible seat allocations of the other customers.

(b) Indicate, in a distinctive way, those elements of the clerk's original matching that are still possible.

(c) Using your answer to part (b) as the initial matching, apply the maximum matching algorithm. You must state your alternating path and your final maximum matching. [E]

8

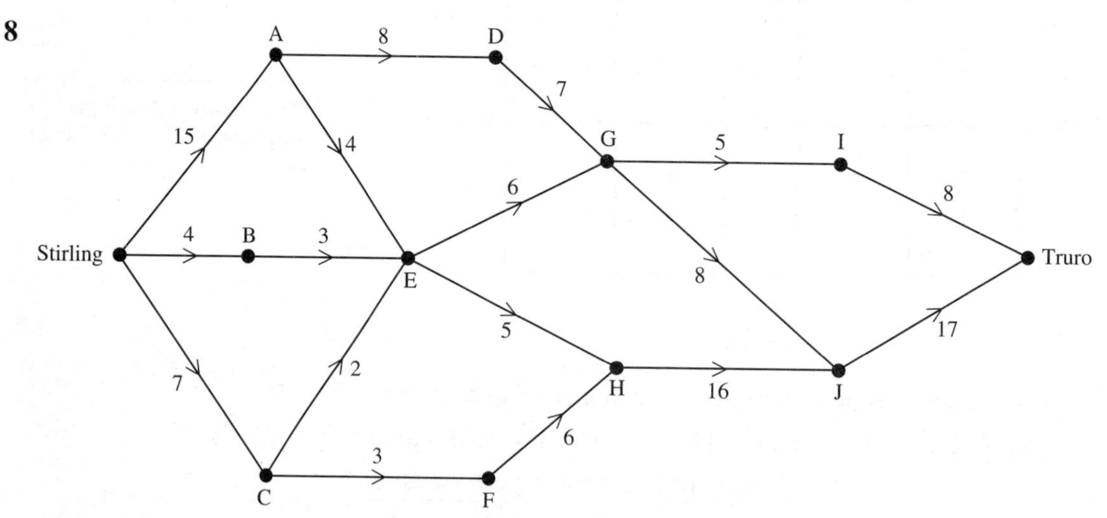

Twenty students wish to travel by bus from Stirling to Truro. The network above shows various routes and the number of free seats available on the coaches that travel on these routes between the two cities. The students agree to travel singly or in small groups and meet up again in Truro.

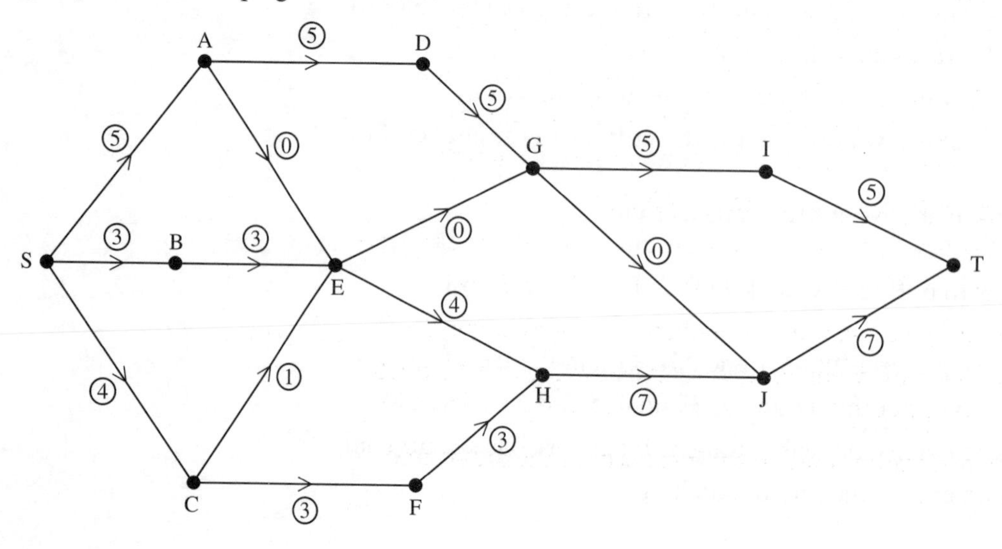

The network at the bottom of page 244 shows how 12 of the students could travel to Truro.

(a) Using this network as your initial flow pattern and listing each flow-augmenting route you use, together with its flow, use the labelling procedure to find the maximum flow.

(b) Draw a network showing the maximum flow.

(c) State how many students can travel from Stirling to Truro along these routes.

(d) Verify your answer using the maximum flow–minimum cut theorem, listing the arcs that your minimum cut passes through. [E]

9

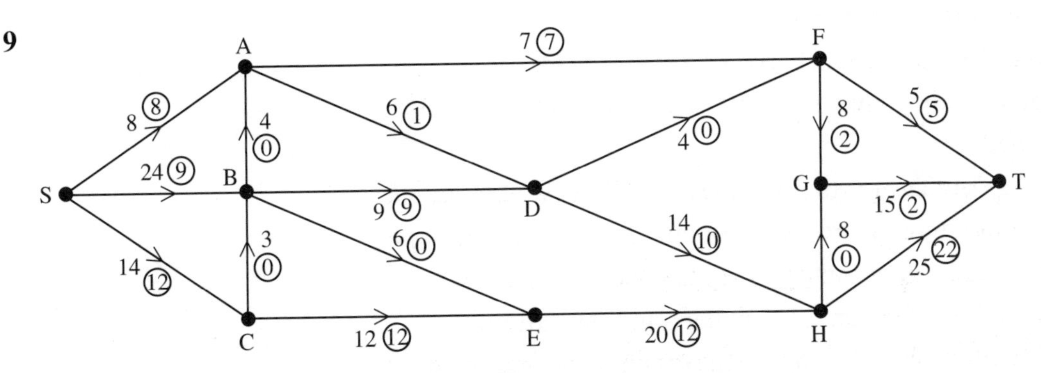

A capacitated directed network is shown above. The number on each arc is the capacity of the maximum flow along that arc.

(a) Describe briefly a situation for which this type of network could be a suitable model.

The numbers in circles show a feasible flow of value 29 from source S to sink T. Take this as the initial flow pattern.

(b) Use the labelling procedure to find the maximum flow through the network from S to T. You must list each flow-augmenting route you use together with its flow.

(c) Indicate on a diagram your maximum flow pattern and state the final flow.

(d) Verify that your answer is a maximum flow by using the maximum flow–minimum cut theorem, listing the arcs through which your cut passes.

(e) For the maximum flow, state a property of the arcs found in (d). [E]

10

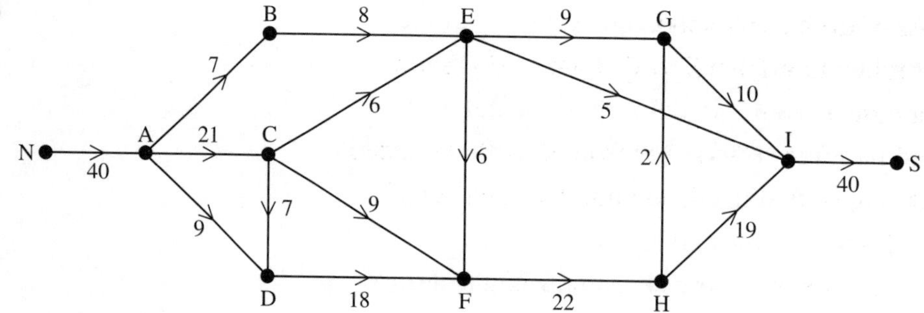

The network above shows the road routes from a bus station, N, on the north side of a town to a bus station, S, on its south side. The number on each arc shows the maximum flow rate, in vehicles per minute, on that route.

(a) State four junctions at which there could be traffic delays, giving a reason for your answer.

Given that AB, AD, CE, CF and EI are saturated:

(b) Show on a diagram a flow of 31 from N to S that satisfies this demand.

(c) Taking your answer to (b) as the initial flow pattern, use the labelling procedure to find the maximum flow. You should list each flow-augmenting route you use together with its flow.

(d) Indicate on a diagram your maximum flow pattern.

(e) Verify your solution using the maximum flow–minimum cut theorem, listing the arcs through which your minimum cut passes.

(f) Show that, in this case, there is a second minimum cut and list the arcs through which it passes. [E]

11

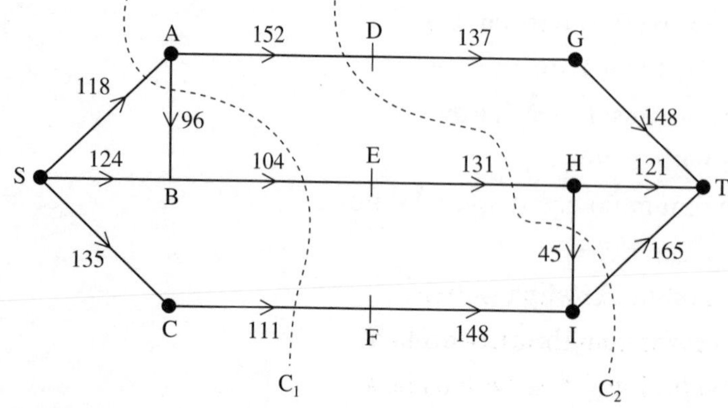

In the network above the number on each arc indicates the maximum flow possible through that arc.

(a) State the values of the cuts C_1 and C_2.

(b) By inspection, find a flow from S to T of value 333. Show your flow on a diagram.

Additional directed arcs from C to E and from B to D are added to the network. The capacity of CE is 123 and the capacity of BD is 95.

(c) Starting with your answer to part (b) as the initial flow pattern, use the labelling procedure to find the new maximal flow. You should list each flow-augmenting route together with its flow.

(d) Show your maximum flow on a diagram.

(e) Prove that your flow is maximal.

(f) Give an example of a practical situation that could have been modelled by the original network. [E]

12

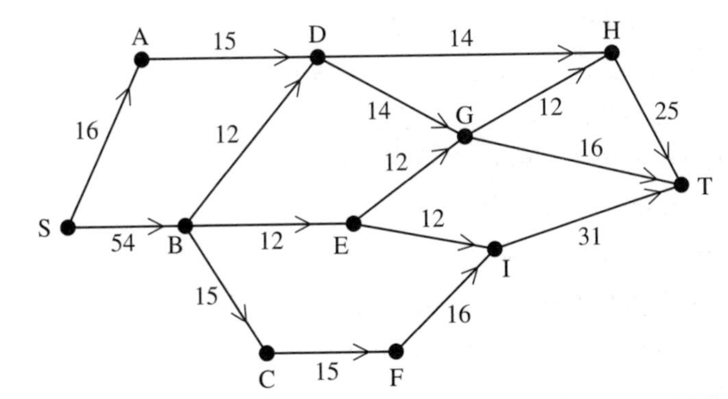

The network above represents a road system through a town. The number on each arc represents the maximum number of vehicles that can pass along that road every minute, i.e. the capacity of the road.

(a) State the maximum flow along:

(i) SBCFIT

(ii) SADHT.

(b) Show these maximum flows on a diagram.

(c) Taking your answer to part (b) as the initial flow pattern, use the labelling procedure to find a maximum flow from S to T. List each flow-augmenting route you find, together with its flow.

(d) Indicate a maximum flow on a diagram.

(e) Prove that your final flow is maximal.

The council has funding to improve one of the roads to increase the flow from S to T. It can choose to increase the flow along one of BE, DH or CF.

(f) Making your reasoning clear, explain which one of these three roads the council should improve, given that it wishes to maximise the flow through the town. [E]

13

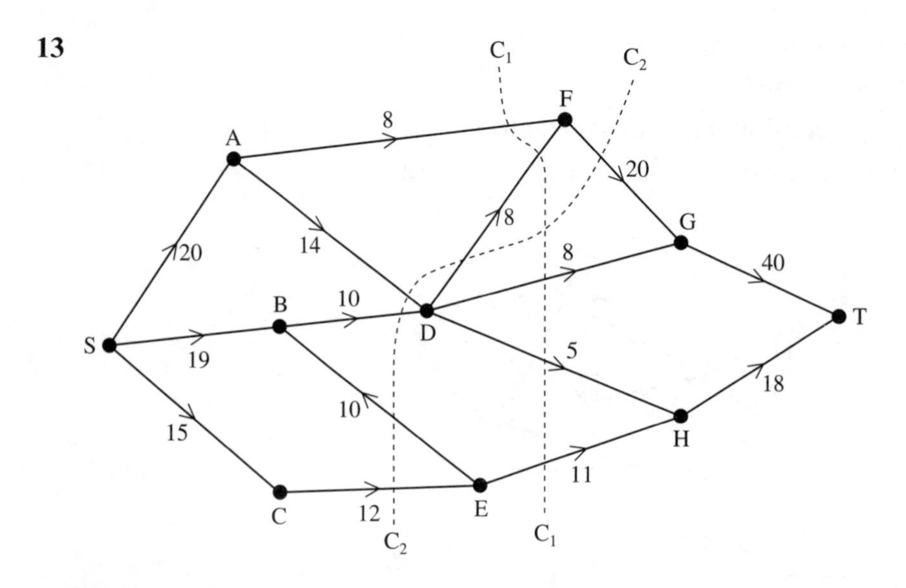

A capacitated, directed network is shown above. The number on each arc indicates the capacity of that arc.

(a) Calculate the values of cuts C_1 and C_2.

Given that one of these cuts is a minimum cut:

(b) State the maximum flow.

(c) Deduce the flow along GT, making your reasoning clear.

(d) By considering the flow into D, deduce that there are only two possible integer values for the flow along SA.

(e) For each of the two values found in part (d), draw a complete maximum flow pattern.

(f) Given that the flow along each arc must be an integer, determine the number of other maximum flow patterns.

Give a reason for your answer. [E]

14

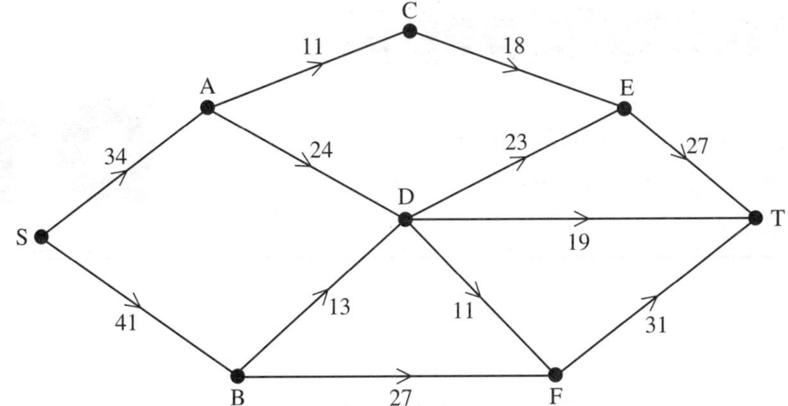

A capacitated network is shown above. The numbers on the arcs indicate the capacities of the arcs.

(a) State the maximum flow along SADT for this network.

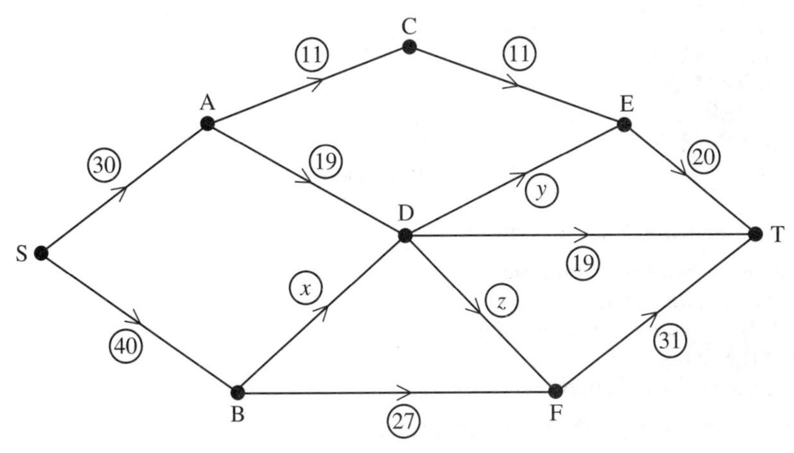

A feasible flow of value 70 through the same network is shown above.

(b) Explaining your reasoning carefully, work out the value of the flows x, y and z.

(c) Explain why 70 is not the maximum flow.

(d) Using this as your initial flow pattern, use the labelling procedure to find the maximum flow through this network. You should list each flow-augmenting route you use, together with its flow.

(e) Draw a network showing the maximum flow.

(f) Verify your answer using the maximum flow–minimum cut theorem, listing the arcs that your minimum cut passes through. [E]

Examination style paper

D1

Answer all questions **Time allowed 90 minutes**

1. Use the binary search algorithm to locate the name JONES in the following alphabetical list:

1 ALLEN
2 BIGGS
3 BROWN
4 CHARLES
5 DIGGENS
6 FIELD
7 GRANT
8 HILL
9 JONES
10 KNOWLES
11 NORMAN
12 PIPER
13 STONER

(6 marks)

2. Five families each wish to rent a particular villa for a month.

The Baileys wish to go in either May or September
The Craigs wish to go in either June or July
The Dales wish to go in July or August or September
The Evans wish to go in either May or June
The Fords wish to go in either July or September

In an initial allocation of months, the Baileys, the Craigs and the Dales are given their earliest preferences.

(*a*) Draw a bipartite graph to model this situation and indicate the initial allocation in a distinctive way. **(2 marks)**

(*b*) Starting from this initial matching use the maximum matching algorithm to find a complete matching. Indicate clearly how the algorithm has been applied and state the complete matching. **(6 marks)**

3.

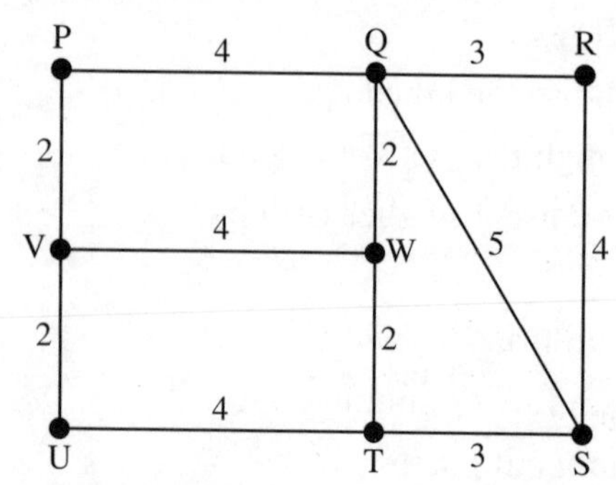

The above logo for HN rugby club is to be machined onto a T-shirt. The numbers on the edges are the lengths, in cm, of the edges.

(*a*) The first machinist starts at V, machines without a break and finishes at V. Use the Chinese postman algorithm to obtain the minimum length of stitching. State this minimum length and a possible route. **(6 marks)**

(*b*) A second machinist is given the same task and produces the logo starting and finishing at different vertices. Determine which vertex she started at and which vertex she finished at if the length of stitching is a minimum. Give the route and the length of this route. **(4 marks)**

4.

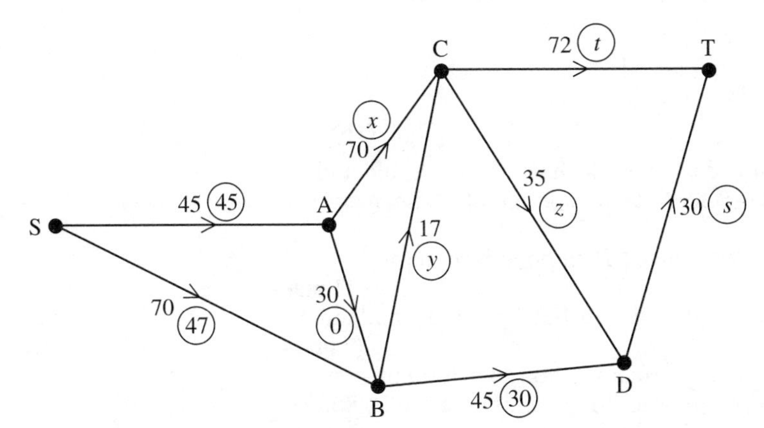

The figure above shows a capacitated network. The unringed numbers on the arcs indicate the capacities of the arcs. The ringed numbers show a feasible flow.

(*a*) Show that $x = 45$ and $y = 17$. **(3 marks)**

(*b*) Explain why $z = 0$ and $s = 30$. **(3 marks)**

(*c*) Deduce the value of t. **(1 mark)**

(*d*) Show that the given flow is a maximum flow from S to T by finding a cut whose capacity has the same value. **(3 marks)**

5.

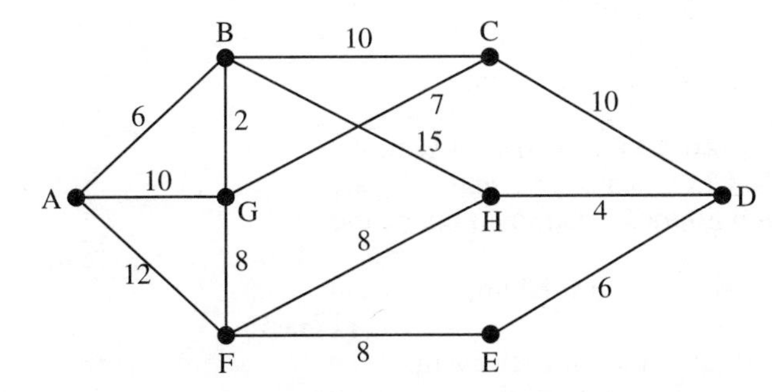

The diagram above shows eight villages, A, B, C, D, E, F, G and H, and the roads connecting them. The distances on the edges are given in miles.

(*a*) Use Dijkstra's algorithm to find a shortest route from A to D. State clearly the order in which the vertices are labelled and show clearly how you determined your shortest path from your labelling. Give its length. **(7 marks)**

Due to road works the road HD is closed.

(*b*) Find the shortest route now and give its length. **(4 marks)**

6. A manufacturer makes two kinds of chairs, A and B, each of which has to be processed in three departments, I, II and III. Chair A has to be processed in Department I for 2 hours, in Department 2 for 1 hour and in Department III for 6 hours. Chair B has to be processed in Department I for 2 hours, in Department II for 5 hours and in Department III for 2 hours. In a particular week there are 24 hours available in Department I, 44 hours available in Department II and 60 hours available in Department III.
Let x be the number of chairs of type A made in this week and y be the number of chairs of type B made in this week.
(*a*) Explain why this situation can be modelled by the inequalities:

$$x + y \leqslant 12$$
$$x + 5y \leqslant 44$$
$$3x + y \leqslant 30$$
$$x \geqslant 0, y \geqslant 0$$

(3 marks)

The manufacturer makes a profit of £6 on each chair of type A sold and a profit of £9 on each chair of type B sold. Assuming that all chairs made are sold:
(*b*) Write down an expression for the profit P, in pounds, in terms of x and y. **(1 mark)**
(*c*) On graph paper display the inequalities satisfied by x and y and indicate clearly the feasible region. **(4 marks)**
(*d*) Use either the vertex method or the ruler method to determine how many of each chair should be made in order to maximise the profit. State the maximum profit. **(5 marks)**

7.

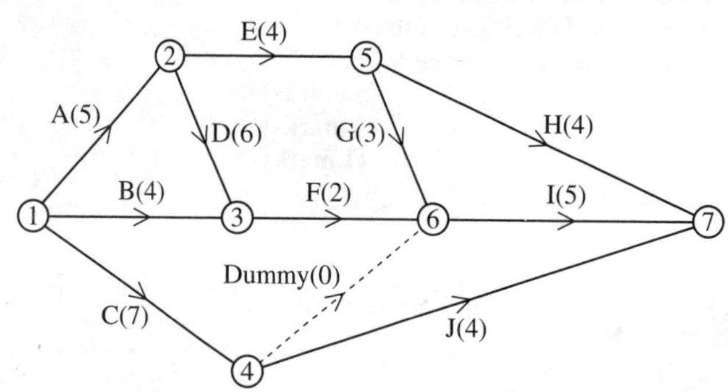

The diagram above shows an activity network for a project. The nodes represent events and the activities A, B, C, . . . , J are represented by arcs. The number in each bracket gives the time in hours needed to complete the activity.
(*a*) Explain what the dummy, represented by the dotted line, means in practical terms. **(1 mark)**
(*b*) (i) Calculate the early time and the late time for each event.
(ii) Hence obtain the critical activities and the length of the critical path. **(7 marks)**
(*c*) Calculate the float for each of the non-critical activities. **(2 marks)**
(*d*) Draw a Gantt (cascade) chart to show the information obtained in parts (*b*) and (*c*). **(3 marks)**
Each activity requires one worker.
(*e*) (i) Show that at least three workers are required to complete the project in the critical time.
(ii) Draw up a possible schedule to show that the project can be completed in the critical time by three workers. **(4 marks)**

Answers

Exercise 1A

1 (a) $x = 2, x = -1$ (b) $x = -\frac{1}{3}, x = 5$

(c) No real solutions

2 (a) 13 (b)18

3 1, 4, 9, 16, 25, 36, 49, 64, 81, 100

All square numbers less than or equal to 100

4 1, 2, 4, 5, 10, 20, 25, 50, 100

Works out all factors of 100

5

1
1 1
1 2 1
1 3 3 1
1 4 6 4 1
1 5 10 10 5 1
1 6 15 20 15 6 1
1 7 21 35 35 21 7 1

6 1, 1, 2, 3, 5, 8, 13, 21, 34, 55, 89, 144

Exercise 1B

1 (a) 42, 45, 50, 55, 68, 70

(b) 70, 68, 55, 50, 45, 42

2 2, 4, 5, 6, 8, 9, 10

3 Jones, Malik, Monro, Shah, Wilson

4 −5, −3, −1, 0, 2, 4, 5

5 (a) & (b) 8.9, 9.0, 9.1, 9.2, 9.6, 9.7, 9.8

(c) quick sort

6 DE(6), CE(7), BE(8), CD(10), BC(12), AD(16), AB(18)

Exercise 1C

1 (a) 5 workers (b) 5 workers (c) No

2 Yes

3 2 workers

4 3 lengths

5 (a) 4 (b) 3 (c) 3

6 (a) 150, 120, 100, 90, 80, 78, 60, 38, 26

(i) Bin 1: 150 + 38

Bin 2: 120 + 80

Bin 3: 100 + 90

Bin 4: 78 + 60 + 26

(ii) 78 could not go in bin 1 because it only had space for 50 available. Bin 2 was full and bin 3 only had space for 10 available. 78 therefore had to go into a new bin, bin 4.

Exercise 1D

1 (a) (3) Orange (b) (2) 25 (c) (4) CLARK

(d) (9) EDNA (e) (8) KNOWLES

4 10.30 a.m.

Exercise 2A

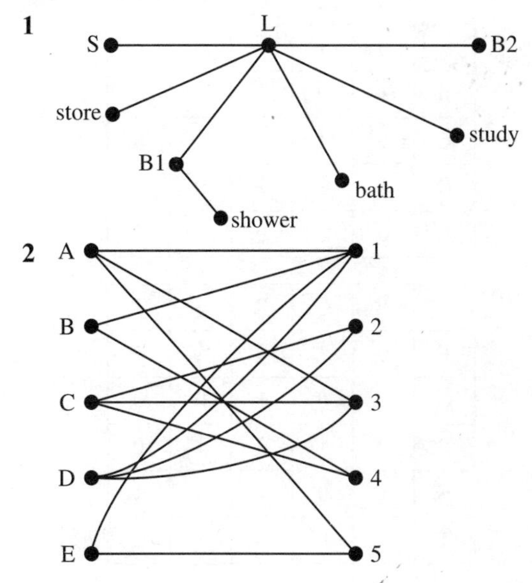

3

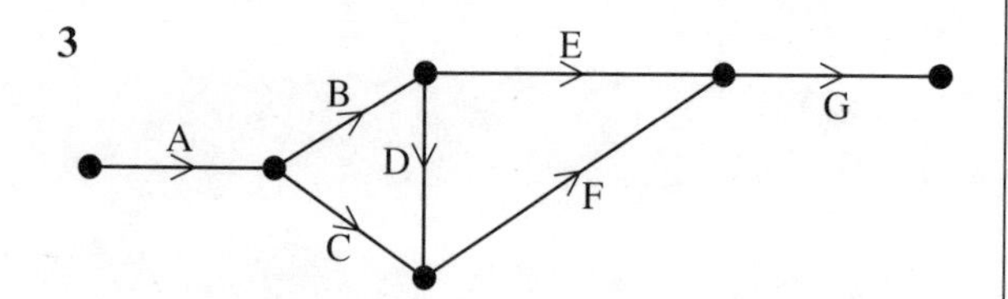

4 (a) eg ABFCD, ABFED, AGFCD, AGFED
(b) A(2), B(3), C(3), D(2), E(3), F(4), G(3)
(c) eg ABCDEGA, ABFCDEGA
(d) eg ABFCDEGA, ABCDEFGA

5 (a) P(4), Q(2), R(4), S(2), T(4), U(2)
(b) eg PQRSTRPTUP
(c) No. P and R now have odd degrees. To obtain an Eulerian cycle all degrees must be even.

6 ABC, ADC, AEC, ADEC

7 Not connected. There is no path between A and B.

Exercise 2B

1 (a) {A, B, C, D, E}
{AB, AD, BC, BD, CD, DE}
(b) {P, Q, R, S, T}
{PQ, PS, PT, QR, RS, RR, ST, TT}

2 (a)

	A	B	C	D	E
A	0	1	0	1	0
B	1	0	1	1	0
C	0	1	0	1	0
D	1	1	1	0	1
E	0	0	0	1	0

(b)

	P	Q	R	S	T
P	0	1	0	1	1
Q	1	0	1	0	0
R	0	1	2	1	0
S	1	0	1	0	1
T	1	0	0	1	2

3

	G	UR	LR	K	D	H	T	L
G	0	1	0	0	0	0	0	0
UR	1	0	1	0	0	0	0	0
LR	0	1	0	1	0	0	0	0
K	0	0	1	0	1	1	0	0
D	0	0	0	1	0	0	0	1
H	0	0	0	1	0	0	1	1
T	0	0	0	0	0	1	0	0
L	0	0	0	0	1	1	0	0

4 To (columns), From (rows)

From \ To	A	B	C	D	E
A	0	1	0	0	1
B	0	0	1	0	0
C	0	0	1	0	0
D	1	1	1	0	0
E	0	0	0	1	0

5 {A, B, C, D, E}
{AB, AE, BC, CC, DA, DB, DC, ED}

6 To (columns), From (rows)

From \ To	P	Q	R	S	T
P	0	1	0	1	1
Q	1	0	1	1	0
R	0	1	0	0	1
S	1	0	1	0	1
T	1	0	1	0	2

7

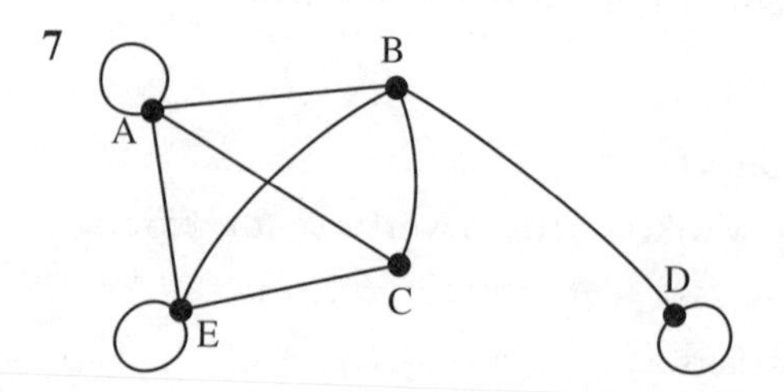

8

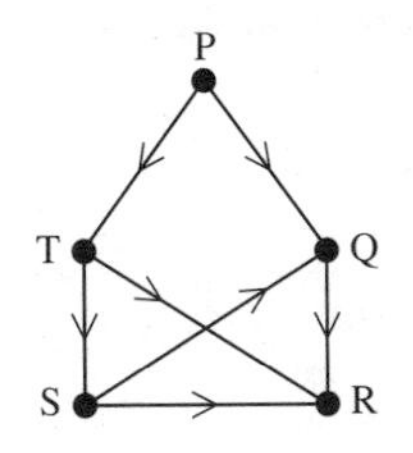

9

		To				
		X	**Y**	**Z**	**V**	**W**
From	**X**	0	0	1	1	1
	Y	1	0	0	0	0
	Z	1	0	0	0	0
	V	1	1	1	0	1
	W	0	0	0	1	2

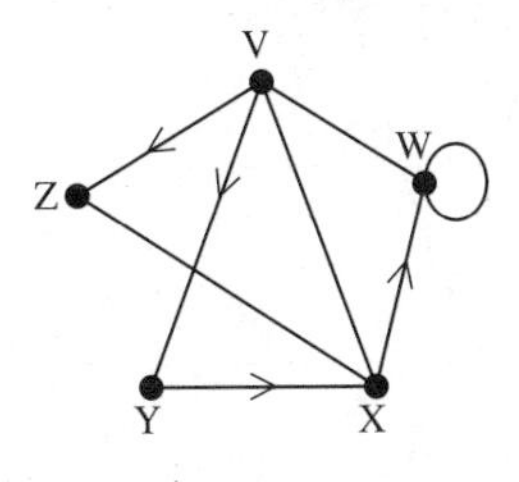

10 (a) Every edge starts and ends at a vertex, so each edge contributes 2 to the sum of the degrees of vertices. If there are n vertices, the sum of the degrees of the vertices is $2n$, which must be even.

(b) As the sum of degrees of vertices is even, so must the sum of the degrees of the odd vertices be even. The number of odd vertices must therefore be even.

(c) Draw a graph with seven vertices representing people. Join with an edge pairs who are friends. If each was friends with five others, the degree of each vertex would be 5 and the total sum of the degrees of the vertices would be $7 \times 5 = 35$. This is an odd number and so contradicts the rule in (a). The contradiction shows it is not possible for each person to be friends with exactly five others.

Exercise 2C

1 (a) Tree (b) Not a tree
(c) Not a tree (d) Tree

2

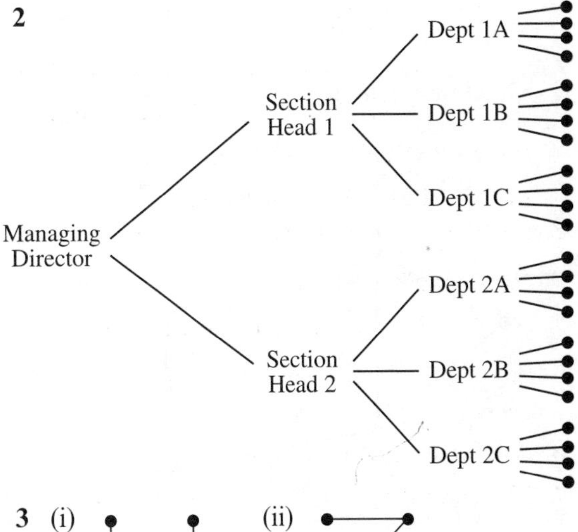

3

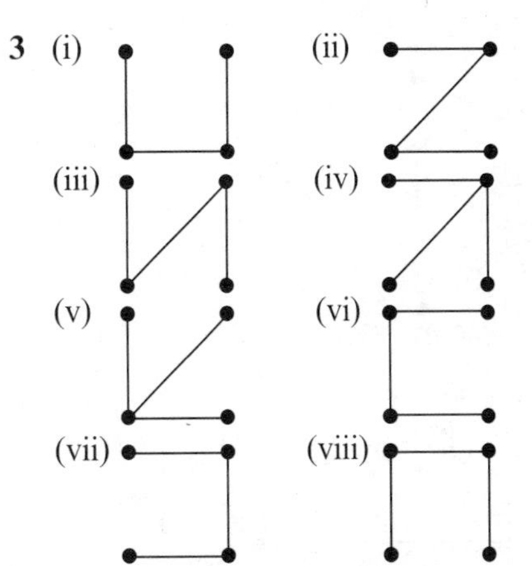

4

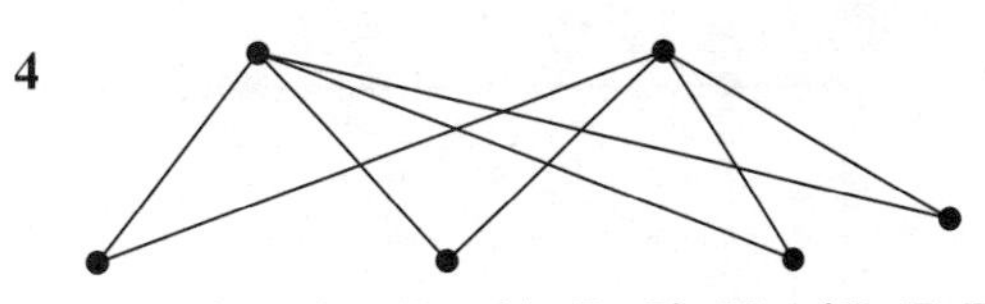

5 (a) Bipartite: X = {A, B, C}; Y = {D, E, F}

(b) Not bipartite: P joined to R, S and T, not Q. Q joined to R, S and T, not P but R is joined to S

6 (a) Satisfies the triangle inequality: in $\triangle ABD$ and $\triangle BCD$ inequality is satisfied

(b) Does not satisfy the triangle inequality: inequality satisfied in 'triangles' SQR and PSR but not in PSQ ($10 + 13 < 25$)

7 (a) (i) 16; (ii) 23; (iii) 19; (iv) 20; (v) 22; (vi) 18; (vii) 17; (viii) 15

(b) Least weight 15;

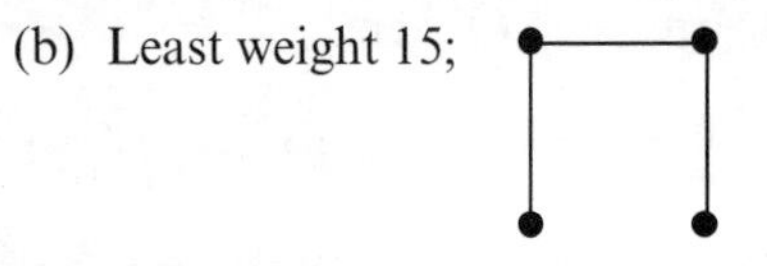

Greatest weight 23;

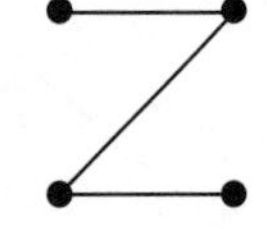

8

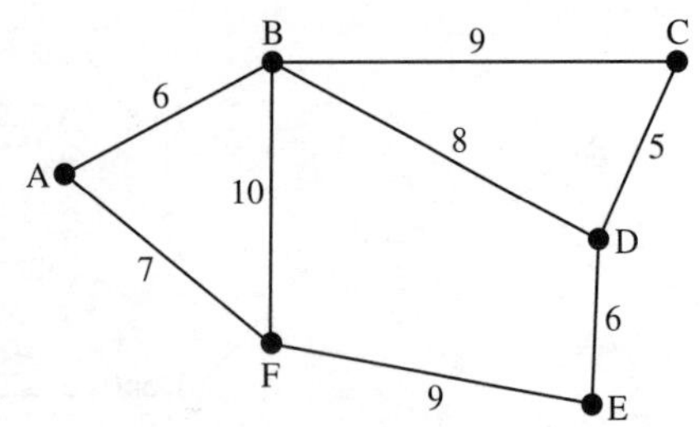

9

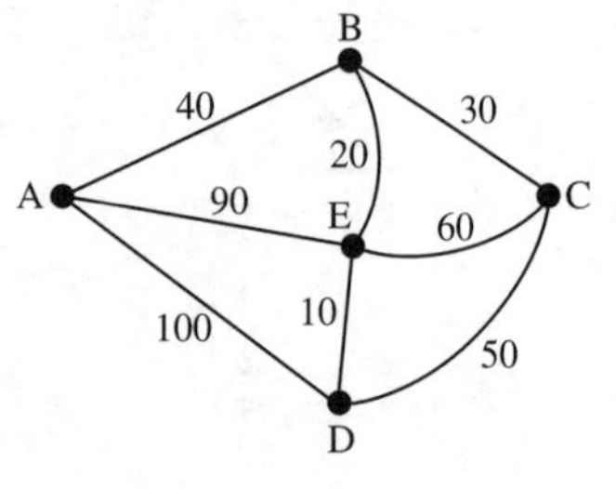

EB + BC = 50 < 60 (EC)

EB + AB = 60 < 90 (AE)

10

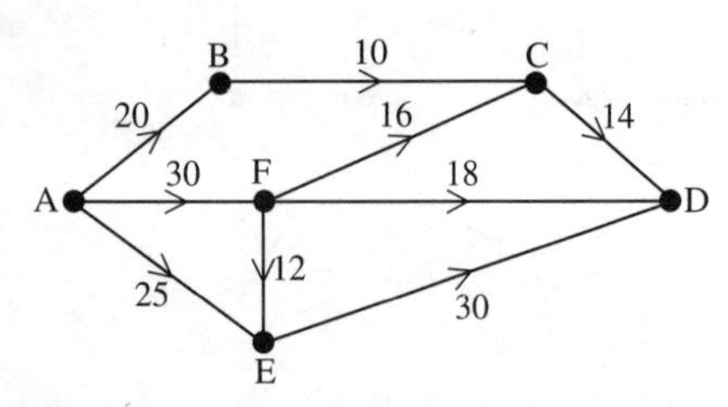

Exercise 3A

1

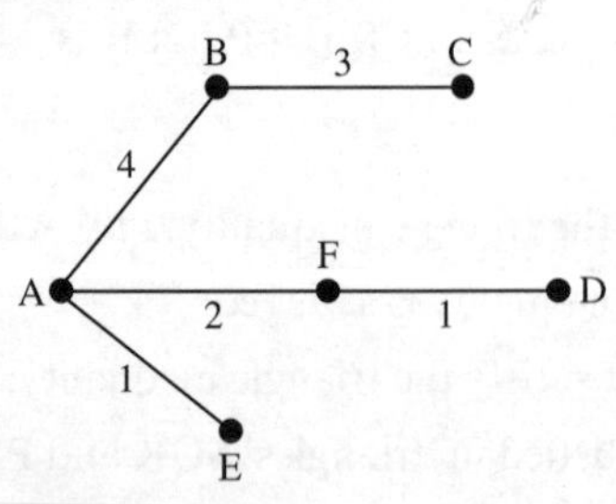

Total weight = 11

2

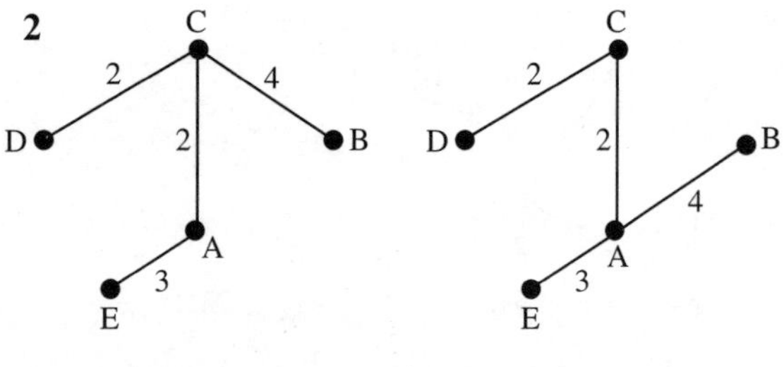

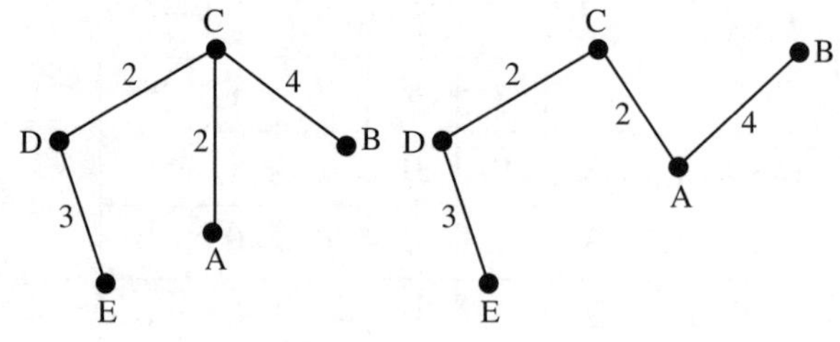

3 (a)

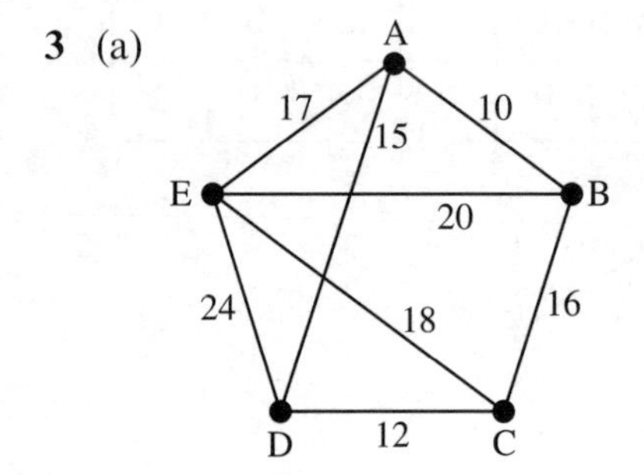

(b)

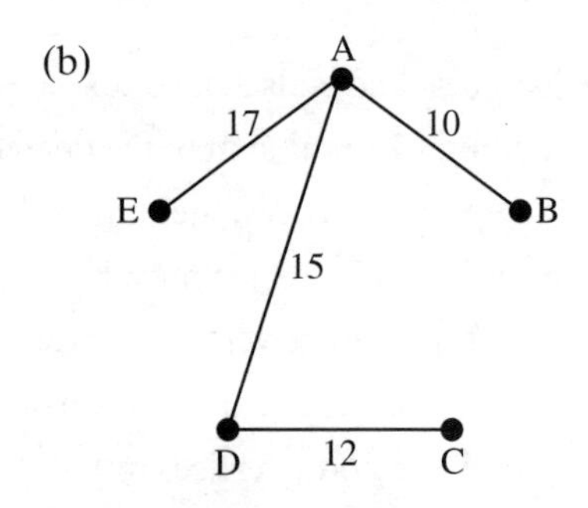

Total weight = 54

4

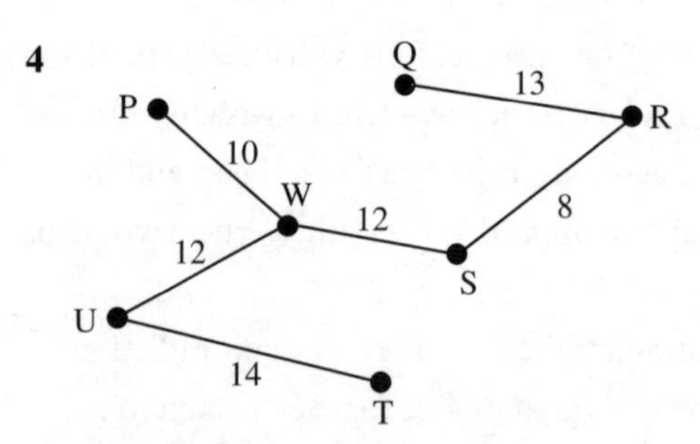

Minimum total length = 69 km

5

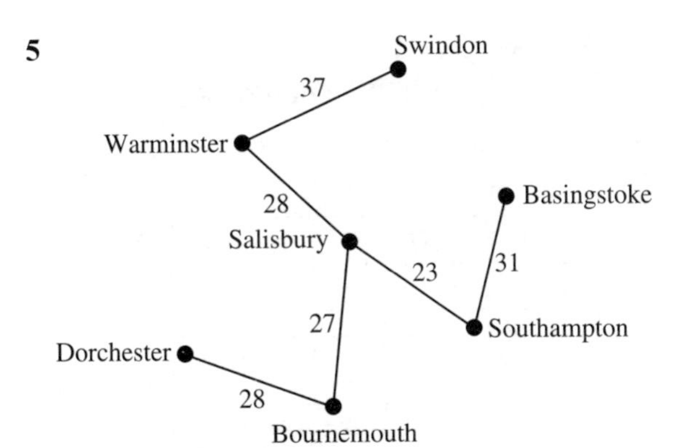

Minimum length of cable:

$23 + 27 + 28 + 28 + 31 + 37 = 174$ miles

Sa–So = 23

Sa–Bo = 27

Bo–Do = 28

Sa–Wa = 28

So–Ba = 31

(Bo–So = 31 cycle)

(Sa–Ba = 36 cycle)

Wa–Sw = 37

All vertices connected

Exercise 3B

1

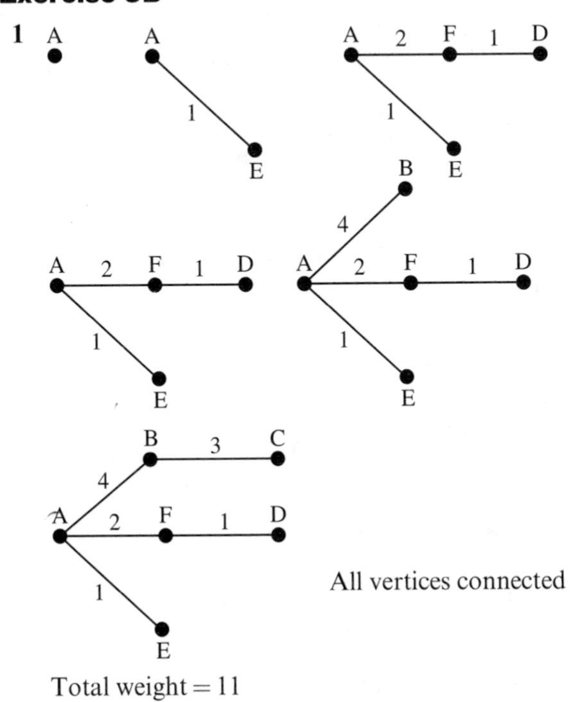

All vertices connected

Total weight = 11

2

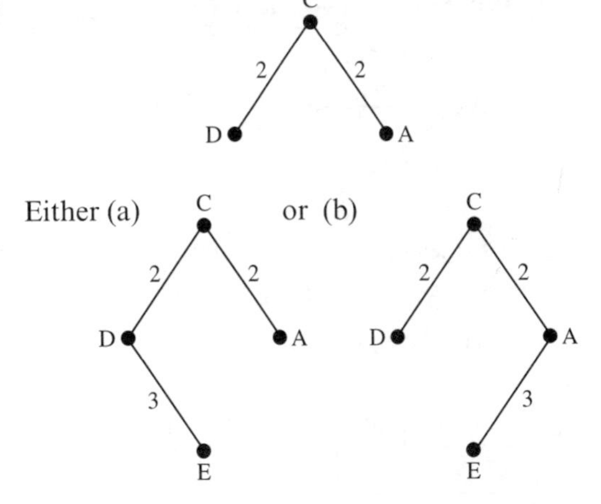

From (a):

Either (c) or (d)

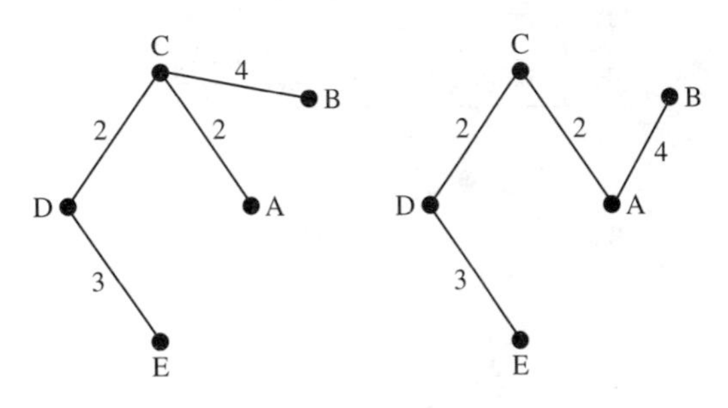

From (b):

Either (e) or (f)

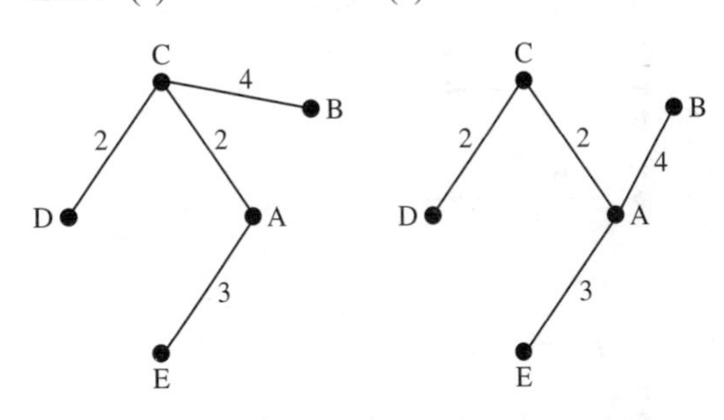

3 (a)

	1	5	6	4	2	3
	A	**B**	**C**	**D**	**E**	**F**
A	—	4	—	—	1	2
B	(4)	—	3	—	—	5
C	—	(3)	—	7	—	6
D	—	—	7	—	3	(1)
E	(1)	—	—	3	—	—
F	(2)	5	6	1	—	—

(b)

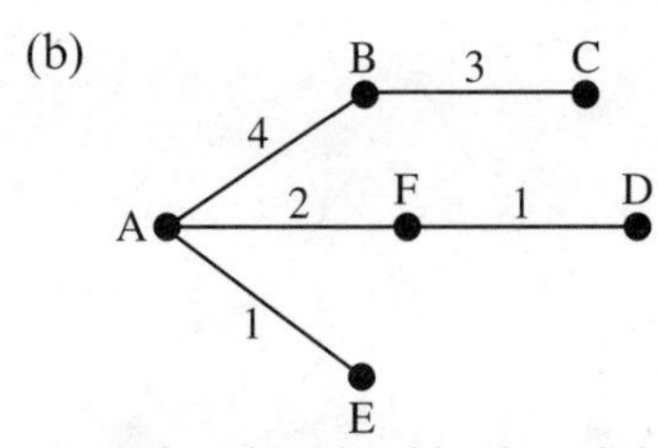

Edges involved in the minimum spanning are AB, AE, AF, BC and FD.

Total weight = 11

4

	1	2	4	3	5
	A	**B**	**C**	**D**	**E**
A	—	10		15	17
B	10	—	16	—	20
C	—	16	—	12	18
D	15	—	12	—	24
E	17	20	18	24	—

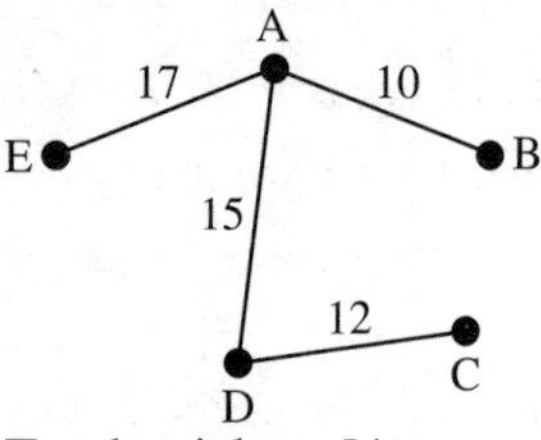

Total weight = 54

5

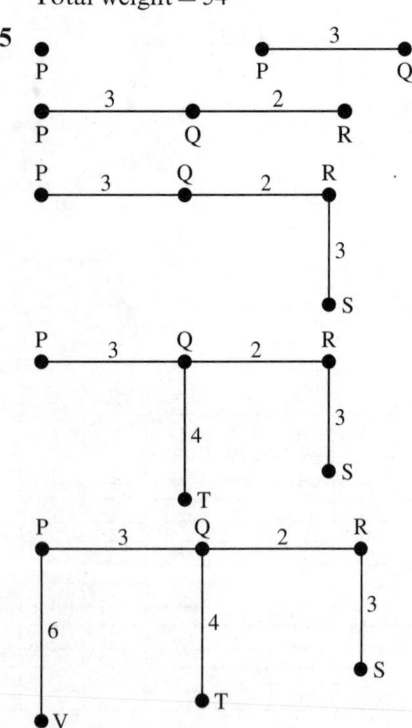

Total weight = 18

Edges used: PQ, PV, QR, QT, RS

Total weight: $3 + 6 + 2 + 4 + 3 = 18$

	1	2	3	4	5	6
	P	**Q**	**R**	**S**	**T**	**V**
P	—	3			5	6
Q	3	—	2	—	4	—
R	—	2	—	3	6	—
S	—	—	3	—	5	—
T	5	4	6	5	—	7
V	6	—	—	—	7	—

6

	1	5	2	3	4	6
	A	**D**	**G**	**L**	**S**	**W**
A	—	78	56	73	71	114
D	78	—	132	121	135	96
G	56	132	—	64	80	154
L	73	121	64	—	144	116
S	71	135	80	144	—	185
W	114	96	154	116	185	—

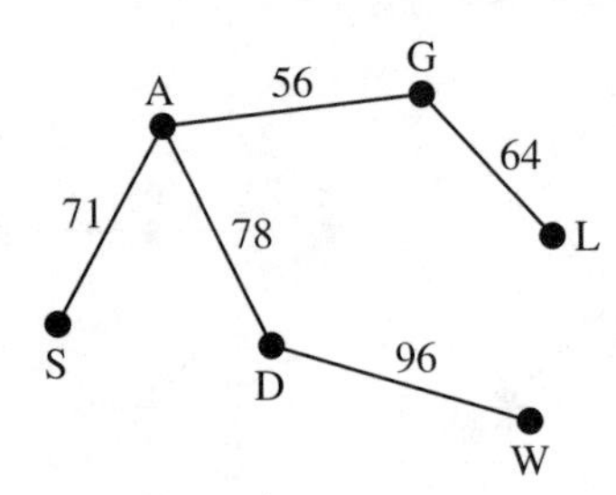

Total length:

$78 + 56 + 71 + 96 + 64 = 365$ miles

7

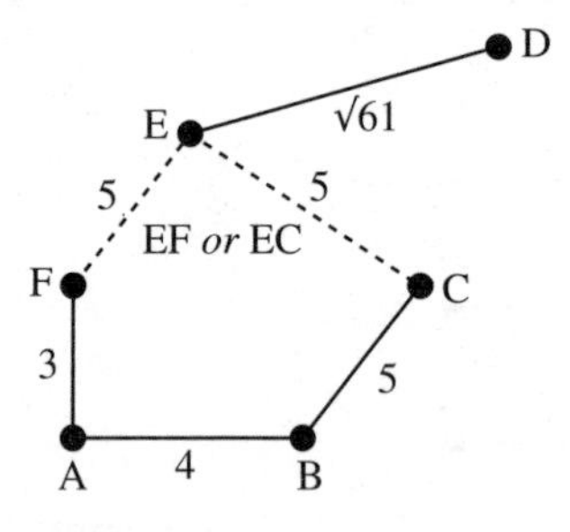

Length: $3 + 4 + 5 + 5 + \sqrt{61}$

$= 17 + \sqrt{61}$

$= 24.8$

8 (a) eg

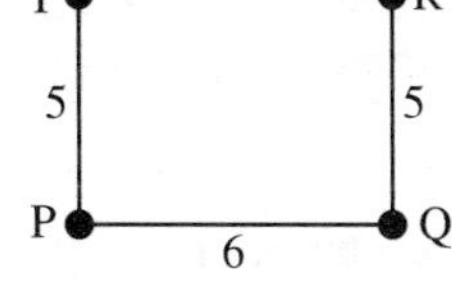

Length = 16

(b)

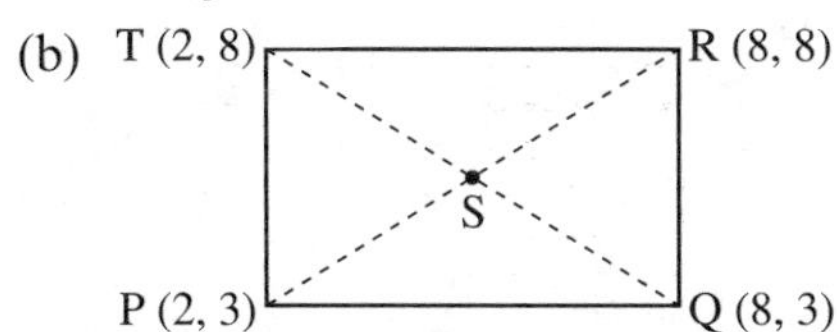

S has coordinates $(5, 5\frac{1}{2})$ and is the mid-point of the diagonals

New connector is

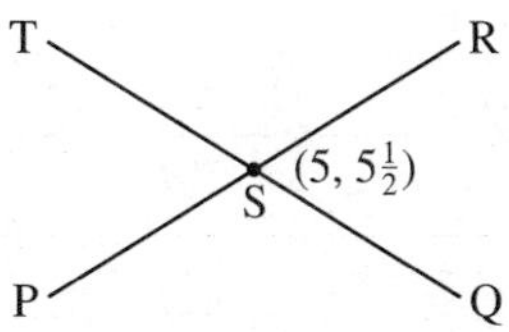

Length PS = length SR = length SQ $= \text{length ST} = \frac{1}{2}\sqrt{[(6)^2 + (5)^2]} = \frac{1}{2}\sqrt{61}$

Total length of connector $= 2\sqrt{61} = 15.6 (< 16)$

Exercise 3C

1 (a) SACT or SABT; 46

(b) SPMT; 24

(c) SGKT; 93

2 eg

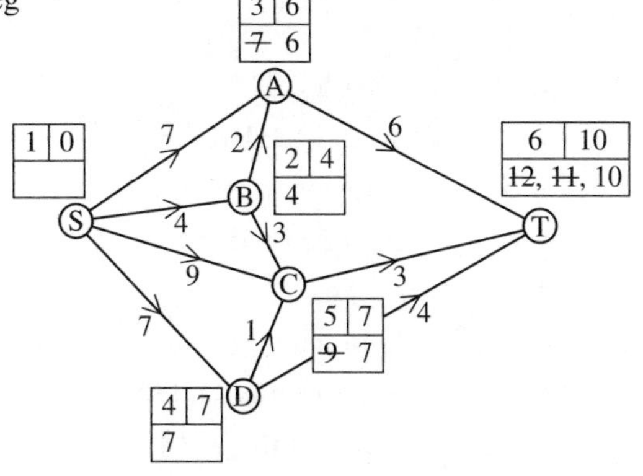

Route includes:

CT, since label T(10) – label C(7) = CT

BC, since label C(7) – label B(4) = BC

SB, since label B(4) – label S(0) = SB

Route is $S \to B \to C \to T$

Cost $(4 + 3 + 3) = 10 \times £100$

$= £1000$

3

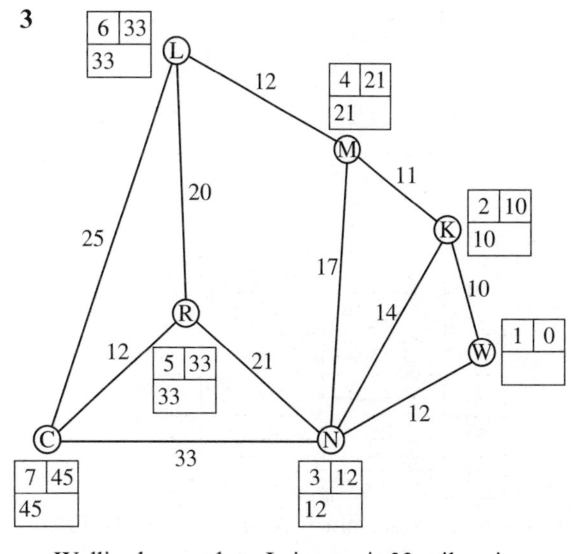

Wellingborough to Leicester is 33 miles via Kettering and Market Harborough.

Wellingborough to Coventry is 45 miles via Northampton.

Mrs Brown will therefore go to Leicester.

4

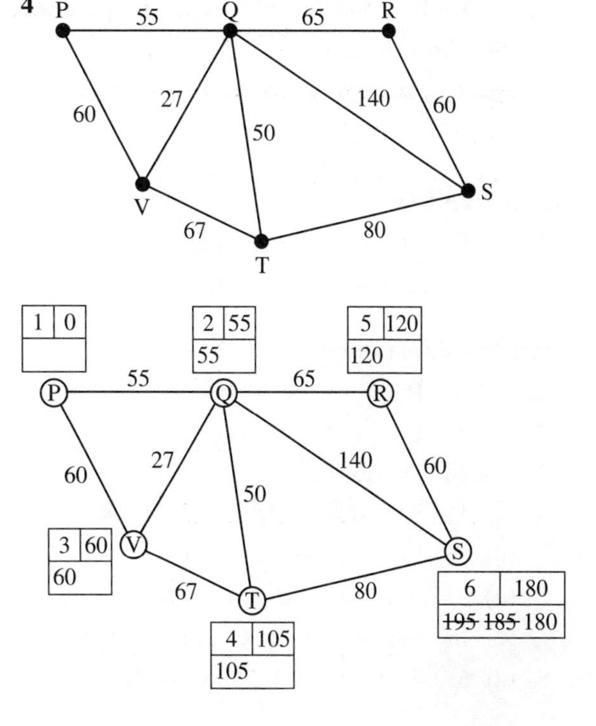

Hence the cheapest route costs 180p and is PQRS, since

label S(180) – label R(120) = RS(60)

label R(120) – label Q(55) = QR(65)

label Q(55) – label P(0) = PQ(55)

5 (a)

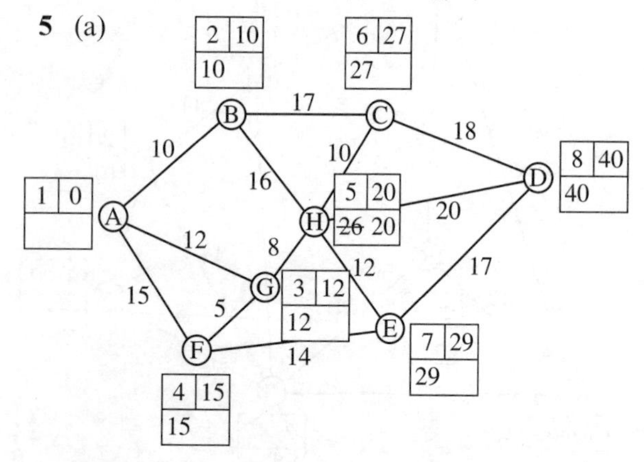

The shortest route is AGHD, of length 40 km.

(b) Convert distances to times in minutes. Either add 5 minutes to all edges incident on G, so ensuring that entering and leaving G collects a delay of 10 minutes, or delete G and compare the fastest time from A to D in the reduced network to 40 + 10 = 50 minutes, which is the quickest time through G.

Quickest route is now ABCD, which takes 45 minutes.

Exercise 3D

2 (a) Planar (b) Planar

(c) Non-planar

Review exercise 1

1 1, 2, 15, 16, 27, 38

2 (b) $5 + 4 + 3 + 2 + 1 = 15$

3 2, 4, 8, 9, 13, 15, 17

4 (a) 115, 111, 103, 98, 93, 82, 81, 77, 68

(b) (i) Five bins required.

(ii) 77 was placed in the fourth bin because it was the first one with space available for 77.

5 (a) 65 minutes (b) Not possible

6 (a) Total lengths = 12 m. Full bin solution not possible

7 (a) 4.8

(b) It selects the number nearest to 5.

(c) It selects the number furthest from 5.

8 (b)

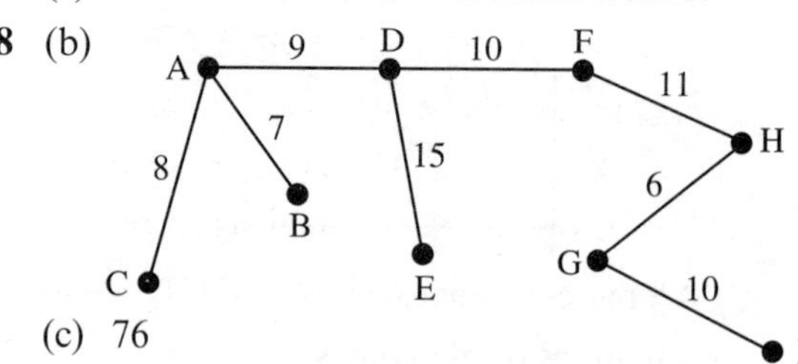

(c) 76

(d) Prim, because it does not require a check for cycles.

9 (a)

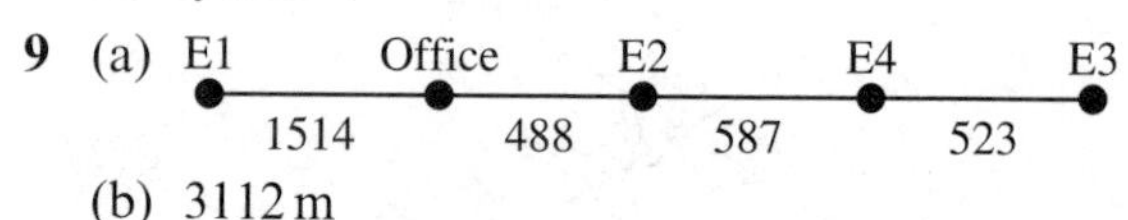

(b) 3112 m

10 (b)

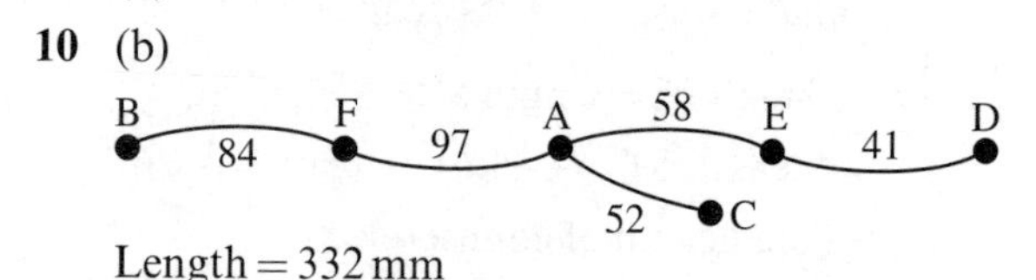

Length = 332 mm

11 Shortest route = SCEFT

Length = 411 miles

12 (a) Shortest route = WIPGYS

Length = 368 miles

(b) Start from S

13 Cheapest route = ABFH

Cost = £120

14 (a) ABCK, length = 44

(b) AHGEIK and AHJIK

(c) vertices = towns, arcs = roads, weights = distances. Find the shortest route between two towns.

15 (a) AFHJ, 43 minutes

(b) Use the algorithm to find shortest path from J to C then add this to CA.

16 (a) (ii) $x = 4$

(b) 18 to 23

(c) Find a minimum cost route from S to T using public transport.

17 (a) Quickest route takes 63 minutes.
(b) AC (or D) EHJ

Exercise 4A

1 Valencies: A(3), B(2), C(3), D(2)
There are two odd vertices and so the graph is semi-Eulerian. Possible route: ABCDAC

2 Valencies: P(3), Q(3), R(3), S(3), T(3), U(5)
There are more than two odd vertices and so the graph is not semi-Eulerian.

3. Valences: A(4), B(4), C(2), D(4), E(4), F(2), G(4)
All the vertices are even and so the graph is Eulerian. Eulerian cycle: ABCDBGDEGAEFA

4 Valencies: U(4), V(4), W(3), X(3), Y(3), Z(3)
Not all the vertices even so the graph is not Eulerian.
Add XZ and YW
or YZ and XW
or WZ and XY

5 **1** No. of edges = 5
sum of valencies = 10 (2×5)
No. of odd vertices = 2 (even)
2 No of edges = 10,
sum of valencies = 20 (2×10)
No. of odd vertices = 6 (even)
3 No. of edges = 12,
sum of valencies = 24 (2×12)
No. of odd vertices = 0 (even)
4 No. of edges = 10,
sum of valencies = 20 (2×10)
No. of odd vertices = 4 (even)

Exercise 4B

1 (a) C(2), S(4), L(2), R(4), W(2), H(4)
(b) As all vertices are even the graph is Eulerian and so is traversable. A possible route is CSLHWRHSRC, of length 223 m.

2 Odd vertices are P(3) and R(3). Repeat edges PQ and QR, therefore weight $= 8 + 5 = 13$.
Possible route: TUPQRQPVRSVUST
Length $= 65 + 13 = 78$

3 Odd vertices are A(3) and C(3).
The shortest route between A and C is AHJC, of length 8.
A possible route:
ABCDEFGAHBJCJDFJGHJHA.
Weight $= 48 + 8 = 56$.

4 Odd vertices are A(3), B(3), C(3) and E(5).
Best pairing is (A with B) and (C with E) *or* (A with E) and (B with C). Repeated edges are AB and CE *or* AE and BC, of length 9.
Possible route: ABCEDCEFAEBA
Length $= 35 + 9 = 44$ cm.

5 (a)

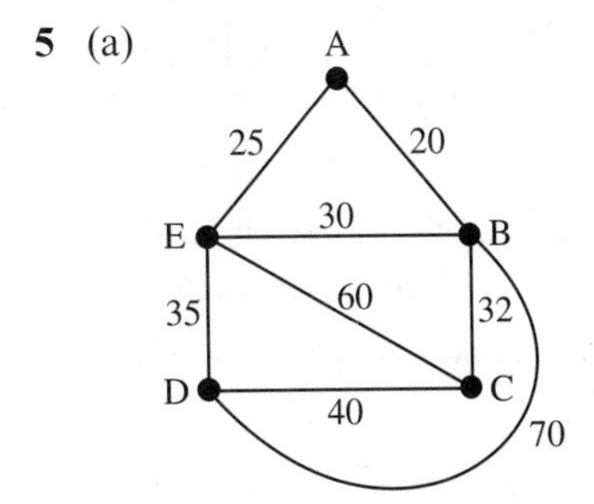

(b) Odd vertices are C(3) and D(3). Repeat edge CD, weight 40.
Possible route: ABCDECDBEA.
Weight $= 312 + 40 = 352$

6 Odd vertices are Sw(3), Ba(3), So(3) and Bo(3).
Best pairing is (Sw with Ba) and (So with Bo).
Edges to be repeated are SwBa(41) and SoBo(31).
Possible route:
SoWSwSaBaSwBaSoBoSaSoBoDSa
Length $= 362 + 72 = 434$ miles.

7 Odd vertices are X(3) and Z(3).
Edges to be repeated are XY(50) and YZ(45).
Possible route: SYZUVWSXYZVYXVS.
Cost $= 927 + 95 =$ £10.22.

Exercise 5A

1

Activity	*Depends on*
A	—
B	—
C	—
D	A
E	B
F	B, C
G	B, D
H	G
I	E, F, H

2

Activity	*Depends on*
A	—
B	—
C	—
D	—
E	—
F	B, D, E
G	C, F
H	A, G
I	H

3

Activity	*Depends on*
A	—
B	—
C	A
D	B
E	C, D

4

Activity	*Depends on*
A	—
B	—
C	—
D	A, B
E	D
F	D
G	E, F
H	F
I	C, G
J	H, I

Exercise 5B

1

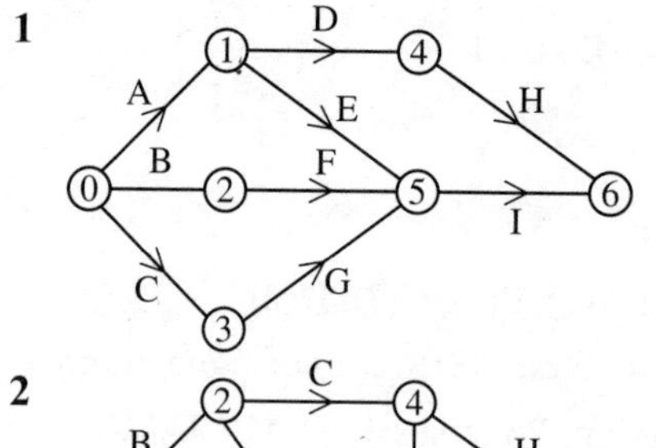

2

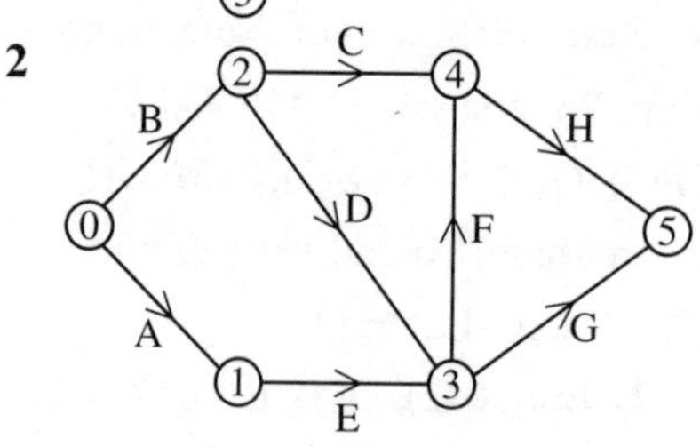

3

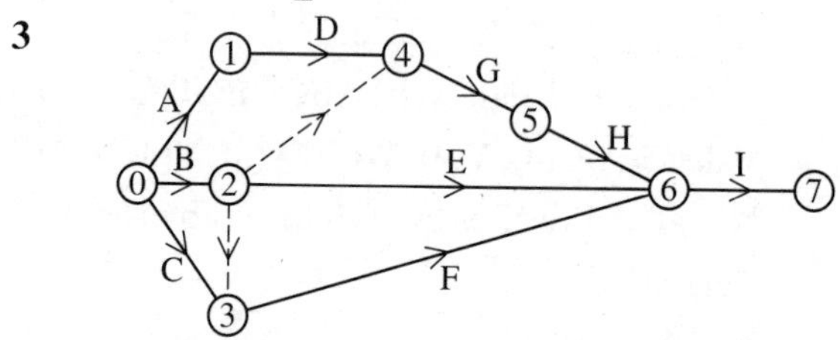

4

Activity	*Depends on*
A	—
B	A
C	A
D	A
E	A
F	A
G	E
H	D, G
I	D, G
J	B
K	C
L	H, J, K
M	F, I
N	L, M

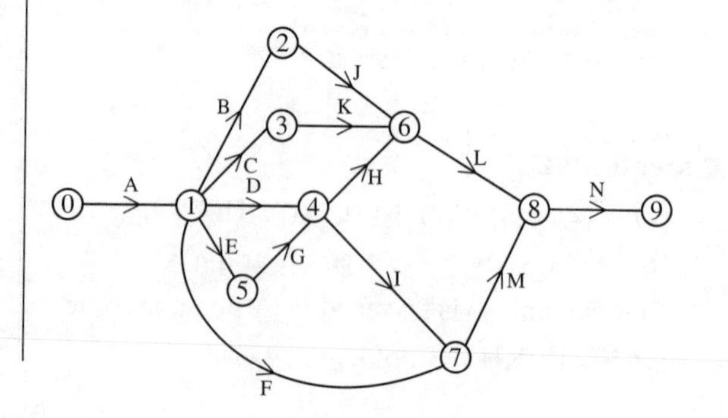

5

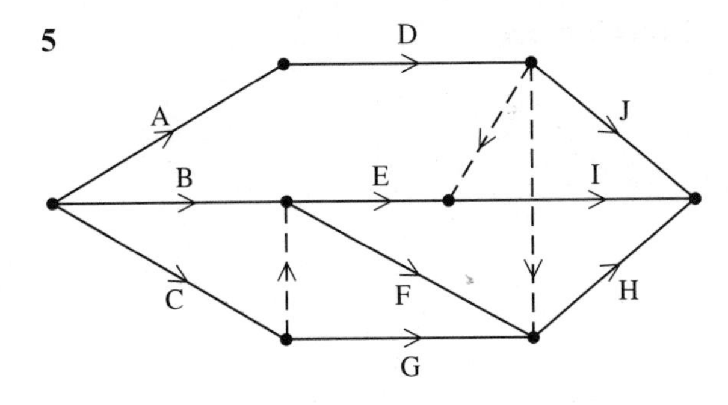

Exercise 5C

1 Completed table is

i	0	1	2	3	4
e_i	0	8	4	10	18
l_i	0	8	10	15	18

2 $e_2 = 7, \quad e_4 = 16, \quad e_5 = 20$

$l_4 = 16, \quad l_3 = 14, \quad l_2 = 7$

3

i	0	1	2	3	4	5	6
e_i	0	2	7	9	11	13	21
l_i	0	2	7	9	18	13	21

Exercise 5D

1 (a) $s_0 = 0, s_1 = 0, s_2 = 6, s_3 = 5, s_4 = 0$

So critical events are ⓪, ① and ④.

(b) Activity A — (0, 1) — 0
Activity B — (0, 2) — 6
Activity C — (2, 3) — 6
Activity D — (1, 3) — 5
Activity E — (1, 4) — 0
Activity F — (3, 4) — 5
So critical activities are A and E.

(c) Hence the critical path is

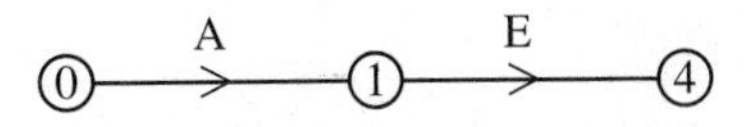

which is of length 18.

2 (a) $s_0 = 0, s_1 = 0, s_2 = 0, s_3 = 8, s_4 = 0,$
$s_5 = 0, s_6 = 0.$

So critical events are ⓪, ①, ②, ④, ⑤ and ⑥.

(b) Activity A — (0, 1) — 0
Activity B — (1, 2) — 0
Activity C — (1, 3) — 8
Activity D — (2, 4) — 0
Activity E — (2, 5) — 12
Activity F — (3, 5) — 8
Activity G — (4, 5) — 0
Activity H — (5, 6) — 0

So critical activities are A, B, D, G and H.

(c) Hence the critical path is

⓪ —A→ ① —B→ ② —D→ ④ —G→ ⑤ —H→ ⑥

which is of length 29.

3 (a) $s_0 = 0, s_1 = 0, s_2 = 0, s_3 = 0, s_4 = 7,$
$s_5 = 0, s_6 = 0.$

So critical events are ⓪, ①, ②, ③, ⑤ and ⑥.

(b) Activity A — (0, 1) — 0
Activity B — (1, 2) — 0
Activity C — (1, 3) — 0
Activity D — (1, 4) — 7
Activity E — (2, 5) — 0
Activity F — (3, 5) — 0
Activity G — (5, 6) — 0
Activity H — (4, 6) — 7

So critical activities are A, B, C, E, F and G.

(c) There are two critical paths:

⓪ —A→ ① —B→ ② —E→ ⑤ —G→ ⑥

⓪ —A→ ① —C→ ③ —F→ ⑤ —G→ ⑥

They are both of length 21.

Exercise 5E

		Start		Finish		
Activity	Duration	Earliest	Latest	Earliest	Latest	Float
A (1,2)	2	0	0	2	2	0
B (1,3)	7	0	0	7	7	0
C (1,4)	10	0	1	10	11	1
D (2,3)	5	2	2	7	7	0
E (3,6)	8	7	8	15	16	1
F (3,5)	6	7	7	13	13	0
G (4,5)	2	10	11	12	13	1
H (5,7)	4	13	13	17	17	0
I (4,7)	3	10	14	13	17	4
J (6,7)	1	15	16	16	17	1

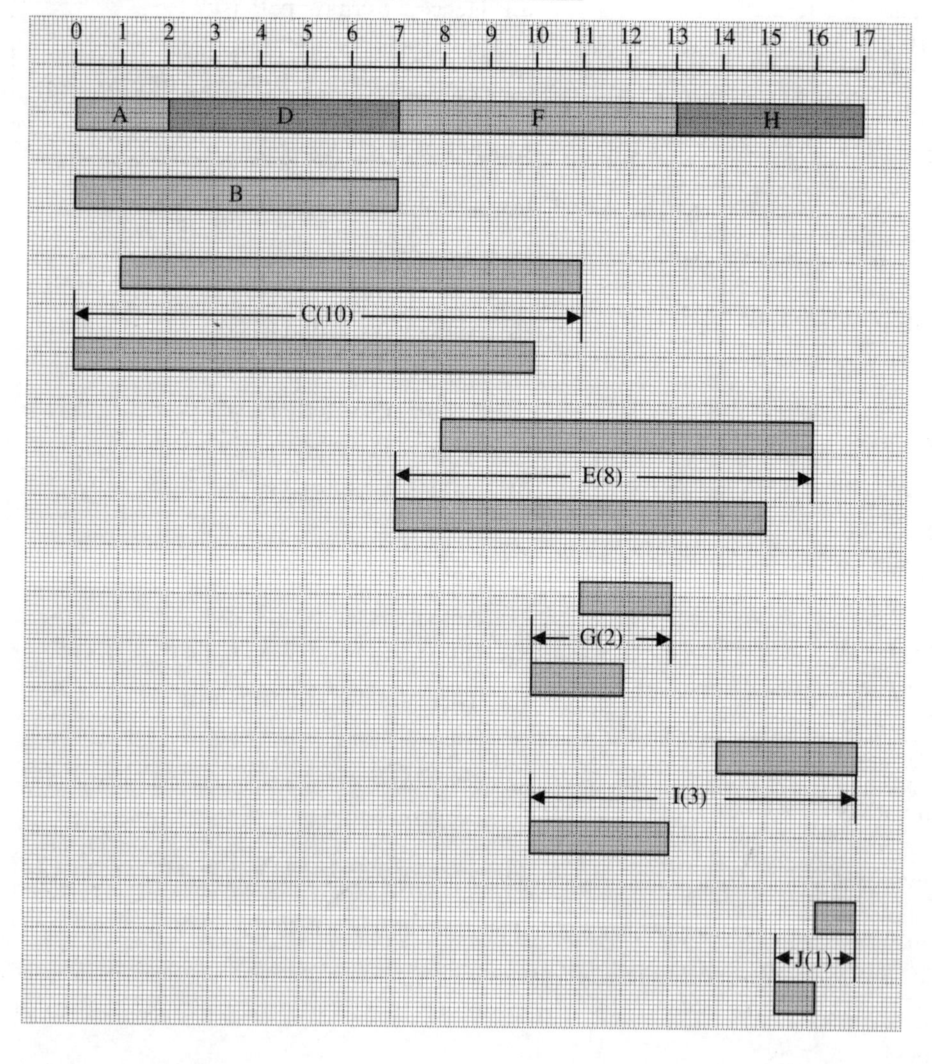

Exercise 5F

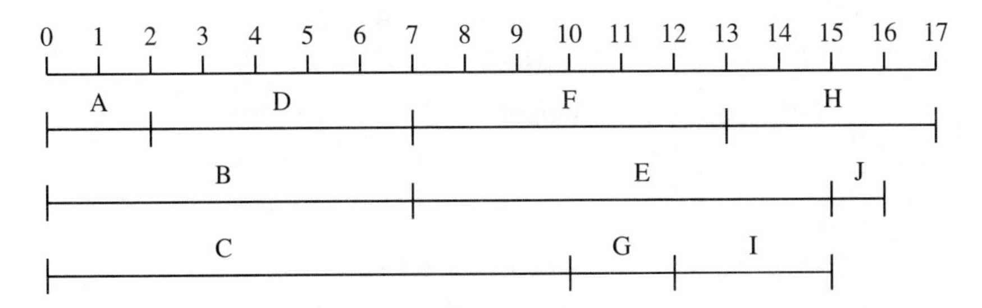

3 workers required

Exercise 5G

1

i	0	1	2	3	4	5
e_i	0	3	4	9	14	16
l_i	0	3	5	9	14	16

Floats A–0, B–1, C–3, D–1,
E–0, F–0, G–2, H–0

Length of critical path 16

Critical events 0,1,3,4,5

Critical activities A,E,F,H

2

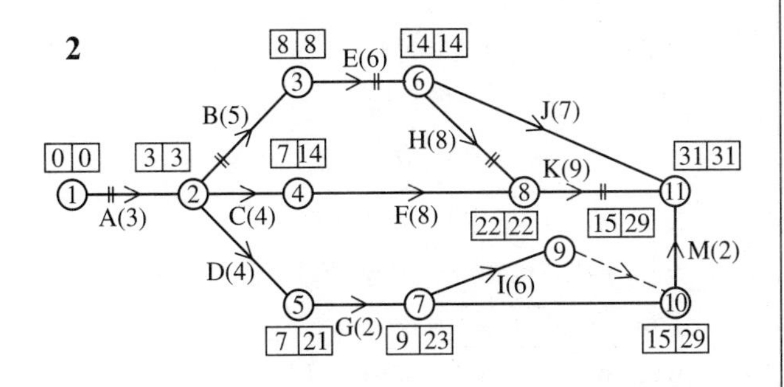

3 (a) The earliest time for event 8 is now increased to 23 days.

(b) This implies that the earliest time for the sink event is now 32. Thus the new project time is 32 days.

The critical path is then A, C, F, K.

4 (a)

i	1	2	3	4	5	6
e_i	0	3	5	9	8	14
l_i	0	3	11	9	12	14

Project time 14 days

Critical path ACF

(b)

		Start		Finish		
Activity	Duration	Earliest	Latest	Earliest	Latest	Float
A (1,2)	3	0	0	3	3	0
B (1,3)	5	0	6	5	11	6
C (2,4)	6	3	3	9	9	0
D (2,5)	5	3	7	8	12	4
E (3,5)	1	5	11	6	12	6
F (4,6)	5	9	9	14	14	0
G (5,6)	2	8	12	10	14	4

(c) Schedule for 2 workers

worker 1		A				C						F		
worker 2			B					D			E		G	

Minimum number of workers required if the project is to be completed in the project time is 2.

5 (a)

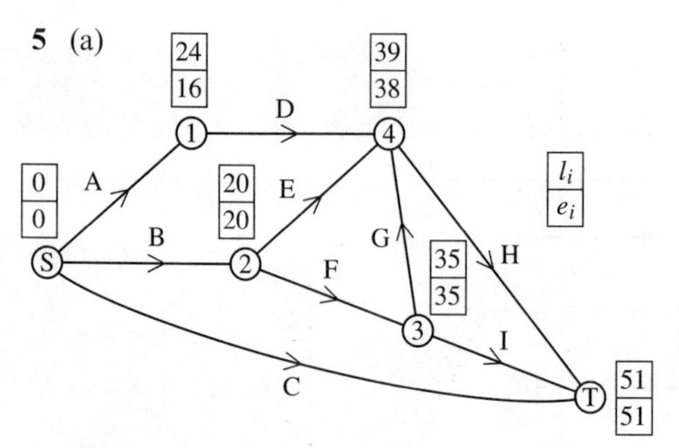

Length of critical path 51

(b) Critical events S, 2, 3, T

Critical activities B, F, I

Critical path S → 2 → 3 → T

(c) (i)

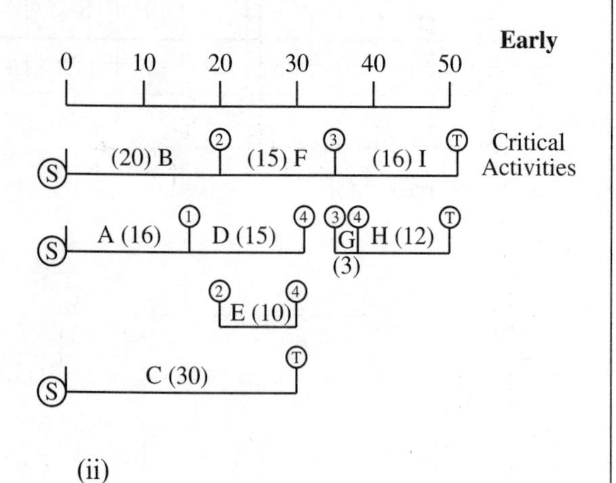

(ii)

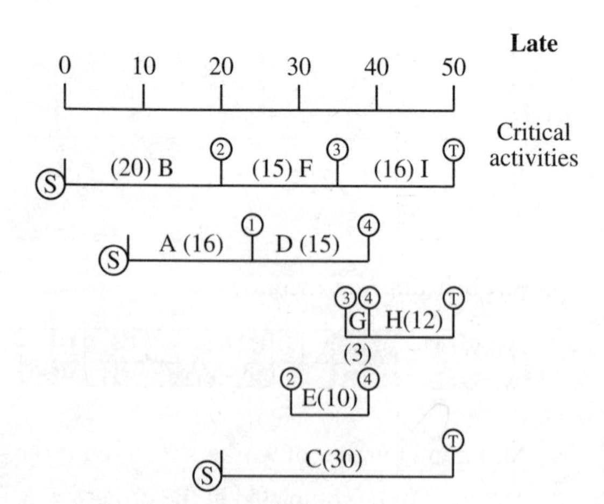

(d) 4 workers required

Exercise 6A

1 x = no. of type A tables

y = no. of type B tables

Maximise $Z = 15x + 17y$

Subject to $20x + 12y \leqslant 480$

$10x + 15y \leqslant 330$

$x \geqslant 0,\ y \geqslant 0$

2 Suppose she buys

x (hundred) grams of α and

y (hundred) grams of β

Minimise $C = 40x + 30y$

Subject to $30x + 10y \geqslant 25$

$20x + 25y \geqslant 30$

$10x + 40y \geqslant 15$

$x \geqslant 0,\ y \geqslant 0$

3 x = no. of type A machines

y = no. of type B machines

Maximise $Z = 75x + 120y$

Subject to $x + 2y \leqslant 40$

$3x + 4y \leqslant 100$

$x \geqslant 0,\ y \geqslant 0$

4 f = no. of full-fare passengers

h = no. of half-fare passengers

Maximise $Z = 10f + 5h$

Subject to $f + h \leqslant 14$

$f + h \geqslant 10$

$f \geqslant 0,\ h \geqslant f$

5 x = no. of small bookshelves

y = no. of medium bookshelves

z = no. of large bookshelves

Maximise $P = 4x + 6y + 12z$

Subject to $4x + 8y + 16z \leqslant 500$

$2x + 4y + 6z \leqslant 400$

$x \geqslant 0,\ y \geqslant 0,\ z \geqslant 0$

6 x = no. of type A made

y = no. of type B made

z = no. of type C made

Maximise $P = 15x + 20y + 25z$

Subject to $9x + 6y + 4z \leqslant 3600$

$2x + 9y + 12z \leqslant 3600$

$18x + 4y + 6z \leqslant 3600$

$6x + 9y + 8z \leqslant 3600$

$x \geqslant 0, \ y \geqslant 0, \ z \geqslant 0$

Exercise 6B

1

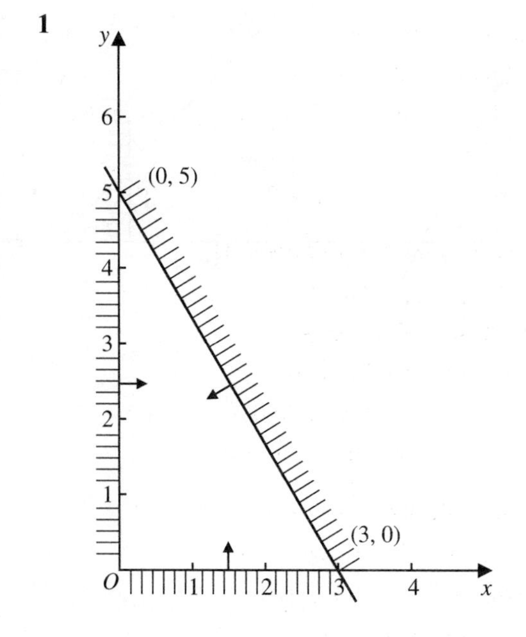

2

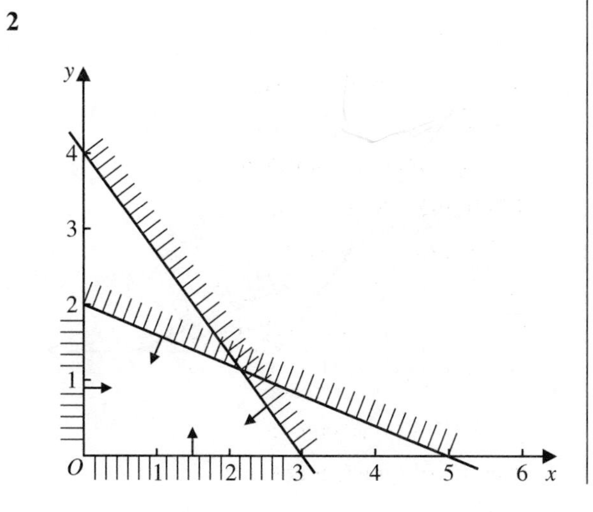

3

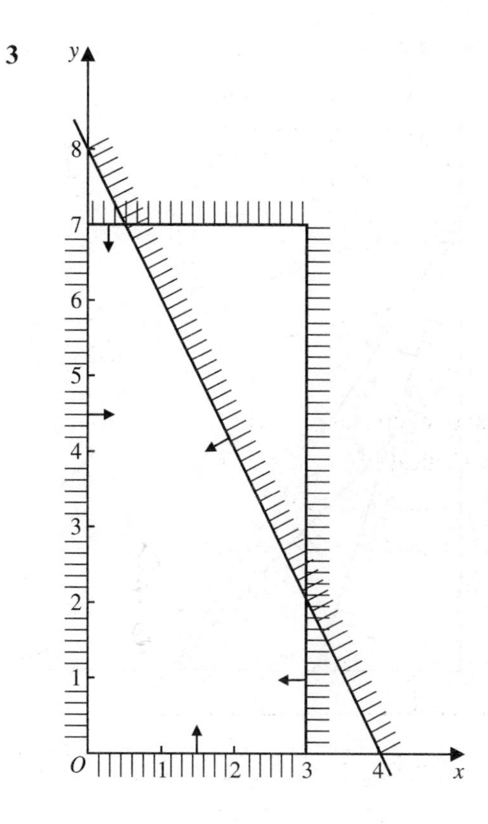

4

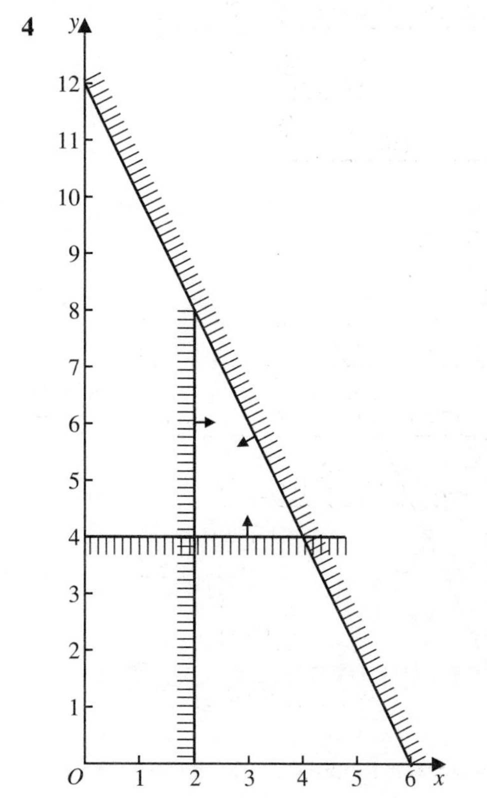

5

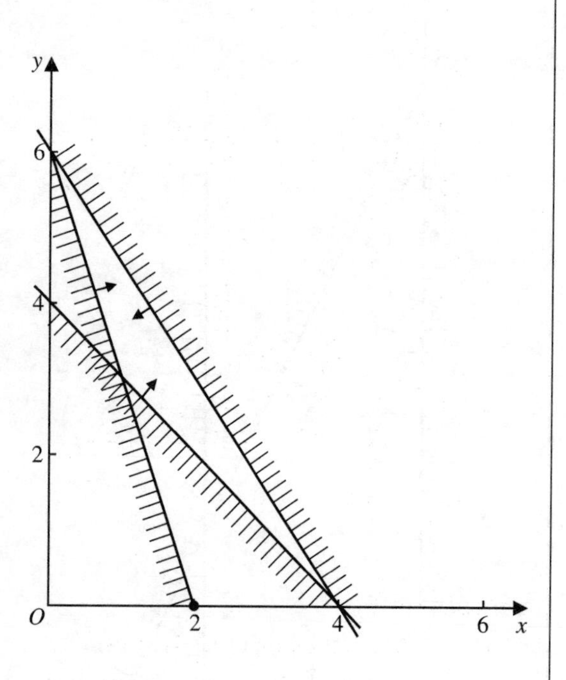

6

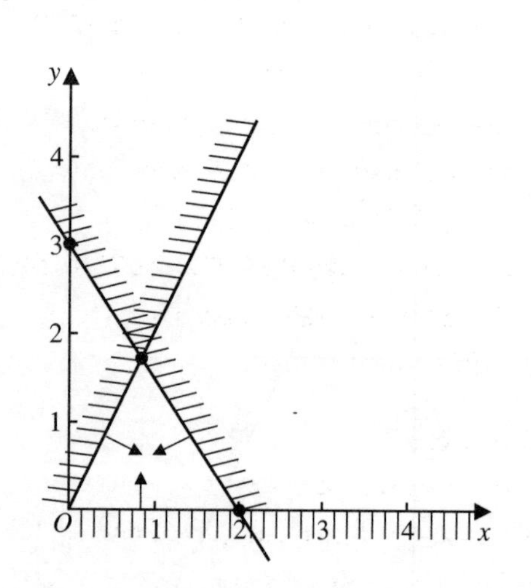

Exercise 6C

1

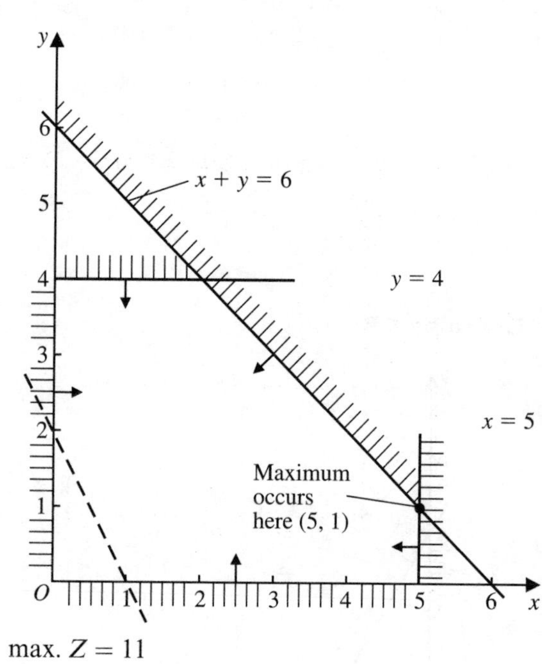

max. $Z = 11$

2

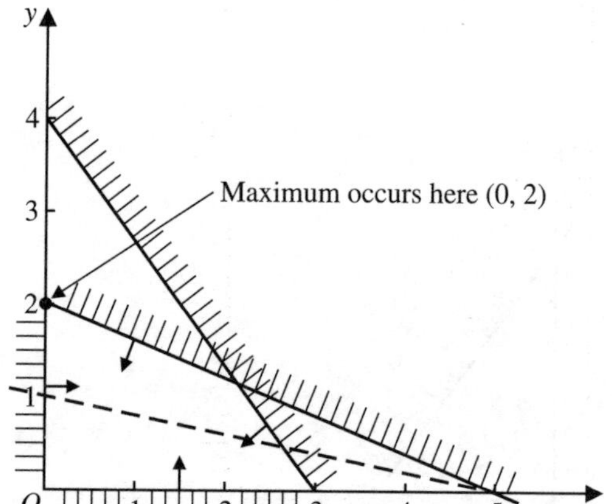

max. $Z = 10$

3

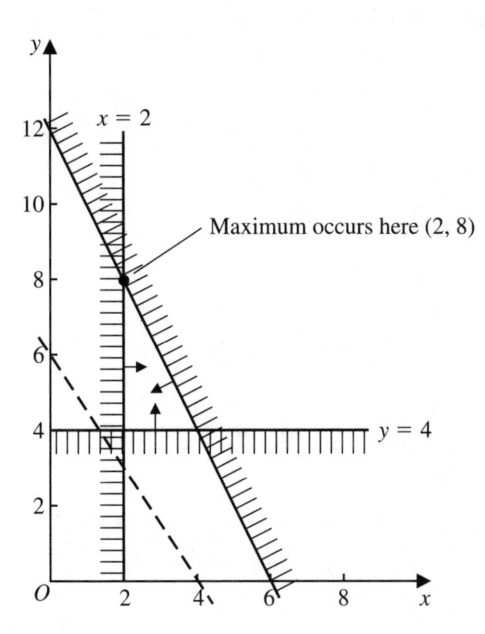

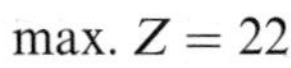
max. $Z = 22$

4

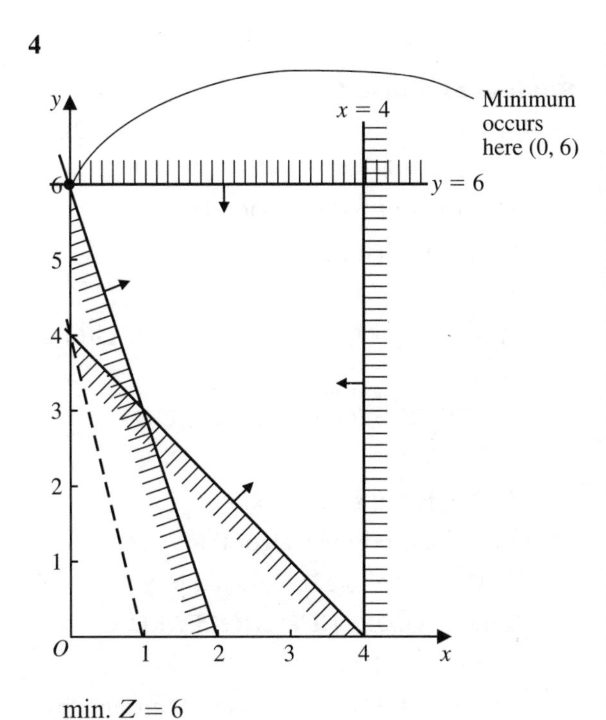

min. $Z = 6$

5

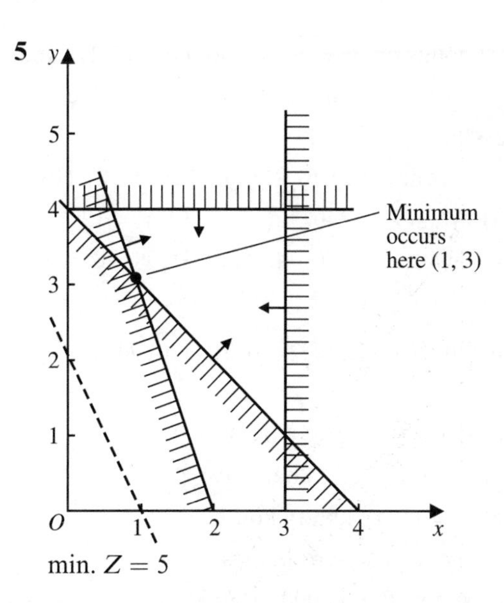

min. $Z = 5$

Exercise 6D

1 (a) $(0, 0), (5, 0), (5, 1), (2, 4), (0, 4)$

(b) Corresponding values of Z are 0, 10, 11, 8, 4.

(c) The maximum value 11 occurs at $(5, 1)$.

2 (a) $(0, 0), (3, 0), \left(2\frac{1}{7}, 1\frac{1}{7}\right), (0, 2)$

(b) Corresponding values of Z are 0, 3, $7\frac{6}{7}$, 10.

(c) The maximum value 10 occurs at $(0, 2)$.

3 (a) $(2, 4), (4, 4), (2, 8)$

(b) Corresponding values of Z are 14, 20, 22.

(c) The maximum value 22 occurs at $(2, 8)$.

4 (a) $(4, 0), (4, 6), (0, 6), (1, 3)$

(b) Corresponding values of Z are 16, 22, 6, 7.

(c) The minimum value of 6 occurs at $(0, 6)$.

5 (a) $(3, 1), (3, 4), \left(\frac{2}{3}, 4\right), (1, 3)$

(b) Corresponding values of Z are 7, 10, $5\frac{1}{3}$, 5.

(c) The minimum value 5 occurs at $(1, 3)$.

Exercise 6E

1 $x = 6, \quad y = 12, \quad Z = 48$

Vertices are $(0, 0), (10, 0), (0, 15)$ and $(6, 12)$.

2 $u = 5, \quad v = 2, \quad W = 80$

Vertices are $(0, 12), (5, 2), (8, 0)$.

3 All points on line segment joining (12, 12) and (15, 8) give maximum Z. Maximum value of Z is 336.

Vertices are (0, 0), (0, 21), (12, 12), (15, 8), (15, 0).

4 (a) Max. $Z = 240$ at $x = 12, y = 12$

(b) Max. $Z = 570$ at $x = 15, y = 8$

5 (a) $x = \frac{4}{5}$, $y = 2\frac{2}{5}$, $Z = 3\frac{1}{5}$

(b) (0, 0), (0, 1), (0, 2), (0, 3), (1, 0), (1, 1), (1, 2), (2, 0);

(0, 3) and (1, 2) give max. Z of 3

(c) (0, 3) and (1, 2) both give $Z = 3$

6 (a) l = no. of luxury houses

s = no. of standard houses

Maximise $P = 12\,000\,l + 8000\,s$

Subject to $l \geqslant 5$, $s \geqslant 10$, $l + s \leqslant 30$

$300\,l + 150\,s \leqslant 6000$

(b) $l = 10$, $s = 20$, max. $P = £280\,000$. That is, 10 luxury homes and 20 standard homes should be built.

7 $l = 8$, $s = 0$ and minimum cost £320

8 $x = £20\,000$, $y = £10\,000$

and maximum yield is £1900.

9 (a) $x \leqslant 3$, $y \leqslant 4$, $x + y \leqslant 5$, $3x + 2y \geqslant 10$

(b)

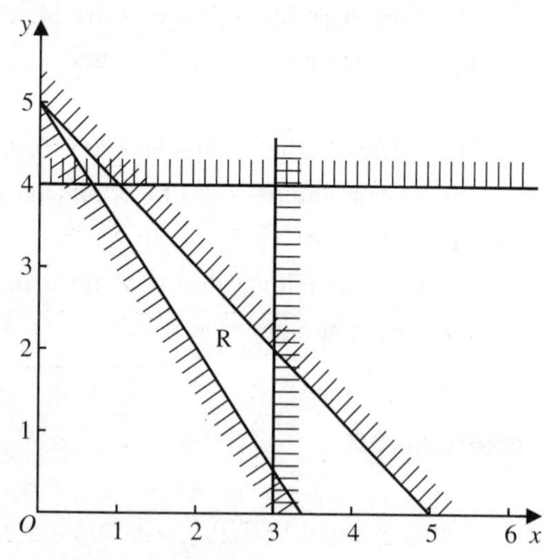

(c) (1, 4) (2, 2) (2, 3) (3, 1) (3, 2)

(d) Cost £C: $C = 50x + 40y$

(1, 4) = 50 + 160 = 210

(2, 2) = 100 + 80 = 180 *

(2, 3) = 100 + 120 = 220

(3, 1) = 150 + 40 = 190

(3, 2) = 150 + 80 = 230

Minimum cost £180; $x = 2$, $y = 2$

Exercise 6F

1 $x = 6$, $y = 12$, max. $Z = 48$

2 $x = 15$, $y = 8$, max. $Z = 570$

3 $x = 4$, $y = 2$, $z = 0$, max. $P = 24$

4 $x = 0$, $y = \frac{1}{3}$, $z = 3\frac{1}{3}$

max. $P = 6\frac{1}{3}$

5 (a) x = no. of lengths of A made

y = no. of lengths of B made

Maximise $P = 12x + 15y$

Subject to $4x + y \leqslant 56$

$5x + 3y \leqslant 105$

$x + 2y \leqslant 56$

$x \geqslant 0$, $y \geqslant 0$

(b) $x = 6$, $y = 25$ and max. $P = £447$

6 max. $P = £56$ when $x = 0$, $y = 9$, $z = 2$

Review exercise 2

1 (a) A(2), B(2), C(2), D(5), E(4), F(3), G(2), H(3), I(3), J(4)

(b) ABFIJGEJHDFIHDCEDA

Repeat DH and FI

(c) 774 m

2 (a) A(2), B(3), C(4), D(3), E(4), F(2), G(6), H(3), I(3)

(b) ABCDEBEGDGFHGHIEIGCA

(c) 61.3 km

(d) Use BD, it saves 0.1 km

3 (a) Chinese postman algorithm

(b) Odd nodes are B, C, E and H

Route is ABDBCHCDEHGEFAEDA

Length = 1179 m

(c) This would decrease his total distance by 18 km.

4 Route is PVUTVRVQRTSTQSQP
Length = 2158

5 (a) H may not start until both E and F have been completed.
(b) (ii) B, E, G, K

6 (c) L, P, Q, U, X, Z
(d) 18 hours
(e) Three workers are needed.

7 (a) 23, 26, 27, 29
(c) B, E, J, M, Q, S
(d) 29 hours

8 (a)

Vertex	1	2	3	4	5	6	7	8	9	10
Early	7	5	3	13	27	24	37	38	49	52
Late	7	6	4	14	27	25	37	39	49	52

(b) A, D, J, N, Q
(c) 52 days

9 (b) Length = 71 minutes, ABEFIJL
(c) 7
(d) ABEFKL, length = 69 minutes

10 (b) AEJKNSP and Q
(c) 106 hours
(d) $3 \times 106 <$ sum of durations $= 323$

11 (b) $5x + 3y \leqslant 48$
(d) $P = 3.5x + 1.5y$
(e) Six cakes and five fruit loaves
(f) £28.50

12 (b) $5x + 4y \leqslant 350$
(e) 43 CD units, 33 cassette units

13 (c) Cost $= 336x + 252y$
(e) Seven large, three small
Minimum total cost = £3108

14 (a) $4x + 5y \leqslant 47$, $y \geqslant 2x - 8$,
$4y - x - 18 \leqslant 0$, $x \geqslant 0$, $y \geqslant 0$
(b) $(6\frac{3}{14}, 4\frac{3}{7})$ (c) (ii) (6, 4)

15 (a) All entries are non-negative in the first row
(c) (i) £840
(ii) A(0), B(56), C(75)

16 (a) Maximise $P = 14x + 20y + 30z$
subject to $5x + 8y + 10z + r = 25\,000$
$5x + 6y + 15z + s = 36\,000$
$x, y, z, r, s \geqslant 0$
(c) (i) $x = 600$, $y = 0$, $z = 2200$
(ii) £744
(iii) Optimal since all numbers in profit row are non-negative.

17 (a) $-x + 4y - z \leqslant 0$
$3x + 5y + 8z \leqslant 100$, $x \geqslant 0$, $y \geqslant 0$, $z \geqslant 0$
(b) $S = 2x + 4y + 6z$
(c) There are more than two variables
(g) All entries in objective row are non-negative.
(h) Small (0), medium (2), large (11)
Cost = £1960

18 (b) (i) $P = 1.5x + 1.75y$
(ii) $x = 125\frac{5}{7}$, $y = 171\frac{3}{7}$
(c) First stage gives A, second stage gives C

19 (b) $P = 12x + 24y + 20z$
(e) All the entries in the profit row are non-negative
(f) A(0), B(15), C(20)
Profit = £760

20 (a) Maximum $P = 14x + 12y + 13z$
subject to $4x + 5y + 3z \leqslant 16$
$5x + 4y + 6z \leqslant 24$
$x \geqslant 0$, $y \geqslant 0$, $z \geqslant 0$

(c)

P	x	y	z	r	s	
1	0	$\frac{11}{2}$	$-\frac{5}{2}$	$\frac{7}{2}$	0	56
0	1	$\frac{5}{4}$	$\frac{3}{4}$	$\frac{1}{4}$	0	4
0	0	$-\frac{9}{4}$	$\frac{9}{4}$	$-\frac{5}{4}$	1	4

(d) From making no lions, move to making four lions.

Exercise 7A

1

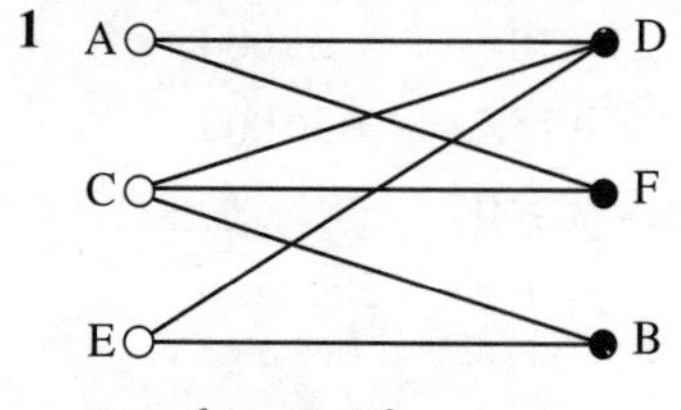

$X = \{A, C, E\}$

$Y = \{B, D, F\}$

2 (a)

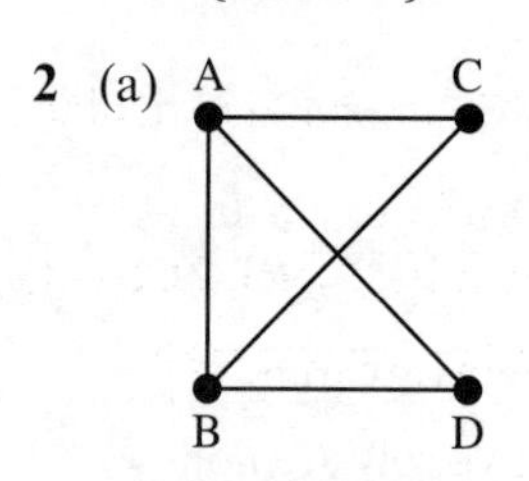

Not bipartite

(b)

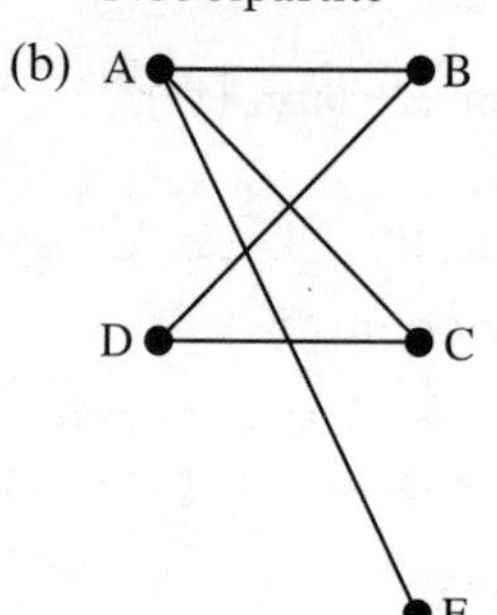

Bipartite: $X = \{A, D\}$, $Y = \{B, C, E\}$

3

4

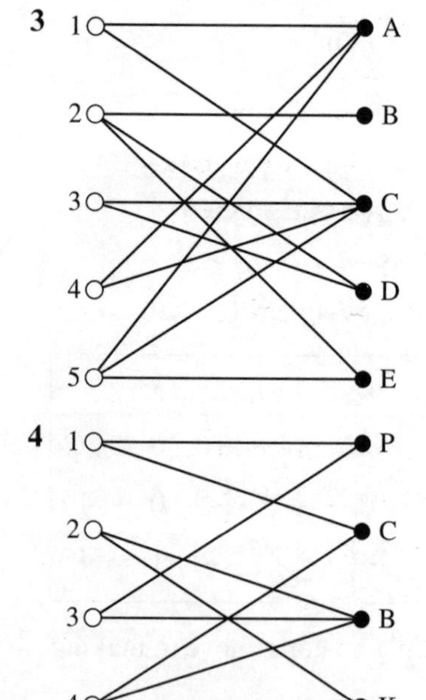

5

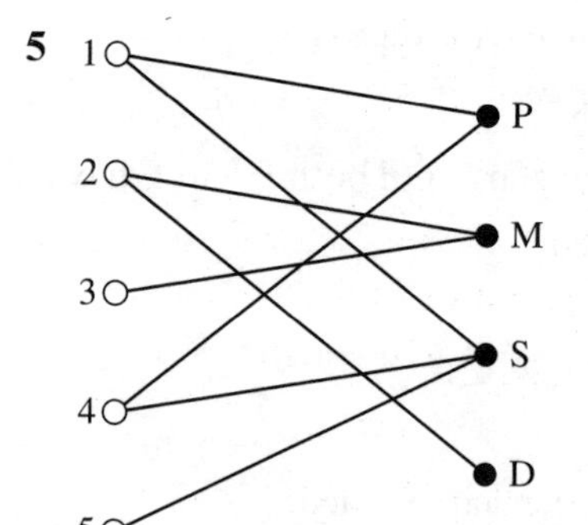

6

	R1	R2	R3	R4
C1				×
C2	×	×	×	×
C3		×	×	
C4	×	×		

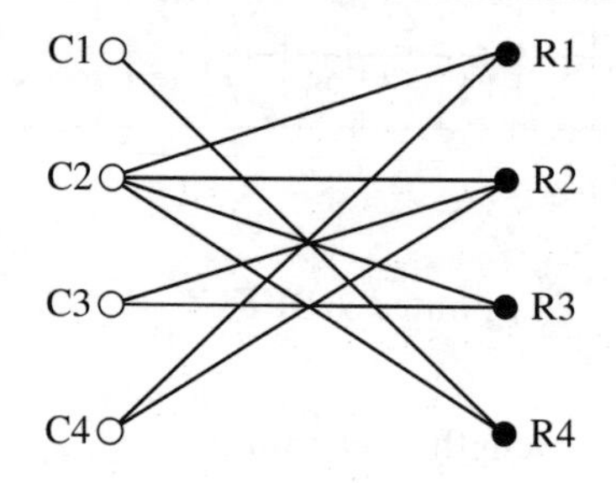

Exercise 7B

1

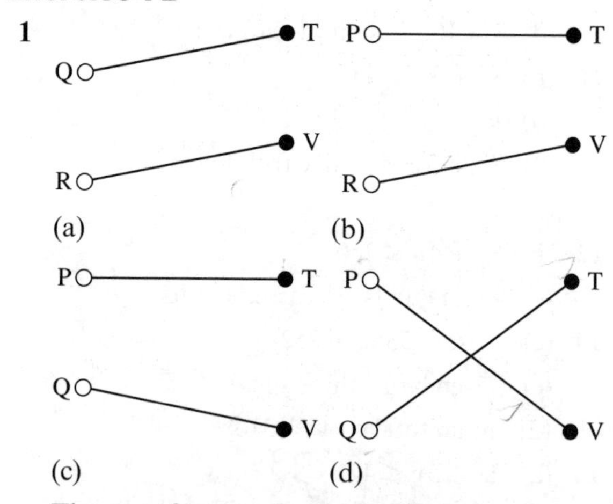

There are four maximal matchings

2 (a) N must pair with 3 (no choice)

E must pair with 1 (3 already paired)

S must pair with 2 (3 already paired)

R must therefore pair with 4

A maximal matching is then

E with 1, S with 2, N with 3, R with 4

(b) N must pair with 3 (no choice)

E must pair with 1 (3 already paired)

S must pair with 2 (3 already paired)

M must therefore pair with 4

A maximal matching is then

E with 1, S with 2, N with 3, R with 4

3 A pairs with X

B pairs with Y

C pairs with Z

X must pair with A (no choice)

B must pair with Y (no choice)

C must pair with Z (Y already paired)

4 1 must pair with A (no choice)

E must pair with 5 (no choice)

2 must pair with B (A already paired)

D must pair with 4 (5 already paired)

C must pair with 3 (4 and 5 already paired)

Hence the only possible complete matching is

A with 1, B with 2, C with 3, D with 4

E with 5

5 Begin with E paired with R

Then either (a) D is paired with Y

or (b) D is paired with B.

(a) If D is paired with Y, then A must be paired with G, which implies C must be paired with B.

Complete matching is

A with G, C with B, D with Y,

E with R

(b) If D is paired with B, then C must be paired with G, which implies that A must be paired with Y.

Complete matching is

A with Y, C with G, D with B,

E with R

6 (a) Only x_4 can be paired with y_1 and y_4, therefore one of y_1 or y_4 must be unpaired.

(b) eg x_2 paired with y_2

x_3 paired with y_3

x_4 paired with y_4

(c) x_3 must be paired with y_3 (no choice)

x_2 must be paired with y_2 (y_3 already paired)

Therefore x_1 cannot be paired (y_2 and y_3 already paired)

Exercise 7C

1 (a) Starting with D, obtain maximal matching (D, 1), (A, 3), (B, 2)

(b) Starting with C, obtain (C, 3), (B, 2), (A, 1)

2 (a) eg R——T, so add edge RT to obtain improved matching

(b) eg S——U━━Q——W

is an alternating path. Complete matching is PV, SU, QW, RT

3 Start with, for example, B2, C3, E5.

Alternating path starting at D is

D——3━━C——1 (breakthrough)

This gives B2, D3, C1, E5

Search for path starting at A:

A——2━━B (no path)

Hence this is a maximum matching and there is no complete matching.

4 (b) eg 1A, 2B, 3C, 5E

(c) eg Using the alternating path

4——C━━3——D (breakthrough)

we obtain the complete matching

1A, 2B, 3D, 4C, 5E

5 eg Initial matching 1P, 3B, 4K

Alternating path

2——K━━4——C (breakthrough)

Maximal matching (complete) 1P, 3B, 2K, 4C

6 (a) Initial matching 1P, 2M, 5S

An alternating path is:

3——M━━2——D

so a maximal matching is

1P, 2D, 3M, 5S

(b) eg Start with 1S, 2D, 3M.
A trivial alternating path is then

giving the matching 1S, 2D, 3M, 4P.

7 The bipartite graph is

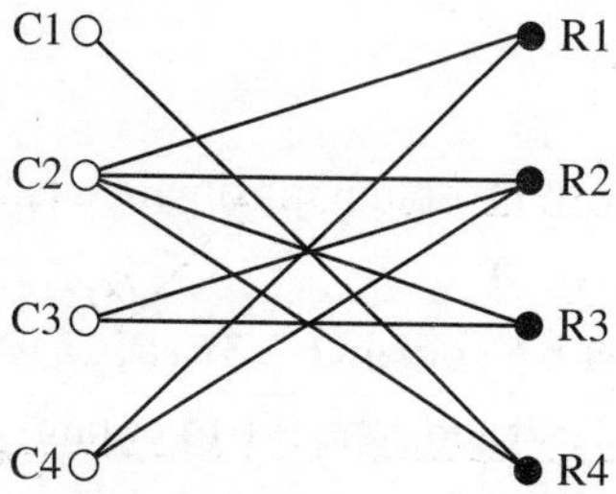

C1 must be paired with R4 (no choice)
C4 must be paired with either R1 or R2
(a) If C4 is paired with R1, then C2 can be paired with either R2 or R3.
(i) If C2 is paired with R2, then C3 can be paired with R3, giving complete matching (C1,R4), (C4, R1), (C2, R2) (C3, R3).
(ii) If C2 is paired with R3, then C3 can be paired with R2, giving complete matching (C1, R4), (C4, R1), (C2, R3), (C3, R2).
(b) If C4 is paired with R2, then C3 can only be paired with R3. This leaves C2 to be paired with R1. This gives the matching (C1, R4), (C4, R2), (C3, R3), (C2, R1).
Hence there are only three complete matchings.

8 (a)

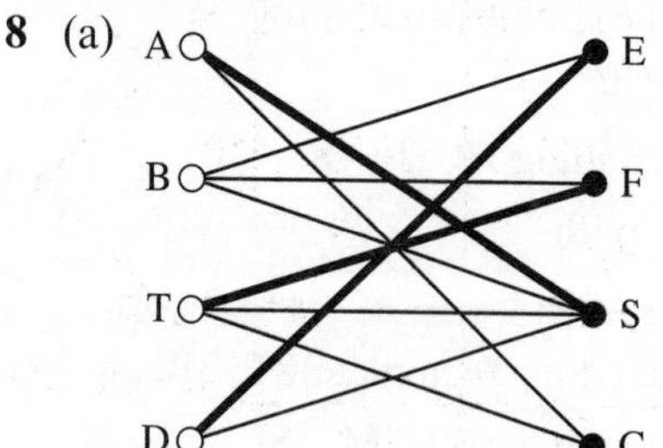

Initial matching shown with thick lines
eg Alternating path:

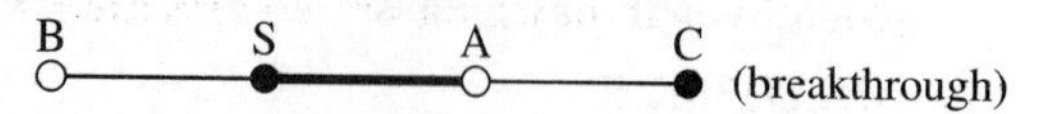

Changing the status of edges:

Complete matching is (T, F), (D, E), (B, S), (A, C), i.e.

Amy	clothing
Bhavana	stationery
Tina	furniture
Dylan	electrical

Exercise 8A

2 $x = 5, y = 2, z = 1, t = 6$

3 (a) (i) minimum (15, 9, 10) = 9
(ii) minimum (10, 8) = 8
(b) SADT, value 6
(c) SBDT, value 2
(d)

4 (a) (i)

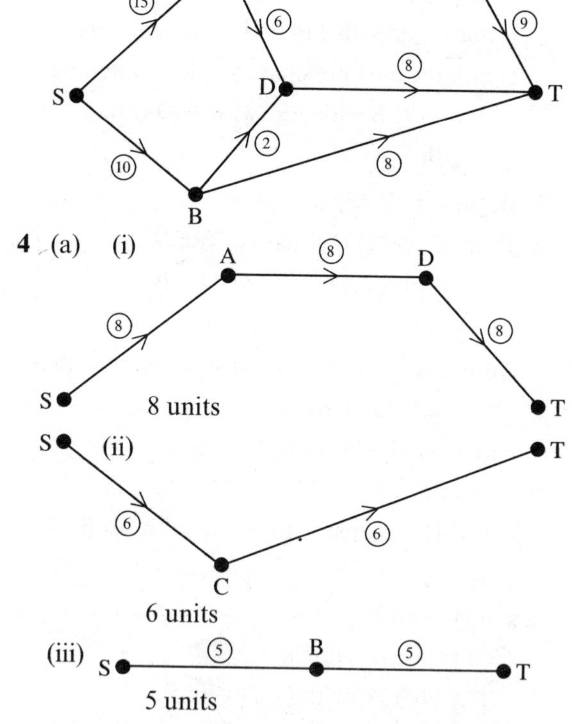

(b) (i) Flow of 8 units, saturated arc AD
(ii) Flow of 6 units, saturated arc CT
(iii) Flow of 5 units, saturated arc SB

(c) As SB is saturated, no further flow is possible on BA, BT and BC.

As AD is saturated, no further flow is possible on DT.

As in addition CT is saturated, there is no further flow possible to T, therefore no further flow is possible.

Total flow is $8 + 6 + 5 = 19$ units

5 Although it is possible for 15 vehicles to pass along arc SA it is only possible for $(8 + 4) = 12$ (i.e. capacity of AE + capacity of AD) to leave A. Similarly, although the capacity of SB is 14 the capacity of BC + capacity of BF is only 13. It is therefore not possible for more than $12 + 13$ vehicles to leave A and B. Thus a flow of 29 from S to T is not possible.

Exercise 8B

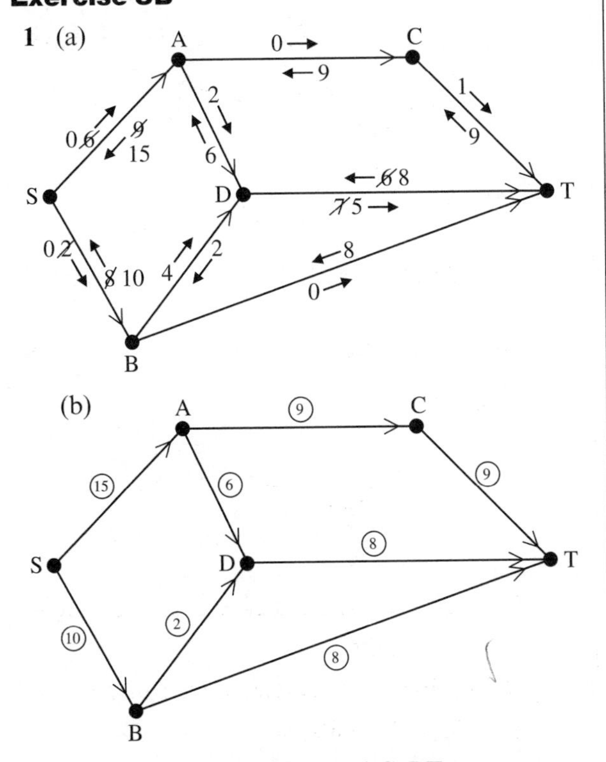

Saturated arcs are SA, SB, AC, BT

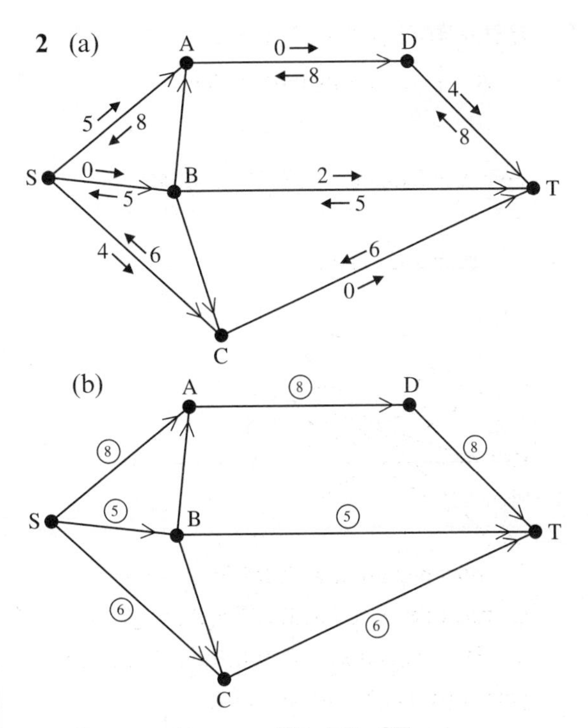

Saturated arcs are SB, AD, CT

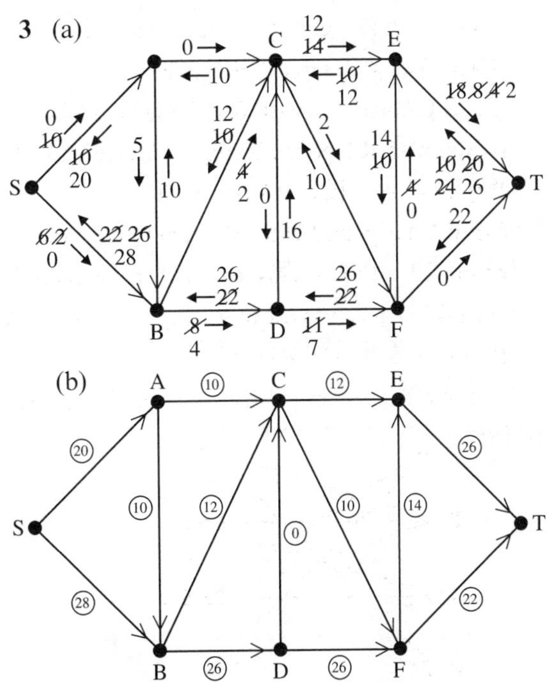

This is a maximum flow as both SA and SB are saturated. The value of this flow is $20 + 28 = 48$ units. Saturated arcs are SA, SB, AC, FE, FT

Exercise 8C

1

Cut	Set X	Set Y	Arcs in cut	Capacity
(i)	{S}	{A, B, C, D, E, T}	SA, SB	35
(ii)	{S, A, B, C, D, E}	{T}	DT, ET	30
(iii)	{S, A, B}	{C, D, E, T}	AC, AD, BD, BE	83
(iv)	{S, A, B, C, D}	{E, T}	BE, DT	44

2

Cut	Set X	Set Y	Arcs in cut	Capacity
(i)	{S}	{A, B, C, D, T}	SA, SB, SC	37
(ii)	{S, C}	{A, B, D, T}	SA, SB, BC, CT	35
(iii)	{S, A, B, C, D}	{T}	BT, CT, DT	43

3 Arcs AC, BC and DT, capacity = 12.

4

Cut	Set X	Set Y	Capacity
SA, SC	{S}	{A, B, C, T}	17
SA, CB, CT	{S, C}	{A, B, T}	21
SA, AB, BT, CT	{S, B, C}	{A, T}	15
SA, AB, SC, CB, BT	{S, B}	{A, C, T}	23
SC, AB, AT	{S, A}	{B, C, T}	30
SC, CB, BT, AT	{S, A, B}	{C, T}	26
AB, AT, CB, CT	{S, A, C}	{B, T}	34
AT, BT, CT	{S, A, B, C}	{T}	18

The minimum cut is SA, AB, BT, CT, capacity = 15.

5 (a) (i) SBET, 10 units (ii) SADFT, 15 units

(b) eg Flow-augmenting paths:

SCEFT, 10 units

SBAEDFT, 5 units

SCET, 2 units

Total maximum flow of value 42.

(c) eg

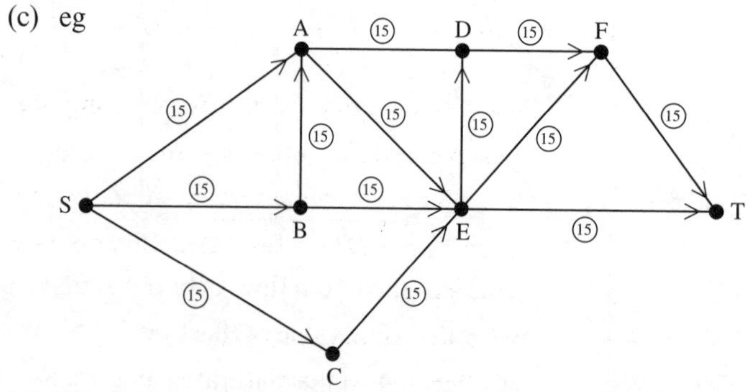

(d) The minimum cut consists of arcs DF, EF and ET, capacity 20 + 10 + 12 = 42 units

6 (a)

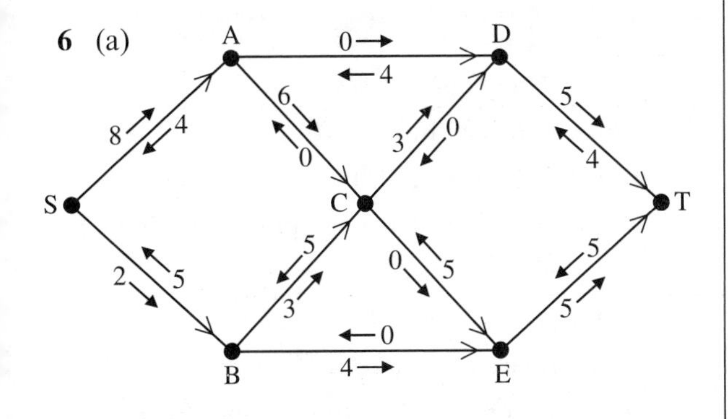

(b) SACDT, 3 units
SBET, 2 units
SACBET, 2 units
Total flow = 16 units

(c)

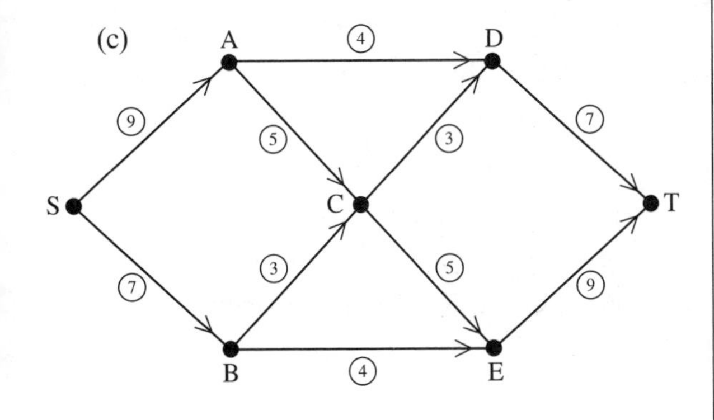

(d) Cut consisting of arcs AD, CD, CE and BE is the minimal cut of capacity $4 + 3 + 5 + 4 = 16$.

Exercise 8D

1 Maximum flow is 19 units
Flow-augmenting paths are eg:
SS_1AT_1T, 8 units (AT_1 and T_1T saturated)
SS_1CET_2T, 7 units (S_2C and ET_2 saturated)
SS_2BDT_2T, 4 units (DT_2 and T_2T saturated)
where S is a supersource and T is a supersink.

Maximum as T_1T and T_2T are saturated.

eg

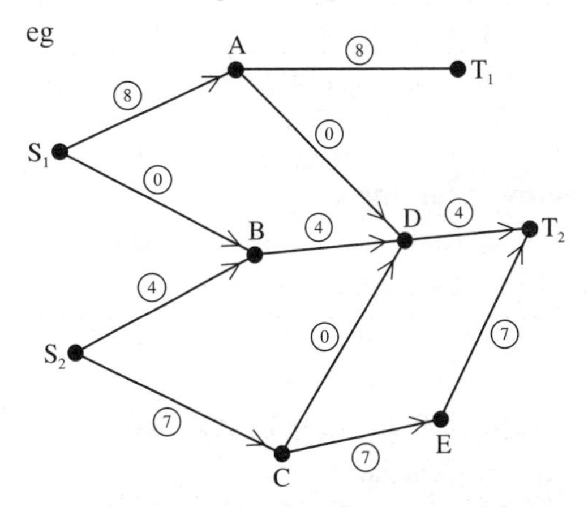

Saturated arcs are AT_1, DT_2, S_2C and ET_2.

2 Maximum flow is 20 units
Flow-augmenting paths are eg:
SS_3DT_2T, 6 units
SS_1BT_1T, 5 units
SS_2ACT_2T, 4 units
SS_1BCT_1T, 5 units

eg

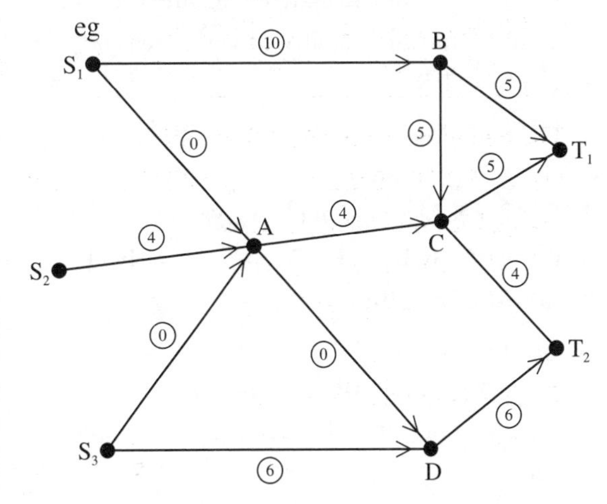

There is a cut of capacity 20 consisting of arcs S_1B, AC and DT_2 $(10 + 4 + 6)$.
As S_1B and AC are saturated no more flow can reach T_1. As DT_2 is saturated no more flow can reach T_2.

(a) flow out of S_1, 10 units
flow out of S_2, 4 units
flow out of S_3, 6 units

(b) flow into T_1, (5) + (5) = 10 units

flow into T_2, (4) + (6) = 10 units

(c) flow through C, (5) + (4) = 9 units

Review exercise 3

1 (b) eg Ruth – French, Steve – Japanese, Tony – German, Ursula – Spanish, Victoria – Italian

2 (b) eg Ann – E, Barry – H, David – C, Gemma – P, Jasmine – M, Nickos – F

(c) Not unique

3 (b) eg Mihi – L, Pat – K, Robert – B, Sarah – L, Tony – A

4 (a) eg A – 6, B – 2, C – 5, D – 3, E – 4, F – 1

(b) Remove E and 2 and all arcs incident to them. Repeat algorithm on reduced diagram.

5 (c) eg A – P, D – T, H – M, K – W, N – S, Y – L

6 (c) eg A – T, B – S, C – U, D – R, F – V

(d) For a complete matching C must run stall U and B must run stall W, then there is no-one to run stall S.

(e) Complete matching now possible eg: A – S, B – W, C – U, D – R, E – T, F – V

7 (a) Mr & Mrs G should be given seats 3 and 4

(c) eg A – 6, B – 5, C – 2, D – 7, E – 8, F – 1

8 (a) Maximum flow = 19

(c) 19

(d) cut DG, AE, BE, CE and CF

9 (a) Flow of traffic through a system of one-way streets.

(b) Maximum flow = 39

(d) Cut edges SA, BA, BD, BE and CE

(e) They are all saturated.

10 (a) A, F, G and H; possible flow in > possible flow out

(c) Maximum flow = 34

(d) GI, EI and HI

(f) AB, CE, HG and HI

11 (a) C_1 has capacity 333

C_2 has capacity 448

(c) Maximum flow = 376

(e) Cut DG, BE, SC has capacity 376

(f) Flow of fluid through pipes with limited capacity.

12 (a) (i) 15 (ii) 14

(c) Maximum flow = 54

(e) Cut through AD, BD, BE and BC *or* CF has capacity 54.

(f) Increase the flow along BE.

13 (a) C_1 has capacity 40, C_2 has capacity 56.

(b) Maximum flow = 40

(c) 24

(d) Flow is either 19 or 20

(f) There are two more.

14 (a) 19 units

(b) $x = 13, y = 9, z = 4$

(c) All cuts have capacity > 70

(d) Maximum flow = 74 units

(f) Cut SA, BD, BF has capacity 74

Examination style paper

1 9 JONES

2 (b)

	or	
Baileys – May		Baileys – September
Craigs – July		Craigs – June
Dales – August		Dales – August
Evans – June		Evans – May
Fords – September		Fords – July

3 (a) 42 cm

eg VWQPVWTSQRSTUV

(b) Start at V and finish at S

eg VWQPVUTWTSRQS

37 cm

4 (b) $z + 30 = s$, but $z \geqslant 0$ and $s \leqslant 30 \Rightarrow z \leqslant 0$

Only solution is therefore $z = 0$ and so $s = 30$

(c) $t = 62$

(d) Maximum flow = 92

Cut consisting of arcs SA(45), BC(17), DT(30) has total capacity 92.

5 (a) Shortest route = AFHD

Vertices labelled in order ABGFCEHD

Length = 24 miles

(b) Shortest route = ABGCD

Length = 25 miles

6 $P = 6x + 9y$

(d) $x = 4, y = 8, P = 96$

7 (a) Activity I cannot start until C, F and G are all completed.

(b) (i)

Event	1	2	3	4	5	6	7
Early time	0	5	11	7	9	13	18
Late time	0	5	11	13	10	13	18

(ii) Critical activities = ADFI

Length of critical path = 18 days

(c) B(7), C(6), E(1), G(1), H(5), J(7)

(e) (i) Total time durations is 44.

$\frac{44}{18} = 2\frac{4}{9} > 2$

Therefore at least three workers are required

(ii) First worker does critical activities.

eg Second worker:

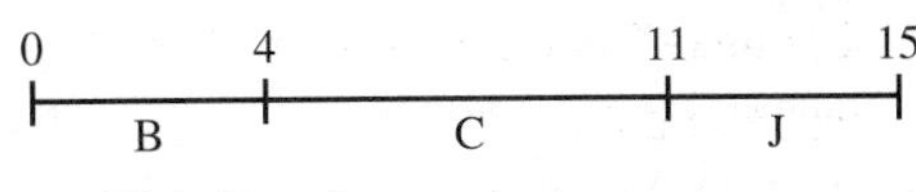

Third worker:

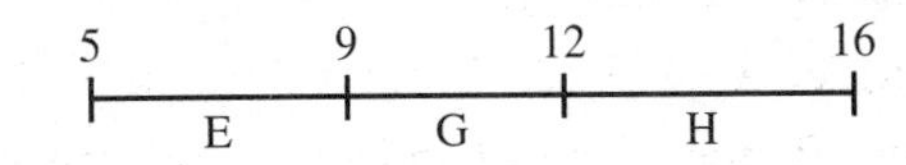

(d)

1 2 3 4 5 6 7 8 9 10 11 12 13 14 15 16 17 18

A D F I

B

C

E

G

H

J

Index